R00937 68477

D1223017

Volume I

MATHEMATICS

People · Problems · Results

Edited by Douglas M. Campbell
and John C. Higgins

Brigham Young University

Wadsworth International
Belmont, California
A Division of Wadsworth, Inc.

To Jill and Seiko

Acquisitions Editor: John Kimmel
Production Editor: Andrea Cava
Copy Editor: Ann Draus
Cover Designer: Catherine Flanders
Interior Designer Leigh McLellan

Printed in the United States of America

1 2 3 4 5 6 7 8 9 10—88 87 86 85 84

ISBN 0-534-02879-9 (hardback)
ISBN 0-534-22884-5 (paperback)

Library of Congress Cataloging in Publication Data
Main entry under title:

Mathematics : people, problems, results.

1. Mathematics—Addresses, essays, lectures.
I. Campbell, Douglas M. II. Higgins, John C., 1935–
QA7.M34466 1984 510 83-17039
ISBN 0-534-02879-9
ISBN 0-534-22884-5 (pbk.)

Preface

LET US BEGIN by stating what these volumes are not. They are not a history of mathematics, not an anthology of famous mathematical lives, not a text, not a survey of mathematical results. They are an introduction to the spirit of mathematics. The purpose of this anthology is to give the nonmathematician some insight into the nature of mathematics and those who create it.

Although these essays were selected for a general audience, it is virtually impossible to write about mathematics without using mathematics somewhere in the discussion. This can lead to problems since the level of mathematical knowledge of our intended audience varies substantially. To the potential reader whose retention of previous mathematical training is nil we have one piece of advice: read on. Ignore the mathematical fine points; they don't affect the narrative and there is much of value to be gained even if the details of the mathematics remain obscure. In almost every case the author's thesis can be grasped no matter what the reader's level of training.

If this anthology has a central purpose it is to help the reader answer the question, Why mathematics? This is an existential *why*, the *why* of an eight year old confronted with fractions. Any such *why* demands a better reply than the usual and transparently artificial applications. No one uses mathematics to slice pies; piggy bank accounts don't earn interest. What the child really wants to know is, Why ought we to care? What is there in the nature of mathematics that demands its perpetuation at such effort? And isn't that precisely the question that those regiments of adults, perplexed and harried by the mandatory mathematics classes of their youth, would ask as well? But to understand why mathematics exists and why it is perpetuated, one must know something of its history and of the lives of the mathematicians and their productions, which brings us to an anthology such as this.

As editors of these volumes we make no claim of uniqueness for this particular set of essays. What we do claim is that we have tried to provide as broad a selection of essays as the limitations of space will allow. We have tried to include some sense of the development of mathematics in time and across various cultures. We have included accounts of, and by, some very famous mathematicians and some not quite so famous. Some might feel that there is a certain aimlessness in this, a failure, if you will, of the editors to get a message across. The truth is that we have no *particular* message to get across. Our intent is to allow the reader to examine mathematics from as many different viewpoints as our space permits. We come bearing no slogans. This is a collection of attitudes and perspectives relative to mathematics that is deliberately untidy. In our view it is a rather untidy universe in which we live, and the world of mathematics is no less so. We would not presume to tell the readers what mathematics is. Let them read the articles, listen to the voices, and form their own opinions.

Finally, acknowledgment of the help received in compiling these volumes is surely in order. Our heartfelt thanks to John Kimmel, Andrea Cava, and Marta Kongsle of Wadsworth. They have carefully shepherded these volumes through an arduous production process and for this we are most grateful. Thanks also to Andrea Thompson and Dianne Johnson of the Department of Mathematics at Brigham Young University for their help in preparing the manuscript. We cannot begin to list by name the editors, secretaries, officers of learned societies, publishers, and others who have so willingly cooperated in allowing these articles to be republished. In all cases we have received cordial and prompt response to our requests and for this we are most appreciative. And lastly, a special thanks must go to the authors of these articles. The volumes are really theirs. We have organized and made comments but the substance of this anthology is that of its numerous authors. We are indebted to their genius.

Douglas M. Campbell
John C. Higgins

Contents

part three The Development of Mathematics

Chronological Table*

Carl B. Boyer

Carl B. Boyer has published extensively on the history of mathematics. He is a professor emeritus of mathematics at Brooklyn College.

Date	Mathematics	Date	General events
		−15,000,000,000	Origin of the sun
		−5,000,000,000	Origin of the earth
		−600,000,000	Beginning of Paleozoic Age
		−225,000,000	Beginning of Mesozoic Age
		−60,000,000	Beginning of Cenozoic Age
		−2,000,000	Origin of man
−50,000	Evidence of counting	−50,000	Neanderthal Man
−25,000	Primitive geometrical designs	−25,000	Paleolithic art; Cro-Magnon Man
		−10,000	Mesolithic agriculture
		−5000	Neolithic civilizations
−4241	Hypothetical origin of Egyptian calendar		
		−4000	Use of metals
		−3500	Use of potter's wheel; writing
−3000	Hieroglyphic numerals in Egypt	−3000	Use of wheeled vehicles
		−2800	Great Pyramid
−2773	Probable introduction of Egyptian calendar		
−2400	Positional notation in Mesopotamia	−2400	Sumerian-Akkadian Empire
−1850	Moscow (Golenishev) papyrus; cipherization		
		−1800	Code of Hammurabi
		−1700	Hyksos domination of Egypt; Stonehenge in England
		−1600	Kassite rule in Mesopotamia; New Kingdom in Egypt

*Dates before −776 are approximations only.

		−1400	Catastrophe in Crete
		−1350	Phonecian alphabet; use of iron; sundial; water clocks
		−1200	Trojan War; Exodus from Egypt
−1100?	Chou-pei		
		−776	First Olympiad
		−753	Traditional founding of Rome
		−743	Era of Nabonassar
		−740	Works of Homer and Hesiod (approx.)
		−586	Babylonian Captivity
−585	Thales of Miletus; deductive geometry (?)		
−540	Pythagorean arithmetic and geometry (approx.) Rod numerals in China (approx.) Indian *Sulvasūtras* (approx.)		
		−538	Persians took Babylon
		−480	Battle of Thermopylae
		−477	Formation of Delian League
		−461	Beginning of Age of Pericles
−450	Spherical earth of Parmenides (approx.)		
−430	Death of Zeno; works of Democritus Astronomy of Philolaus (approx.) *Elements* of Hippocrates of Chios (approx.)	−430	Hippocrates of Cos (approx.) Atomic doctrine (approx.)
		−429	Death of Pericles; plague at Athens
−428	Birth of Archytas; death of Anaxagoras		
−427	Birth of Plato		
−420	Trisectrix of Hippias (approx.) Incommensurables (approx.)		
		−404	End of Peloponnesian War
		−399	Death of Socrates; *Anabasis* of Xenophon
−369	Death of Theaetetus		
−360	Eudoxus on proportion and exhaustion (approx.)		
−350	Menaechmus on conic sections (approx.) Dinostratus on quadratrix (approx.)		
		−347	Death of Plato
−335	Eudemus: *History of Geometry* (approx.)		
		−332	Alexandria founded
−330	Autolycus: *On the Moving Sphere* (approx.)		
		−323	Death of Alexander
		−322	Deaths of Aristotle and Demosthenes

−320 Aristaeus: *Conics* (approx.)

−300 Euclid's *Elements* (approx.)

−260 Aristarchus' heliocentric astronomy (approx.)

−230 Seive of Eratosthenes (approx.)
−225 *Conics* of Apollonius (approx.)
−212 Death of Archimedes

−180 Cissoid of Diocles (approx.)
Conchoid of Nicomedes (approx.)
Hypsicles and 360° circle (approx.)

−150 Spires of Perseus (approx.)

−140 Trigonometry of Hipparchus (approx.)

−60 Geminus on parallel postulate (approx.)

+75 Works of Heron of Alexandria (approx.)

100 Nicomachus: *Arithmetica* (approx.)
Menelaus: *Spherics* (approx.)

125 Theon of Smyrna and Platonic mathematics
150 Ptolemy: *The Almagest* (approx.)

250 Diophantus: *Arithmetica* (approx.?)

320 Pappus: *Mathematical Collections* (approx.)

390 Theon of Alexandria (fl.)
415 Death of Hypatia

470 Tsu Ch'ung-chi's value of π (approx.)
476 Birth of Aryabhata
485 Death of Proclus

−311 Beginning of Seleucid Era in Mesopotamia
−306 Ptolemy I (Soter) of Egypt

−283 Pharos at Alexandria
−264 First Punic War opened

−232 Death of Asoka, the "Buddhist Constantine"

−210 Great Chinese Wall begun

−166 Revolt of Judas Maccabaeus

−146 Destruction of Carthage and Corinth

−121 Gaius Gracchus killed
−75 Cicero restored tomb of Archimedes
−60 Lucretius: *De rerum natura*
−44 Death of Julius Caesar

+79 Death of Pliny the Elder at Vesuvius

116 Trajan extends Roman Empire
122 Hadrian's Wall in Britain begun

180 Death of Marcus Aurelius

286 Division of Empire by Diocletian

324 Founding of Constantinople
378 Battle of Adrianople

455 Vandals sack Rome

476 Traditional "fall" of Rome

520 Anthemius of Tralles and Isidore of Miletus
524 Death of Boethius
529 Closing of the schools at Athens
560 Eutocius' commentaries on Archimedes (approx.)
628 Brahma-sphuta-siddhânta
662 Bishop Sebokht mentioned Hindu numerals
735 Death of the Venerable Bede
775 Hindu works translated into Arabic
830 Al-Khowarizmi: *Algebra* (approx.)
901 Death of Thabit ibn-Qurra
998 Death of abu'l-Wefa
1037 Death of Avicenna
1039 Death of Alhazen
1048 Death of al-Biruni
1114 Birth of Bhaskara
1123 Death of Omar Khayyam
1142 Adelard of Bath translated Euclid
1202 Fibonacci: *Liber abaci*
1260 Campanus' trisection (approx.)
Jordanus Nemorarius: *Arithmetica* (approx.)
1270 Wm. of Moerbeke translated Archimedes (approx.)
1274 Death of Nasir Eddin

496 Clovis adopted Christianity
526 Death of Theodoric
529 Founding of the monastery at Monte Cassino
532 Building of Hagia Sophia by Justinian
590 Gregory the Great elected pope
622 Hejira of Mohammed
641 Library at Alexandria burned
732 Battle of Tours
814 Death of Charlemagne
910 Benedictine abbey at Cluny
987 Accession of Hugh Capet
999 Gerbert became Pope Sylvester II
1028 School of Chartres
1066 Battle of Hastings
1096 First Crusade
1100 Henry I of England crowned
1170 Murder of Thomas à Becket
1204 Crusaders sack Constantinople
Death of Maimonides
1215 Magna Carta
1265 "First" parliament in England
1271 Travels of Marco Polo; mechanical clocks (approx.)

1303 Chu Shi-kié and the Pascal triangle
1328 Bradwardine: *Liber de proportionibus*
1336 Death of Richard of Wallingford

1360 Oresme's latitude of forms (approx.)

1436 Death of al-Kashi

1464 Death of Nicholas of Cusa
1472 Peurbach: *New Theory of the Planets*

1476 Death of Regiomontanus
1482 First printed Euclid

1484 Chuquet: *Triparty*

1489 Use of + and − by Widmann
1492 Use of decimal point by Pellos
1494 Pacioli: *Summa*

1525 Rudolff: *Coss*
1526 Death of Scipione dal Ferro
1527 Apian published the Pascal triangle

1543 Tartaglia published Moerbeke's *Archimedes*
Copernicus: *De revolutionibus*
1544 Stifel: *Arithmetica integra*
1545 Cardan: *Ars magna*

1557 Recorde: *Whetstone of Witte*

1564 Birth of Galileo

1572 Bombelli: *Algebra*
1579 Viète: *Canon mathematicus*

1585 Stevin: *La disme*
Harriot's report on "Virginia"

1595 Pitiscus: *Trigonometria*

1603 Death of Viète
1609 Kepler: *Astronomia nova*
1614 Napier's logarithms

1286 Invention of eyeglasses (approx.)

1348 The Black Death

1364 Death of Petrarch
1431 Joan of Arc burned

1440 Invention of printing
1453 Fall of Constantinople

1473 Sistine Chapel

1483 Murder of the princes in the Tower

1485 Henry VII, first Tudor

1492 Discovery of America by Columbus

1498 Execution of Savonarola
1517 Protestant Reformation
1520 Field of the Cloth of Gold

1534 Act of Supremacy
1543 Vesalius: *De fabrica*
Ramus: *Reproof of Aristotle*

1553 Servetus burned at Geneva

1558 Accession of Elizabeth I
1564 Birth of Shakespeare; deaths of Vesalius and Michelangelo
1572 Massacre of St. Bartholomew

1584 Assassination of William of Orange

1588 Drake defeated the Spanish Armada

1598 Edict of Nantes
1603 Deaths of Wm. Gilbert and Elizabeth I
1609 Galileo's telescope

1620 Bürgi's logarithms

1629 Fermat's method of maxima and minima
1631 Harriot: *Artis analyticae praxis*
Oughtred: *Clavis mathematicae*
1635 Cavalieri: *Geometria indivisibilibus*

1637 Descartes: *Discours de la méthode*
1639 Desargues: *Bruillon projet*
1640 *Essay pour les coniques* of Pascal
1642 Birth of Newton; death of Galileo

1647 Deaths of Cavalieri and Torricelli

1655 Wallis: *Arithmetica infinitorum*
1657 Neil rectified his parabola
1658 Huygens' cycloidal pendulum clock

1667 Gregory: *Geometriae pars universalis*
1668 Mercator: *Logarithmotechnia*
1670 Barrow: *Lectiones geometriae*
1672 Assassination of De Witt
1678 Ceva's theorem

1684 Leibniz' first paper on the calculus

1687 Newton: *Principia*

1690 Rolle: *Traité d'algèbre*
1696 Brachistochrone (the Bernoullis)
L'Hospital's rule

1706 Use of π by William Jones

1715 Taylor: *Methodus incrementorum*
1718 De Moivre: *Doctrine of Chances*
1722 Cotes: *Harmonia mensurarum*
1730 Stirling's formula
1731 Clairaut on skew curves

1616 Deaths of Shakespeare and Cervantes
1620 Landing of the Pilgrims
1626 Deaths of Francis Bacon and Willebrord Snell
1628 Harvey: *De motu cordis et sanguinis*

1636 Harvard College founded

1643 Accession of Louis XIV
1644 Torricelli's barometer

1649 Charles I beheaded
1651 Hobbes: *Leviathan*
Von Guericke's air pump

1660 The Restoration
1662 Royal Society founded
1666 Académie des Sciences founded

1679 Writ of Habeas Corpus
1682 *Acta eruditorum* founded
1683 Seige of Vienna

1685 Revocation of the Edict of Nantes

1689 The Glorious Revolution

1699 Death of Racine
1702 Opening of Queen Anne's War

1711 Birth of Hume

1718 Fahrenheit's thermometer

1730 Rèaumur's thermometer

1733	Saccheri: *Euclid Vindicated*		
1734	Berkeley: *The Analyst*		
		1737	Linnaeus: *Systema naturae*
		1738	Daniel Bernoulli: *Hydrodynamica*
		1740	Accession of Frederick the Great
1742	Maclaurin: *Treatise of Fluxions*	1742	Centigrade thermometer
1743	D'Alembert: *Traité de dynamique*		
1748	Euler: *Introductio;* Agnesi: *Istituzioni*		
		1749	Volume I of Buffon's *Histoire naturelle*
1750	Cramer's rule; Fagnano's ellipse		
		1751	Volume I of Diderot's *Encyclopédie*
		1752	Franklin's kite
1759	*Die freye Perspektive* of Lambert		
		1767	Watt's improved steam engine
1770	Hyperbolic trigonometry		
		1774	Discovery of Oxygen (Priestley, Scheele, Lavoisier)
		1776	American Declaration of Independence
1777	Buffon's needle problem		
1779	Bézout on elimination		
		1781	Discovery of Uranus by Herschel
		1783	Composition of water (Cavendish, Lavoisier)
1788	Lagrange: *Mécanique analytique*		
		1789	French Revolution
1794	Legendre: *Eléments de géométrie*	1794	Lavoisier guillotined
1795	Monge: *Feuilles d'analyse*	1795	École Polytechnique; École Normale
1796	Laplace: *Système du monde*	1796	Vaccination (Jenner)
1797	Lagrange: *Fonctions analytiques* Mascheroni: *Geometria del compasso* Wessel: *Essay on . . . direction* Carnot: *Métaphysique du calcul*		
		1799	Metric system
		1800	Volta's battery
1801	Gauss: *Disquisitiones arithmeticae*	1801	Ceres discovered
1803	Carnot: *Géométrie de position*	1803	Dalton's atomic theory
		1804	Napoleon crowned emperor
1810	Volume I of Gergonne's *Annales*		
		1814	Fraunhofer lines
1815	"The Analytical Society" at Cambridge	1815	Battle of Waterloo
1817	Bolzano: *Rein analytischer Beweis*	1817	Optical transverse vibrations (Young and Fresnel)
		1820	Oersted discovered electromagnetism
1822	Poncelet: *Traité;* Fourier series; Feuerbach's theorem		
1826	Crelle's *Journal* founded Principle of Duality (Poncelet, Plücker, Gergonne) Elliptic functions (Abel, Gauss, Jacobi)	1826	Ampère's work in electrodynamics

1827	Homogeneous coordinates (Möbius, Plücker, Feuerbach)	1827	Ohm's law
	Cauchy: *Calculus of Residues*		
1828	Green: *Electricity and Magnetism*	1828	Synthesis of urea by Wöhler
1829	Lobachevskian geometry	1829	Death of Thomas Young
	Death of Abel at age 26		
1830	Peacock: *Algebra*	1830	Lyell: *Principles of Geology*
			Comte: *Cours de philosophie positive*
		1831	Faraday's electromagnetic induction
1832	Bolyai: *Absolute Science of Space*	1832	Babbage's analytical engine
	Death of Galois at age 20		
1834	Steiner became professor at Berlin		
1836	Liouville's *Journal* founded	1836	First telegraph
1837	*Cambridge and Dublin Mathematical Journal*		
		1842	Conservation of energy (Mayer and Joule)
1843	Hamilton's quaternions		
1844	Grassmann: *Ausdehnungslehre*		
		1846	Discovery of Neptune (Adams and Leverŕier)
			Use of anesthesia
1847	Von Staudt: *Geometrie der Lage*		
		1848	Marx: *Communist Manifesto*
		1850	Dickens: *David Copperfield*
1852	Chasles: *Traité de géométrie supérieure*		
1854	Riemann's Habilitationschrift		
	Boole: *Laws of Thought*		
1855	Dirichlet succeeded Gauss at Göttingen		
		1858	Atlantic cable
		1859	Darwin: *Origin of Species*
			Chemical spectroscopy (Bunsen and Kirchhoff)
1863	Cayley appointed at Cambridge		
1864	Weierstrass appointed at Berlin		
		1868	Cro-Magnon caves discovered
		1869	Opening of Suez Canal
			Mendeleef's periodic table
1872	Dedekind: *Stetigkeit und irrationale Zahlen*		
	Heine: *Elemente*		
	Méray: *Nouveau preçis*		
	Klein's Erlanger Programm		
1873	Hermite proved e transcendental	1873	Maxwell: *Electricity and Magnetism*
1874	Cantor's *Mengenlehre*		
		1876	Bell's telephone
1877	Sylvester appointed at Johns Hopkins		
1881	Gibbs: *Vector Analysis*		
1882	Lindemann proved π transcendental		

1884	Frege: *Grundlagen der Arithmetik*		
		1887	Discovery of herzian waves
1888	Beginnings of American Mathematical Society	1888	Pasteur Institute founded
1889	Peano's axioms		
1895	Poincaré: *Analysis situs*	1895	Discovery of X-rays (Roentgen)
1896	Prime number theorem proved (Hadamard and De la Vallée-Poussin)	1896	Discovery of radioactivity (Becquerel)
		1897	Discovery of electron (J. J. Thomson)
		1898	Discovery of radium (Marie Curie)
1899	Hilbert: *Grundlagen der Geometrie*		
1900	Hilbert's problems Volume I of Russell and Whitehead: *Principia*	1900	Freud: *Die Traumdeutung*
		1901	Planck's quantum theory
1903	Lebesgue integration	1903	First powered air flight
		1905	Special relativity (Einstein)
1906	Functional calculus (Fréchet)		
1907	Brouwer and intuitionism		
1914	Hausdorff: *Grundzüge der Mengenlehre*	1914	Assassination of Austrian Archduke
		1915	Panama Canal opened
1916	Einstein's general theory of relativity		
1917	Hardy and Ramanujan on theory of numbers	1917	Russian Revolution
1923	Banach spaces		
		1927	Lindbergh flew the Atlantic
		1928	Fleming discovered penicillin
1930	Weyl succeeded Hilbert at Göttingen		
1931	Gödel's theorem		
1933	Weyl resigned at Göttingen	1933	Hitler became chancellor
1934	Gelfond's theorem		
1939	Volume I of Bourbaki: *Eléments*		
		1941	Pearl Harbor
		1945	Bombing of Hiroshima
		1946	First meeting of U.N.
1955	Homological algebra (Cartan and Eilenberg)		
1963	Paul J. Cohen on continuum hypothesis	1963	Assassination of President Kennedy
		1965	Death of Sir Winston Churchill
1966	15th International Congress of Mathematicians (Moscow)		
		1967*	Six Day War—MidEast
		1969	People land on the moon
1970	Death of Bertrand Russell		
		1972	Nixon visits China

*Entries from 1967 to the present have been added by the editors of this anthology.

1974	Deligne's proof of the modified Riemann hypothesis		
		1975	Apollo/Soyuz meet in space
1976	Proof of the 4-color conjecture by Appel and Haken		
1978	Death of Kurt Gödel		
		1979	Shah of Iran deposed
1980	Classification of finite simple groups		
1983	Proof of Mordell's theorem by Faltings		

part one

Historical Sketches

THE HISTORY OF mathematics, while somewhat less colorful than that of humanity, is very nearly as long. The construction of regular geometric figures (such as circles, triangles, and ellipses), counting by means of tally sticks, and a wide variety of other mathematical activities predate recorded history nearly everywhere. Peoples as diverse as the builders of Stonehenge in southwest England and the Anasazi Indians of New Mexico and Arizona left physical monuments that hint of substantial mathematical insights. The independent discoveries of the Pythagorean and other theorems in Babylon, China, and elsewhere suggest that mathematical discovery may be a fundamental human activity—a way for humanity to impose order and stability on a capricious universe, a practice as widespread and ancient as religion itself. The essays in this first part provide a number of insights into specific aspects of the recorded history of mathematics.

The accounts begin with the earliest documented mathematical activities in Egypt and Babylon. While scholarly debate constantly rearranges the skyline at the farthest edge of recorded history, it can be stated with some confidence that about 5,000 years ago, that is, 3000 B.C., *the Egyptians possessed a serviceable and surprisingly sophisticated arithmetic. As their monumental history unfolded, this body of knowledge was refined, expanded, and developed to eventually include the rudiments of what we would call geometry. The Babylonian civilizations—Mesopotamian might be more accurate—were essentially contemporary with the Egyptian culture, but the Babylonians apparently developed independently an even more useful and advanced arithmetic than that of the Egyptians. The Babylonians computed a variety of square roots, solved second degree equations, solved systems of first degree equations, knew the Pythagorean theorem* ($a^2 + b^2 = c^2$ *for the lengths* a, b, c *of a right triangle), and knew that the distance around a circle was at least 3-1/8 times the distance across it. They may have understood a great deal more; our knowledge of their civilization is limited by the lack of relevant documents.*

From its recorded beginnings in Egypt and Mesopotamia the development of mathematics in the West has had a relatively continuous trace whose gross outline can be discerned. The mathematical lore of the major Near Eastern civilizations was widely transmitted throughout the Mediterranean area. That this lore provided seed for the amazing development of mathematics in Greece beginning about the sixth century B.C. *can hardly be doubted although direct links are difficult to find. Whatever their initial inspiration, the Greeks created an abstract discipline of power, beauty, subtlety, and rigor such that mathematics as we now know it may fairly be said to have started then and there. For centuries after Greece's golden age, the study of mathematics in Europe was almost exclusively the rediscovery of bits and fragments of those sublime Greek achievements.*

The mathematical heritage of the world outside western Europe and the Hellenistic world is less well focused. It is well known that the highly developed civilizations of both ancient India and China had in their possession a substantial body of mathematical knowledge. However, the precise outlines of this body of knowledge and its development and dissemination are much less clear. The reasons for this are various and well beyond the scope of this introduction. The mathematics of the pre-Columbian Americas is also presently unknown although existing physical ruins would suggest a level of knowledge at least on a par with that of ancient Egypt.

In contrast to our rather clouded view of the Far East, the development of mathematics in the Islamic world and its essential contribution to the development of math in Europe are well documented. Much of what western Europe learned of Greek mathematics came not directly from Greek documents but from Islamic sources. This is not to suggest that Islamic cultures served only as a conduit for Greek achievements. They created much of their own—for example, our present system of writing numbers is an Islamic creation. But the Islamic civilization, too, was heavily indebted to Greek writings for much of its mathematics. That they were able to add significantly to what they inherited from the Greeks is a tribute both to their genius and to the universality of mathematics as a creative art.

The civilization of western Europe and its modern offspring in the Americas, eastern Europe, and Japan have, over the last 500 years, expanded their Greek/Islamic heritage into the amazing structure that is modern mathematics. While it is difficult to find any major western European contribution to mathematics prior to the fourteenth century, it is equally difficult to find any mathematical accomplishment after that time that was not either begun or duplicated in western Europe. From the time Cardano published the general solution for cubic equations in 1545, nothing that has been done outside of Western civilization has had any significance at all in the development of mathematics. For example, although Chinese mathematics predated Newton and Leibniz in discussing some aspects of calculus, this event had absolutely no impact on the development of mathematics anywhere in the world, including China. The history of mathematics after 1500 is the history of mathematics in western Europe, and our emphasis on this area is the result not of cultural bias but of historical fact.

One cannot hope to gain from the vignettes presented here an accurate view of the history and development of mathematics. What we hope these articles will accomplish is arousal of the reader's curiosity to learn more; in learning more the reader may appreciate better the monument of ideas that is mathematics.

Egyptian Mathematics and Astronomy

Otto Neugebauer

Otto Neugebauer is the world's leading authority on the history of the exact sciences in Egypt and Babylon. Born and educated in Germany, he currently resides in the United States.

Neugebauer may disconcert the reader when he states, "The fact that Egyptian mathematics did not contribute positively to the development of mathematical knowledge does not imply that it is of no interest to the historian." Hearing a man announce that he has come to bury Caesar and not to praise him, one might be tempted to just ignore his discourse entirely. But Neugebauer's interest is really in astronomy, and he wants to explain why the Egyptians did not develop a competent astronomy. As a historian Neugebauer knows of the Egyptians' interest in the field and he knows of the great achievements made by Greek and late Babylonian astronomers. Thus he presents a history of primitive mathematical operations, done for routine problems, lacking an overall theory of proof or abstractness.

It is of interest to compare in Neugebauer's and Gillings's articles their choice of examples and the development of each discussion from the standpoint of what each author is trying to accomplish, since they have such different attitudes toward Egyptian mathematics.

Neugebauer also presents a satirical letter from a papyrus of the New Kingdom. One can easily imagine it being updated and read on the floor of the U.S. Senate as an example of the uselessness of the new math now taught in some grammar schools. Apparently some things never change.

OF ALL THE civilizations of antiquity, the Egyptian seems to me to have been the most pleasant. The excellent protection which desert and sea provide for the Nile valley prevented the excessive development of the spirit of heroism which must often have made life in Greece hell on earth. There is probably no other country in the ancient world where cultivated life could be maintained through so many centuries in peace and security. Of course not even Egypt was spared from severe outside and interior struggles; but, by and large, peace in Mesopotamia or Greece must have been as exceptional a state as war in Egypt.

It is not surprising that the static character of Egyptian culture has often been emphasized. Actually there was as little innate conservatism in Egypt as in any other human society. A serious student of Egyptian language, art, religion, administration, etc. can clearly distinguish continuous change in all aspects of life from the early dynastic periods until the time when Egypt lost its independence and eventually became submerged in the Hellenistic world.

The validity of this statement should not be contested by reference to the fact that mathematics and astronomy played a uniformly insignificant role in all periods of Egyptian history. Otherwise one should deny the development of art and architecture during the Middle Ages on the basis of the invariably low level of the sciences in

Source: Excerpt reprinted from Otto Neugebauer, *The Exact Sciences in Antiquity*, by permission of University Press of New England. Second edition published 1957 by Brown University Press. (Some references omitted.)

Western Europe. One must simply realize that mathematics and astronomy had practically no effect on the realities of life in the ancient civilizations. The mathematical requirements for even the most developed economic structures of antiquity can be satisfied with elementary household arithmetic which no mathematician would call mathematics. On the other hand the requirements for the applicability of mathematics to problems of engineering are such that ancient mathematics fell far short of any practical application. Astronomy on the other hand had a much deeper effect on the philosophical attitude of the ancients in so far as it influenced their picture of the world in which we live. But one should not forget that to a large extent the development of ancient astronomy was relegated to the status of an auxiliary tool when the theoretical aspects of astronomical lore were eventually dominated by their astrological interpretation. The only practical applications of theoretical astronomy may be found in the theory of sun dials and of mathematical geography. There is no trace of any use of spherical astronomy for a theory of navigation. It is only since the Renaissance that practical aspects of mathematical discoveries and the theoretical consequences of astronomical theory have become a vital component in human life.

The fact that Egyptian mathematics did not contribute positively to the development of mathematical knowledge does not imply that it is of no interest to the historian. On the contrary, the fact that Egyptian mathematics has preserved a relatively primitive level makes it possible to investigate a stage of development which is no longer available in so simple a form, except in the Egyptian documents.

To some extent Egyptian mathematics has had some, though rather negative, influence on later periods. Its arithmetic was widely based on the use of unit fractions, a practice which probably influenced the Hellenistic and Roman administrative offices and thus spread further into other regions of the Roman empire, though similar methods were probably developed more or less independently in other regions. The handling of unit fractions was certainly taught wherever mathematics was included in a curriculum. The influence of this practice is visible even in the works of the stature of the Almagest, where final results are often expressed with unit fractions in spite of the fact that the computations themselves were carried out with sexagesimal fractions. Sometimes the accuracy of the results is sacrificed in favor of a nicer appearance in the form of unit fractions. And this old tradition doubtless contributed much to restricting the sexagesimal place value notation to a purely scientific use.

There are two major results which we obtain from the study of Egyptian mathematics. The first consists in the establishment of the fact that the whole procedure of Egyptian mathematics is essentially additive. The second result concerns a deeper insight into the development of computation with fractions. We shall discuss both points separately.

What we mean by the "additivity" of Egyptian mathematics can easily be explained. For ordinary additions and subtractions nothing needs to be said. It simply consists in the proper collection and counting of the marks for units, tens, hundreds, etc., of which Egyptian number signs are composed. But also multiplication and division are reduced to the same process by breaking up any higher multiple into a sum of consecutive duplications. And each duplication is nothing but the addition of a number to itself. Thus a multiplication by 16 is carried out by means of four consecutive duplications, where only the last partial result is utilized. A multiplication by 18 would add the results for 2 and for 16 as shown in the following example

	1	25
/	2	50
	4	100
	8	200
/	16	400
	total	450

In general, multiplication is performed by breaking up one factor into a series of duplications. It certainly never entered the minds of the Egyptians to ask whether this process will always work. Fortunately it does; and it is amusing to see that modern computing machines have again made use of this "dyadic" principle of multiplication. Division is, of course, also reducible to the same method because one merely asks for a factor which is needed for one given number in order to obtain the second given number. The division of 18 by 3 would simply mean to double 3 until the total 18 can be reached

	1	3
/	2	6
/	4	12
	total	18,

and the result is 2 + 4 = 6. Of course, this process might not always work so simply and fractions must be introduced. To divide 16 by 3 one would begin again with

/	1	3
	2	6
/	4	12

and thus find 1 + 4 = 5 as slightly below the requested solution. What is still missing is obviously 16 − 15 = 1, and to this end the Egyptian computer would state

	$\bar{\bar{3}}$	2
/	$\bar{3}$	1

which means that 2/3 of 3 is 2, 1/3 of 3 is 1 and thus he would find 5 $\bar{3}$ as the solution of his problem.

Here we have already entered the second problem, operations with fractions. . . . Egyptian fractions are always "unit fractions," with the sole exception of 2/3 which we always include under this name in order to avoid clumsiness of expression. The majority of these numbers are written by means of the ordinary number signs below the hieroglyph ⬭ "r" meaning something like "part." We write therefore $\bar{5}$ for the expression "5th part" = 1/5. For 2/3 we write $\bar{\bar{3}}$ whereas the Egyptian form would be "2 parts" meaning "2 parts out of 3", i. e., 2/3. There exist special signs for 1/2 and 1/4 which we could properly represent by writing "half" and "quarter" but for the sake of simplicity we use $\bar{2}$ and $\bar{4}$ as for all other unit fractions.

We shall not go into the details of the Egyptian procedures for handling these fractions. But a few of the main features must be described in order to characterize this peculiar level of arithmetic. If, e.g., $\bar{3}$ and $\overline{15}$ should be added, one would simply leave $\bar{3}$ $\overline{15}$ as the result and never replace it by any symbol like 2/5. Again $\bar{\bar{3}}$ forms an exception in so far as the equivalence of $\bar{2}$ $\bar{6}$ and $\bar{\bar{3}}$ is often utilized.

Every multiplication and division which involves fractions leads to the problem of how to double unit fractions. Here we find that twice $\bar{2}$, $\bar{4}$, $\bar{6}$, $\bar{8}$, etc. are always directly replaced by 1, $\bar{2}$, $\bar{3}$, $\bar{4}$, respectively. For twice $\bar{3}$ one has the special symbol $\bar{\bar{3}}$. For the doubling of $\bar{5}$, $\bar{7}$, $\bar{9}$, . . . however, special rules are followed which are explicitly summarized in one of our main sources, the mathematical Papyrus Rhind. One can represent these rules in the form of a table which gives for every odd integer n the expression for twice $\bar{n}$.

This table has often been reproduced and we may restrict ourselves to a few lines at the beginning:

n	twice $\bar{n}$
3	$\bar{2}$ $\bar{6}$
5	$\bar{3}$ $\overline{15}$
7	$\bar{4}$ $\overline{28}$
9	$\bar{6}$ $\overline{18}$
etc.	

The question arises why just these combinations were chosen among the infinitely many possibilities of representing $2/n$ as the sum of unit fractions.

I think the key to the solution of this problem lies in the separation of all unit fractions into two classes, "natural" fractions and "algorithmic" fractions, combined with the previously described technique of consecutive doubling and its counterpart, consistent halving. As "natural" fractions I consider the small group of fractional parts which are singled out by special signs or special expressions from the very beginning, like $\bar{\bar{3}}$, $\bar{3}$, $\bar{2}$ and $\bar{4}$. These parts are individual units which are considered basic concepts on an equal level with the integers. They occur everywhere in daily life, in counting and measuring. The remaining fractions, however, are the unavoidable consequence of numerical operations, of an "algorism," but less deeply rooted in the elementary concept of numerical entities. Nevertheless there are "algorithmic" fractions which easily present themselves, namely, those parts which originate from consistent halving. This process is the simple analogue to consistent duplication upon which all operations with integers are built. Thus we obtain two series of fractions, both directly derived from the "natural" fractions by consecutive halving. One sequence is $\bar{\bar{3}}$, $\bar{3}$, $\bar{6}$, $\overline{12}$, etc., the other $\bar{2}$, $\bar{4}$, $\bar{8}$, $\overline{16}$, etc. The importance of these two series is apparent everywhere in Egyptian arithmetic. A drastic example has already been quoted above on p. 5 where we found that $\bar{3}$ of 3 was found by stating first that $\bar{\bar{3}}$ of 3 is 2 and only as a second step $\bar{3}$ of 3 is 1. This arrangement $\bar{\bar{3}} \rightarrow \bar{3}$ is standard even if it seems perfectly absurd to us. It emphasizes the completeness of the first sequence and its origin from the "natural" fraction $\bar{\bar{3}}$.

If one now wishes to express twice a unit fraction, say $\bar{5}$, as a combination of other fractional parts, then it seems natural again to have recourse to these two main sequences of fractions. Thus one tries to represent twice $\bar{5}$ as the sum of a natural fraction of $\bar{5}$ and some other fraction which must be found in one way or another. At this early stage, some trials were doubtless made until the proper solution was found. I think one may reconstruct the essential steps as follows. We operate with the natural fraction $\bar{3}$, after other experiments (e.g., with $\bar{2}$) have failed. Two times $\bar{5}$ may thus be represented as $\bar{3}$ of $\bar{5}$ or $\overline{15}$ plus a remainder which must complete the factor 2 and which is 1 $\bar{\bar{3}}$. The question of finding 1 $\bar{\bar{3}}$ of $\bar{5}$ now arises. This is done in Egyptian mathematics by counting the thirds and writing their number in red ink below the higher units, in our case

1 $\bar{\bar{3}}$ (written in black)
3 2 (written in red).

This means that 1 contains 3 thirds and $\bar{\bar{3}}$ two thirds. Thus the remaining factor contains a total of 5 thirds. This is the amount of which $\bar{5}$ has to be taken. But 5 fifths are one complete unit and this was a third of the original higher unit. Thus we obtain for the second part simply $\bar{3}$ and thus twice $\bar{5}$ is represented as $\bar{3}$ $\overline{15}$. This is exactly what we find in the table.

For the modern reader it is more convenient to repeat these clumsy conclusions with modern symbols though we must remember that this form of expression is totally unhistorical. In order to represent 2/5 in the form of $\frac{1}{m} + \frac{1}{x}$ we chose $\frac{1}{m}$ as a natural fraction of 1/5, in this case $1/3 \cdot 1/5 = 1/15$. For the remaining fraction we have

$$\frac{1}{x} = \left(1 + \frac{2}{3}\right)\frac{1}{5} = \frac{5}{3} \cdot \frac{1}{5} = \frac{1}{3}.$$

Thus we have the representation

$$\frac{2}{5} = \frac{1}{15} + \frac{1}{3}$$

of the table. In general we have

$$\frac{2}{n} = \frac{1}{3} \cdot \frac{1}{n} + \frac{5}{3} \cdot \frac{1}{n}$$

and the second term on the right-hand side will be a unit fraction when and only when n is a multiple of 5. In other words a trial with the natural fraction 1/3 will only work if n is a multiple of 5. This is indeed confirmed in all cases available in the table of the Papyrus Rhind which covers all expressions for $\frac{2}{n}$ from $n = 3$ to $n = 101$.

We may operate similarly with the natural fraction 1/2. Then we have

$$\frac{2}{n} = \frac{1}{2} \cdot \frac{1}{n} + \frac{3}{2} \cdot \frac{1}{n}$$

which shows that we obtain a unit fraction on the right-hand side if n is divisible by 3. For $n = 3$ we obtain

$$\frac{2}{3} = \frac{1}{6} + \frac{1}{2}$$

and this is the relation $\bar{\bar{3}} = \bar{2}\,\bar{6}$ which we quoted at the beginning. All other cases in the table for $\bar{n}$'s which are multiples of 3 show the same decomposition operating with $\frac{1}{2n}$ as one term.

It is clear that one can proceed in the same manner by operating with $\bar{4}$, $\bar{8}$, etc. or with $\bar{6}$, $\overline{12}$, etc. In this way, more and more cases of the table can be reached and it seems to me there is little doubt that we have found in essence the procedure which has led to these rules for the replacement of $2/n$ by sums of unit fractions.

For our present purposes it is not necessary to discuss in detail all steps in the structure of Egyptian fractional arithmetic. I hope, however, to have made clear the two leading principles, the strict additivity and the extensive use of the "natural fractions."

A few historical remarks must be added. The Papyrus Rhind is not our only document for the study of Egyptian arithmetic. The other large text, the Moscow papyrus, agrees with rules known from the Papyrus Rhind. We have, however, an ostracon from the early part of the New Kingdom where the duplication of $\overline{7}$ is given as $\overline{6}\ \overline{14}\ \overline{21}$ instead of $\overline{4}\ \overline{28}$ of the standard rule. Much more material is available from Demotic and Greek papyri of the Hellenistic period. Here again, deviations from the earlier rules can be observed, though the main principle remains the same. In other words we cannot assume that once and forever a system of fractional tables was computed and then rigidly maintained. Obviously several equivalent forms were slowly developed but without ever seriously transgressing the original methods. This latter fact is of great historical importance. The handling of fractions always remained a special art in Egyptian arithmetic. Though experience teaches one very soon to operate quite rapidly within this framework, one will readily agree that the methods exclude any extensive astronomical computations comparable to the enormous numerical work which one finds incorporated in Greek and late Babylonian astronomy. No wonder that Egyptian astronomy played no role whatsoever in the development of this field.

It would be quite out of proportion to describe Egyptian geometry here at length. It suffices to say that we find in Egypt about the same elementary level we observed in contemporary Mesopotamia. The areas of triangles, trapezoids, rectangles, etc. are computed, and for the circle a rule is used which we can transcribe as $A = \left(\frac{8}{9}d\right)^2$ if d denotes the diameter. Corresponding formulae for the elementary volumes were known, including a correct numerical computation for the volume of a truncated pyramid. This, as well as the relatively accurate value 3.16 for π resulting from the above formula, give Egyptian geometry a lead over the corresponding arithmetical achievements. It has even been claimed that the area of a hemisphere was correctly found in an example of the Moscow papyrus, but the text admits also of a much more primitive interpretation which is preferable.

A vivid description of the main topics of Egyptian mathematics is given in a papyrus of the New Kingdom, written for school purposes. It is a satirical letter in which an official ridicules a colleague. The section on mathematics runs as follows:* "Another topic. Behold, you come and fill me with your office. I will cause you to know how matters stand with you, when you say 'I am the scribe who issues commands to the army.'

"You are given a lake to dig. You come to me to inquire concerning the rations for the soldiers, and you say 'reckon it out.' You are deserting your office, and the task of teaching you to perform it falls on my shoulders.

"Come, that I may tell you more than you have said: I cause you to be abashed (?) when I disclose to you a command of your lord, you, who are his Royal Scribe, when you are led beneath the window (of the palace, where the king issues orders) in respect of any goodly (?) work, when the mountains are disgorging great monuments for Horus (the king), the lord of the Two Lands (Upper and Lower Egypt). For see, you are the clever scribe who is at the head of the troops. A (building-) ramp is to be constructed, 730 cubits long, 55 cubits wide, containing 120 compartments, and filled with reeds and beams; 60 cubits high at its summit, 30 cubits in the middle, with a batter of twice 15 cubits and its pavement 5 cubits. The quantity of bricks needed for it is asked of the generals, and the scribes are all asked together, without one of them knowing anything. They all put their trust in you and say, 'You are the clever scribe, my friend! Decide for us quickly! Behold your name is famous; let none be found in this place to magnify the other thirty! Do not let it be said of you that there are things which even you do not know. Answer us how many bricks are needed for it?

"See, its measurements (?) are before you. Each one of its compartments is 30 cubits and is 7 cubits broad."

*From Erman, A., The Literature of the Ancient Egyptians: Poems, Narratives, and Manuals of Instruction from the Third and Second Millenia B.C., E. P. Dutton, New York, 1927.

On the whole, one can repeat here what we have already said for Babylonian geometry. Problems concerning areas or volumes do not constitute an independent field of mathematical research but are only one of the many applications of numerical methods to practical problems. There is no essential difference between the determination of the acreage of a field in special measures and the distribution of beer to temple personnel according to different ratings. This is a state of affairs which holds to a large extent even in the Hellenistic period and far beyond it. In Arabic mathematics the "inheritance" problems play an important role, while similar examples are found already in Old-Babylonian texts. The geometrical writings of Heron, whether authentic or merely ascribed to him, contain whole chapters on units, weights, measurements, etc. Of course, since the Hellenistic period, even the writings of Heron and related documents show the influence of scientific Greek geometry. But, by and large, one has to distinguish two widely separate types of "Greek" mathematics. One is represented by the strictly logical approach of Euclid, Archimedes, Apollonius, etc.; the other group is only part of general Hellenistic mathematics, the roots of which lie in the Babylonian and Egyptian procedures. The writings of Heron and Diophantus and works known only from fragments or from papyrus documents form part of this Oriental tradition which can be followed into the Middle Ages both in the Arabic and in the western world. "Geometry" in the modern sense of this word owes very little to the modest amount of basic geometrical knowledge which was needed to satisfy practical ends. Mathematical geometry got one of its most important stimuli from the discovery of irrational numbers in the 4th or 5th century B.C. and remained rather stagnant from the second century B.C. onwards, except for those additions of spherical geometry and descriptive geometry which were introduced by their astronomical importance. On the other hand, geometrical theory had a negative effect on the algebraic and numerical methods which were part of the Oriental background of Hellenistic science. A real insight into the mutual relations between all these fields was not reached before modern times. The role of Egyptian mathematics is probably best described as a retarding force upon numerical procedures.

Mathematics in the Time of the Pharaohs

Richard J. Gillings

Richard J. Gillings was born in England and educated in Australia. He is the author of over 100 books and articles in the areas of mathematics and the history of mathematics.

We are all familiar with the story of the Rosetta stone. Most people are less familiar with the deciphering of Egyptian mathematical symbols. Gillings gives a clear account of addition, subtraction, multiplication, and division, using hieroglyphics and hieratic and demotic scripts.

One might be tempted to dismiss an account of the four arithmetic operations as hopelessly ancient and of questionable relevance. Aware of such a reader response, Gillings provides an account of how to multiply two numbers as the procedure was taught a mere 400 years ago. The painful ugliness of that account, contrasted with the smoothness and generality of the Egyptian account, will cause the reader to read on with greater sympathy.

We have included Gillings's section "The Nature of Proof" because it provides a fairly successful refutation of the idea that the Egyptians were without a formal proof procedure. Too often all credit for the idea of mathematical proof is given to the Greeks.

Finally, notice that in the footnote to the very first sentence Gillings tells us that he wrote this article with Neugebauer's The Exact Sciences in Antiquity *in mind. It is interesting to compare Neugebauer's extract (see previous article) with Gillings's excerpts here; note the subtle changes in emphasis. Keep in mind that Gillings has a pro-Egyptian bias, and be sure to compare their treatments of Egyptian fractions.*

The Four Arithmetic Operations

Addition and Subtraction. Historians of Egyptian mathematics have seldom committed themselves regarding the methods of addition and subtraction employed by the Egyptian scribes; they mostly take for granted the scribes' ability to add or subtract fairly large numbers.[1] Now we of the twentieth century can count to a million and well past a million, yet *we* have difficulties in addition and subtraction, and we use desk calculators and other aids to computation to overcome them. And we have only to open a modern textbook on the teaching of arithmetic to be surprised at the many different methods for subtraction now being taught. Mankind has always had trouble with these two operations. That is why the abacus was invented centuries ago, and is still being used today in many Asian countries.

In the mathematical papyri that have come down to us, there are many problems proposed and solved which require all four fundamental operations for their solution, and we can quite well deduce the scribes' methods for multiplication and division. But for addition and subtraction methods there are hardly any clues. It would seem that these operations were performed and checked elsewhere by the scribes, and the answers

Source: Excerpts from Richard J. Gillings, *Mathematics in the Time of the Pharaohs* (MIT Press, 1972), pp. 11–23, 232–234.

inscribed on the papyri afterwards. A scribal error in this part of their arithmetic is such a rarity that one can be excused for drawing the conclusion that they had tables for additions (and consequently for subtractions), from which they merely read off the answers. If such tables ever existed, however, no copies have come down to us, so that the idea remains purely a conjecture insofar as integers are concerned. For fractions, of course, many varied tables *must* have existed. The EMLR[2] is itself very good evidence for the existence of such tables for addition and subtraction of fractions. Today most nations write from *left to right,* and our numbers are so written also; but the values of the digits in our "Hindu-Arabic" decimal system increase in place value from *right to left.* So if we have to perform an addition or a subtraction, we begin with the units column on the right, and work toward the left through the tens, the hundreds, the thousands, and so on. In these calculations, including multiplication, we reverse our direction of writing.

Conversely, as we have seen, the Egyptians wrote both their words and numbers from *right to left.* Of necessity, however, the Egyptian arithmeticians, like ourselves, had to start adding in the opposite direction to that in which they were accustomed to write, so the place value of the Egyptians' digits increases from *left to right,* and the Egyptian system therefore runs widdershins to ours.

Figure 1 gives some simple examples of addition comparing Hindu-Arabic, hieroglyphics, and hieratic. In Hindu-Arabic addition, the number combinations 3 + 4 = 7, 6 + 7 = 13 and 6 + 9 = 15, are learned by heart, "look-and-say," so that the addition (4 + 3) is not done by starting with 4, and then counting 5, 6, 7, and then stopping. This counting method offers itself naturally to one performing additions in hieroglyphics. In hieratic, however, counting does not so lend itself, and we ask ourselves:

a. Did they have an addition table?
b. Did they add by simple counting?
c. Did they learn number combinations?

If tables such as that shown in Figure 2 were prepared, they would be equally useful in subtraction, because I am sure the Egyptians did not say to themselves, "From 12 take away 7, answer 5," but rather, "Seven, how many more to make 12? It needs 5," that is, the table would supply the answer.

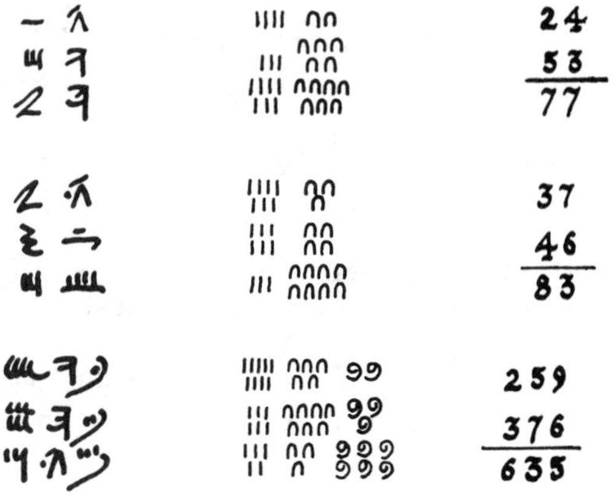

Figure 1 Examples of additions. *Left,* hieratic; *center,* hieroglyphic; *right,* Hindu-Arabic.

It is easy to say that multiplication by 10 was simply performed by changing each 𓏺 to 𓎆, each 𓎆 to 𓍢, each 𓍢 to 𓆼, each 𓆼 to 𓂭, and so on. But this would be true only of hieroglyphics, and in ordinary everyday business, where these calculations would be needed, hieratic—and later, the quicker demotic—was used by the clerk or scribe, and in such writing this simple transfer of signs does not apply.

Again, multiplication by continual doubling, performed by simple duplication of each sign of a number, seems simple enough, even allowing for "carrying," but is feasible only in hieroglyphs, which, as we have seen, were seldom used. We are forced to the conclusion that reference tables were used for the four fundamental operations (which we indicate by +, −, ×, and ÷); and it is probable that portions of these tables were memorized much as we do today. When we add a column of figures, 9 and 5 immediately suggest 4, 8 and 5 suggest 3, 7 and 5 suggest 2, 6 and 5 suggest 1, and in subtraction these particular pairs suggest the same numbers, so that a kind of elementary theory of numbers begins to arise. There are very many examples of the four operations with unit fractions to be found in the pa-

2 9 11	2 8 10	2 7 9	2 6 8	2 5 7	2 4 6	2 3 5
3 9 12	3 8 11	3 7 10	3 6 9	3 5 8	3 4 7	
4 9 13	4 8 12	4 7 11	4 6 10	4 5 9		
5 9 14	5 8 13	5 7 12	5 6 11			
6 9 15	6 8 14	6 7 13				
7 9 16	7 8 15					
8 9 17						

Figure 2 An addition table in hieratic script that could have been constructed by the scribes, and its Hindu-Arabic translation.

pyri, but very few show actual addition and subtraction with integers. In Figure 3 are displayed two additions from RMP[3] Problem 79, the controversial problem thought by some to be the prototype of the Mother Goose rhyme beginning, "As I was going to St. Ives, / I met a man with seven wives. . . ." In the first addition, by chance the units digits are more or less in a ver-

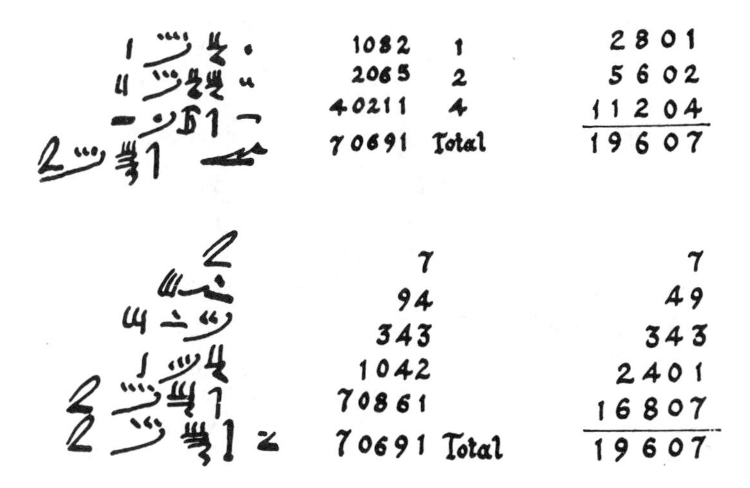

1082 1
2065 2
40211 4
70691 Total
2801
5602
11204
19607
7
94
343
1042
70861
70691 Total
7
49
343
2401
16807
19607

Figure 3 Two additions from Problem 79 of RMP. *Left,* the hieratic; *center,* the transliteration; *right,* in translation.

tical line, and (1 + 2 + 4) = 7 simply enough. There are no tens digits. Then for the hundreds there are (800 + 600 + 200), also more or less in line, which must be written as 6 hundreds with one thousand to be carried. What the scribe's thought process for this step was is the point we are doubtful about. Then for the thousands he had (1,000 + 2,000 + 5,000 + 1,000) = 9,000 with nothing to carry, and of course the one ten thousand was merely written down in the total, the whole of which was pushed toward the left, because the symbol for "total," , being written first, required more space. His multiplier 7, written by the scribe as (1 + 2 + 4), was not recorded in the last line.

In the second addition, all digits are out of alignment, units, tens, hundreds, etc., but the alternative symbol for "total," , has not pushed the answer out toward the left. In reading this addition greater care is necessary. The scribe had (7 + 9 + 3 + 1 + 7), which, however he arrived at it, is 27, so he put down 7 and carried the 2 tens. Notice the working is from left to right. For the tens he had, (20 + 40 + 40) = 100, so that there are no tens in the answer. The Egyptians had no sign for zero, nor did they even leave a space to indicate "no tens." For the hundreds, he had to add (100 + 300 + 400 + 800), which he finds is 1,000 and 600, (however he did it), so he put down 600 and carried the 1,000. Then he had (1,000 + 2,000 + 6,000) = 9,000 with nothing to carry, which goes into the total with the 10,000.

What we have not enough evidence to decide is whether in adding (100 + 300 + 400 + 800) he counted (on his fingers so to speak), one hundred; two, three, four hundreds; five, six, seven, eight hundreds; then eight and eight hundreds makes sixteen hundreds; or whether he thought of units, adding (1 + 3 + 4 + 8) as 16, and then calling them hundreds. Or did he merely read off the answer from handy prepared tables?

Multiplication. It is not uncommon in histories of mathematics to read that Egyptian multiplication was clumsy and awkward, and that this clumsiness and awkwardness was due to the Egyptians' very poor arithmetical notation.

> In Ahmes' treatment of multiplication, he seems to have relied on repeated additions. *Jourdain*[4]

> It is remarkable that the Egyptians, who attained so much skill in their arithmetic manipulations, were unable to devise a fresh notation and less cumbersome methods. *Newman*[5]

> The limitations of this notation made necessary the use of special tables. With such a cumbrous system of fractional notation, calculation was a lengthy process, frequently involving the use of very small fractions. *Sloley*[6]

> Such a calculus with fractions gave to Egyptian mathematics an elaborate and ponderous character, and effectively impeded the further growth of science. *Struik*[7]

If Egyptian multiplication was so clumsy and difficult, how did it come about that these same techniques were still used in Coptic times, in Greek times, and even up to the Byzantine period, a thousand or more years later? No nation, over a period of more than a millennium, was able to improve on the Egyptian notation and methods. The fact is that, despite their notation, the scribes were adept at solving arithmetic problems and were in fact quite skillful in devising ingenious methods of attack on algebraic and geometric problems as well, so that their successors remained content with what came down to them.

How far have we progressed in multiplication since the times of the ancient Egyptians, or even since Greek and Roman times? What are our grounds for being so critical of Egyptian multiplication, in which it was only necessary to use the twice-times tables? In English-speaking countries, at least, as late as the sixteenth century, it was not part of the school curriculum to learn any multiplication tables at all. Samuel Pepys, the famous diarist, educated at St. Paul's School and at Magdalen College, Cambridge University, "an able man of business,"[8] was secretary to the British Admiralty, and in that position must surely have needed to know how to calculate. But note this entry:

> July 4. 1662. . . . Up by 5 o'clock. . . . Comes M. Cooper of whom I intend to learn mathematics, and do, being with him to-day. After an houres being with him at arithmetique, my first attempt being to learn the multiplication table.[9]

If a graduate of Cambridge University was just beginning in his thirtieth year to learn the multiplication tables, what are we to suppose the average schoolchild knew of them? In 1542 the Welshman Robert Recorde published *The Grounde of Artes. Teachyng the Worke and Practice of Arithmetike,*[10] in which he shows how to multiply two numbers between 5 and 10.

MULTIPLY 8 BY 7

First set your digits one over the other.

8
7

Then from the uppermost downwards, and from the nethermost upwards, draw straight lines, so that they make a St. Andrew's cross.

8 ╳
7

Then look how many each of them lacketh of 10, and write that against each of them at the end of the line, and that is called the difference.

8 ╳ 2
7 ╳ 3

I multiply the two differences, saying, "two times three make six," that must I ever set down under the differences.

8 ╳ 2
7 ╳ 3
6

Take from the other digit, (not from his own), as the lines of the cross warn me, and that that is left, must I write under the digits. If I take 2 from 7, or 3 from 8, (which I will, for all is lyke), and there remaineth 5, and then there appeareth the multiplication of 8 times 7 to be 56. A chylde can do it.

8 ╳ 2
7 ╳ 3
5 6

Compare this technique of the sixteenth century A.D. with that of an apprentice scribe of the Hyksos period of the Middle Kingdom, more than 3,000 years earlier.

MULTIPLY 8 BY 7

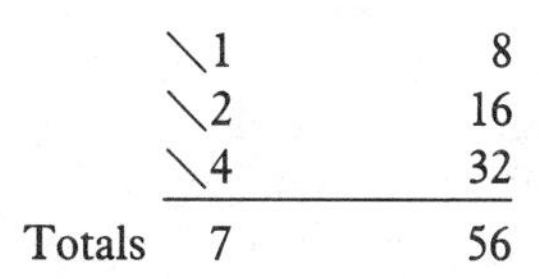

When the Egyptian scribe needed to multiply two numbers, he would first decide which would be the multiplicand, then he would repeatedly multiply this by 2, adding up the intermediate multipliers until they summed to the original multiplier. For example, to multiply 13 and 7, assume the multiplicand to be 13, doubling thus:

1	13
2	26
4	52

Here he would stop doubling, for he would note that a further step would give him a multiplier of 8 which is bigger than 7. In this case he would note that $1 + 2 + 4 = 7$. So he put check marks alongside these multipliers to indicate this:

	\1	13
	\2	26
	\4	52
Totals	7	91

Adding together those numbers in the right-hand column opposite check marks, the scribe would thus obtain the final answer. Had he chosen 7 as the multiplicand, and 13 as the multiplier, his sum would appear as follows:

1	7
2	14
4	28
8	56

Again he would cease doubling at 8, for a further doubling would give 16 which is past 13; then he would note that $1 + 4 + 8 = 13$, and so he would put check marks alongside these multipliers:

	\1	7
	2	14
	\4	28
	\8	56
Totals	13	91

Since 2 is not checked, he took care not to add in the 14 of the right-hand column, where he has 7 + 28 + 56 = 91. The scribe's mental arithmetic had to be pretty accurate, especially for large multipliers, but of course he could keep a check of his totals on a papyritic scribbling pad. These additions were made easier for the scribe by virtue of a special property of the series

1, 2, 4, 8, 16, 32, . . .;

for any integer can be uniquely expressed as the sum of some of its terms. Thus, for example, 19 = 1 + 2 + 16; 31 = 1 + 2 + 4 + 8 + 16; and 52 = 4 + 16 + 32. We do not know whether or not the scribes were explicitly aware of this but they certainly used it, just as do the designers of a modern electronic computer, and this is surely a somewhat sobering thought.

Division. For the Egyptian scribe, the process of division was closely allied to his method of multiplication. Suppose that he wished to divide 184 by 8. The scribe did not say to himself, "I will divide 8 into 184." He said, "By what must I multiply 8 to get 184?" Thus he had,

1	8
2	16
4	32
8	64
16	128

At this state he stops multiplying by 2, for his next doubling would give 256, which is well past 184. Now he must do some mental arithmetic or use his memo pad to locate which numbers of the right-hand column will add up to 184. Finding that 8 + 16 + 32 + 128 = 184, he would place a check mark beside each of these numbers:

1	8/
2	16/
4	32/
8	64
16	128/
Totals 23	184

Then he must add the multipliers corresponding to the checked numbers,

1 + 2 + 4 + 16 = 23,

which is his quotient.

Fractions. When the Egyptian scribe needed to compute with fractions he was confronted with many difficulties arising from the restrictions of his notation. His method of writing numbers did not allow him to write such simple fractions as 3/5 or 5/9, because all fractions had to have unity for their numerators (with one exception[11]). This was because a fraction was denoted by placing the hieroglyph ("r," an open mouth) over any integer to indicate its reciprocal. Thus the number 12, written in hieroglyphs as , became the fraction 1/12 when written as . In the hieratic or handwritten form, in which the scribe used a reed brush and ink, the open mouth became merely a dot, and 1/12 would look like . The dot being so much smaller than the "mouth," it was placed over the first digit of the number (here it is 10), and so in reading numbers in hieratic papyri care must be taken not to think that, say, is (1/10 + 2) instead of 1/12. Such a mistake is even more likely with hundreds or thousands.

Thus all hieroglyphic and hieratic fractions are *unit fractions (stammbruchen)*, and have unit numerators in translation. The fraction 3/4 was written by the Egyptians as 1/2 + 1/4, 6/7 was written as 1/2 + 1/4 + 1/14 + 1/28, etc., for all fractions of the form p/q. To a modern arithmetician this seems unnecessarily complicated, but we shall see that the Egyptian scribes devised means and rules to meet the difficulties of the method as they arose. The exception to the unit-numerator usage—the fraction 2/3—was denoted by a special sign: in hieroglyphics, and in hieratic. There is no doubt that the Egyptians knew that 2/3 was the reciprocal of 1-1/2, as the hieroglyph suggests,[12] for there are many instances, particularly in the RMP, where this relation is specifically shown. Thus:

RMP 33

Since $\overline{\overline{3}}$ of 42 = 28,
then $\overline{28}$ of 42 = 1 $\overline{2}$.

RMP 20

$\overline{\overline{3}}$ of $\overline{24}$ = $\overline{1\,\overline{2}\times 24}$,
= $\overline{36}$.

The numerator 1 of each fraction is here omitted, and $\overline{\overline{3}}$ is 2/3. In multiplying and dividing fractions, the scribes used the same setting out as they did for integers, but they needed to use various techniques for the different problems that arose. In RMP 2, the fraction $\overline{5}$ is explicitly multiplied by 10 (more often multiplications by 10 were written down at once):

1.1	Do it thus	1	$\overline{5}$
1.2		\2	\ $\overline{3}$ $\overline{15}$
1.3		4	$\overline{\overline{3}}$ $\overline{10}$ $\overline{30}$
1.4		\8	\1 $\overline{3}$ $\overline{5}$ $\overline{15}$
1.5	Totals	10	1 $\overline{\overline{3}}$ $\overline{5}$ $\overline{15}$ $\overline{15}$
1.6			2

In line 1.2 the product $2 \times \overline{5}$ is seen to have been recognized by the scribe as equal to the division $2 \div 5$, for the answer ($\overline{3}$ $\overline{15}$) is that given in the RMP Recto[13] Table, which lists such divisors of 2. . . . Lines 1.2 and 1.3 give the double of $\overline{3}$ as $\overline{\overline{3}}$, and $2 \times \overline{15}$ as ($\overline{10}$ $\overline{30}$), which, again from the Recto Table, is the result of the division $2 \div 15$. Lines 1.3 and 1.4 give the double of $\overline{\overline{3}}$ as (1 $\overline{3}$), the double of $\overline{10}$ as $\overline{5}$, and the double of $\overline{30}$ as $\overline{15}$. Check marks were put on lines 1.2 and 1.4,[14] and ($\overline{3}$ $\overline{15}$) + (1 $\overline{3}$ $\overline{5}$ $\overline{15}$) read off as the answer. The scribe would have recognized this quantity to be equal to 2, by some papyritic jottings on his scribbling pad, or merely by referring to his unit fraction tables of addition.

The constant necessity to double fractions in all multiplications gave rise to the construction of the RMP Recto table . . . where unit fraction equivalents of 2 divided by the odd numbers are recorded for the scribe's easy reference. Many of the simpler equalities were no doubt committed to memory, just as were some of the additions of unit fractions, as shown in the EMLR.

In RMP 9, $\overline{2}$ $\overline{14}$ is multiplied by 1 $\overline{2}$ $\overline{4}$.

1.1	\1	\ $\overline{2}$ $\overline{14}$
1.2	\ $\overline{2}$	\ $\overline{4}$ $\overline{28}$
1.3	\ $\overline{4}$	\ $\overline{8}$ $\overline{56}$
1.4	Totals 1 $\overline{2}$ $\overline{4}$	$\overline{2}$ $\overline{4}$ $\overline{8}$ $\overline{14}$ $\overline{28}$ $\overline{56}$
1.5		1

Line 1.2 shows the halving of fractions by merely doubling the denominator numbers, and line 1.3 shows the same. Line 1.4 shows the totals of both columns, and in line 1.5, the scribe wrote at once the answer 1. This final addition of six unit fractions was quite possibly done mentally, for $\overline{14}$ $\overline{28}$ $\overline{56}$ = $\overline{8}$ was a commonplace equality to the scribes,[15] and the steps

$$
\begin{aligned}
\overline{2}\ \overline{4}\ (\overline{8}\ \overline{8}) &= \overline{2}\ (\overline{4}\ \overline{4}) \\
&= (\overline{2}\ \overline{2}) \\
&= 1
\end{aligned}
$$

would have been quite easy for a competent scribe, even though in writing the details it appears rather long. In more difficult multiplications involving fractions, as in 7 $\overline{2}$ $\overline{4}$ $\overline{8}$ × 12 $\overline{\overline{3}}$ (RMP 70), which the scribe showed to be equal to 99 $\overline{2}$ $\overline{4}$, he had to refer to his two-thirds table for integers and fractions, to his rule given in RMP 61B, to the Recto Table, and to his 2-term unit fraction tables, and he did it accurately in six lines. It is instructive to calculate 7-7/8 × 12-2/3, as we would do it today, and then compare the modern working with that of the ancient scribe. It can be quite an enlightening comparison.

The Nature of Proof

Those historians who have adversely criticized the mathematics of the ancient Egyptians confine themselves generally to the lack of formal proof in the Egyptian methods and to the apparent absence of what they call "the scientific attitude of mind" in the Egyptians' treatment of mathematical problems. Thus these commentators have written,

> The fact that the Egyptians had evolved no better means of stating a formula than that of giving three or four examples of its use is hardly a tribute to the scientific nature of their mathematics.
>
> The table [of the RMP Recto] is in itself a monument to the lack of the scientific attitude of mind in the Egyptians.
>
> That they did not reach the conception of scientific mathematics and its dependence on cogent *a priori* demonstration is merely another instance of the vast debt which the world owes to the Greeks. *Peet*[16]

> There are no theorems (in the RMP) properly so called; everything is stated in the form of problems, not in general terms, but in distinct numbers. *Jourdain*[17]

All available texts point to an Egyptian mathematics of rather primitive standards. *Struik*[18]

Egyptian geometry is not a science in the Greek sense of the word, but merely *applied arithmetic*.

The Greeks may also have taken from the Egyptians the rules for the determination of areas and volumes. But for the Greeks, such rules did not constitute mathematics; they merely led them to ask; how does one prove this? *Van der Waerden*[19]

Perhaps a more discerning assessment of the situation was expressed by J. R. Newman when he wrote,

A sound appraisal of Egyptian mathematics depends upon a much broader and deeper understanding of human culture than either Egyptologists or historians of science are wont to recognise.[20]

Let us see if we can develop and amplify Newman's statement somewhat, and perhaps dispel some of the more depressing opinions of some of the less charitable commentators.

It is true that the Egyptians did not show exactly how they established their rules or formulas, nor how they arrived at their methods of dealing with specific values of the variable. But they nearly always proved that the numerical solution to the problem at hand was indeed correct for the particular value or values they had chosen. For them, this constituted both method and proof, so that many of their solutions concluded with sentences like the following:

The producing of the same. (RMP 4)

The manner of the reckoning of it. (RMP 41)

The correct procedure for this [type of] problem. (MMP 9)

Manner of working out. (RMP 43, 44, 46)

Behold! Does one according to the like for every uneven fraction which may occur. (RMP 61B)

Thus findest thou the area. (RMP 55)

These are the correct and proper proceedings. (MMP 6)

The doing as it occurs. [or] That is how you do it. (RMP 28 and 23 other problems)

Shalt do thou according to the like in relation to what is said to thee, all like example this. (RMP 66)

Twentieth-century students of the history and philosophy of science, in considering the contributions of the ancient Egyptians, incline to the modern attitude that an argument or logical proof must be *symbolic* if it is to be regarded as rigorous, and that one or two specific examples using selected numbers cannot claim to be scientifically sound. But this is not true! A nonsymbolic argument or proof *can* be quite rigorous when given for a particular value of the variable; the conditions for rigor are that the particular value of the variable should be *typical,* and that a further generalization to *any* value should be *immediate.* In any of the topics mentioned in this book where the scribes' treatment follows such lines, both these requirements are satisfied, so that the arguments adduced by the scribes are already *rigorous;* the concluding proofs are really not necessary, only confirmatory. The rigor is implicit in the method.

We have to accept the circumstance that the Egyptians did not think and reason as the Greeks did. If they found some exact method (however they may have discovered it), they did not ask themselves *why* it worked. They did not seek to establish its universal truth by an a priori symbolic argument that would show clearly and logically their thought processes. What they did was to explain and define in an ordered sequence the steps necessary in the proper procedure, and at the conclusion they added a verification or proof that the steps outlined did indeed lead to a correct solution of the problem. This was science as they knew it, and it is not proper or fitting that we of the twentieth century should compare too critically their methods with those of the Greeks or any other nation of later emergence, who, as it were, stood on their shoulders. We tend to forget that they were a people who had no plus, minus, multiplication, or division signs, no equals or square-root signs, no zero and no decimal point, no coinage, no indices, and no means of writing even the common fraction p/q; in fact, nothing even approaching a mathematical notation, nothing beyond a very complete knowledge of a twice-times table, and the ability to find two-thirds of any number, whether integral or fractional. With these restrictions they reached a relatively high level of mathematical sophistication.

Notes

1. For example, see O. Neugebauer, "For ordinary additions and subtractions nothing needs to be said" (*The Exact Sciences in Antiquity*, Harper Torchbooks, Harper, New York, 1962, p. 73). Peet has dismissed the subject by remarking that "people who could count beyond a million had no difficulty about the addition and subtraction of whole numbers" ("Mathematics in Ancient Egypt," *Bulletin of the John Rylands Library*, Vol. 15, No. 2 (Manchester, 1931), p. 412).
2. Egyptian Mathematical Leather Roll. British Museum, London.
3. Rhind Mathematical Papyrus. British Museum.
4. Philip E. B. Jourdain, in *The World of Mathematics*, James R. Newman, ed., Vol. 1, p. 12, Simon and Schuster, New York, 1956.
5. *The World of Mathematics*, Vol. 1, p. 172.
6. R. W. Sloley, in *The Legacy of Egypt*, S. R. K. Glanville, ed., pp. 168f., Oxford University Press, London, 1942 (reprinted 1963).
7. Dirk J. Struik. *A Concise History of Mathematics*, pp. 19f., Dover, New York, 1948.
8. Charles J. Finger, *Pepys' Diary*, p. 5.
9. *Ibid.*, p. 10.
10. See *The Mathematical Gazette*, London, Vol. XIV, No. 195 (July 1928), pp. 196f.
11. The fraction 2/3. There is some evidence that a special hieroglyph for 3/4 existed.
12. In earlier times, the two vertical strokes were often drawn the same length, but this may have been merely lack of care.
13. The Recto of the RMP is the first portion, dealing with the division of 2 by the odd numbers 3 to 101. . . . The remainder of the RMP is called the Verso.
14. Only on one column by the scribe. They are on both columns here for ease of reading.
15. For example, EMLR 1. 12 is $\overline{7}\ \overline{14}\ \overline{28} = \overline{4}$, and simple doubling gives $\overline{14}\ \overline{28}\ \overline{56} = \overline{8}$.
16. "Mathematics in Ancient Egypt," *Bulletin of the John Rylands Library*, Vol. 15, No. 2 (Manchester, 1931), pp. 439, 440, 441.
17. "The Nature of Mathematics," in *The World of Mathematics*, James R. Newman, editor, Vol. 1, Simon and Schuster, New York, 1956, p. 12.
18. *A Concise History of Mathematics*, Dover, New York, 1948, p. 23.
19. *Science Awakening*, Arnold Dresden, translator, Noordhoff, Groningen, 1954, pp. 31, 36.
20. *The World of Mathematics*, Vol. 1, p. 178.

Greek Mathematical Philosophy

Edward Maziarz and Thomas Greenwood

Edward Maziarz was educated in Canada and is currently professor of philosophy at Loyola University in Chicago. He is the author of a number of works on philosophy, religion, and the philosophy of mathematics.

Thomas Greenwood was a professor of philosophy at Montreal University in Quebec until his death. Maziarz first studied under Greenwood and then became a close personal friend. Greenwood's professional writings focused on his feelings about the closeness of mathematics to philosophy.

For modern mathematicians, the history of mathematics began with the Greeks. There is little evidence that until the flowering of Hellenistic civilization in the fifth or sixth century B.C. *any earlier civilization approached the subject with sufficient rigor or clarity of thought for us to characterize their efforts as mathematics. While it is true that Egyptian, Babylonian, and Indian civilizations with much mathematical lore predate the Greeks, those cultures produced little that mathematicians today would call real mathematics. The Greeks, in contrast, thought about, taught, and created mathematics in a way that is essentially the same as ours today.*

That we pursue mathematics as the Greeks did is not so surprising when looked at in the proper historical perspective. Until quite recently the mathematics of both western Europe and the Muslim world consisted of scraps and fragments preserved and handed down from the earlier and more enlightened Greek civilization. For centuries mathematics "research" was the discovery of some sliver of insight gleaned from a chance encounter with a newly found Greek manuscript.

This essay explains how the Greeks thought about mathematics, how they created new mathematics, and how the discovery of a discomforting paradox embedded in what they thought was a perfect science changed their view of mathematics.

A brief explanation of the terms used in the article may be helpful. In the vocabulary of mathematics the numbers 1, 2, 3, . . . (and so on) are called the whole numbers. *Numbers expressed as ratios of whole numbers, such as 1/2, 3/4, 15/9, are called* rational numbers. *An* irrational number *is one that can never be written as a ratio of whole numbers no matter how large the numbers selected for the ratio may be. An example of an irrational number is* $\sqrt{2}$*, that is, the number* a *such that* a *times* a *is equal to 2* ($a^2 = 2$)*. The Greeks proved that* $\sqrt{2}$ *is irrational and to their dismay this number occurs often in a very natural way. For example, if you construct a square of one unit on each side (foot, yard, meter—the unit does not matter) and measure the distance between opposite corners in the same units, then that distance is* $\sqrt{2}$ *units.*

The concept of irrational numbers did not receive a full resolution until 1872 with the publication of J. Dedekind's book, Continuity and Irrational Numbers.

Source: Excerpts from Edward Maziarz and Thomas Greenwood, *Greek Mathematical Philosophy*,

Representations of Numbers

The assimilation of number and figure in a rational method of investigating nature called for a practical way of combining arithmetic and geometry. The initial step was a systematic representation of numbers, which the early Greeks accomplished in two ways. The easiest was the method of disposing dots or alphas (units) along straight lines which formed geometrical patterns; the more technical was the construction of straight lines proportional in length to their corresponding numbers. The Pythagoreans are credited with the discovery and use of both methods.

The first method is illustrated by the theory of the *figured numbers,* which assumes that any number can be represented by a rectilinear segment. Except for primes, all numbers could be expressed by straight lines drawn in two or three dimensions. Because prime numbers can be represented in one dimension only, Thymaridas called them supremely rectilinear. According to Speusippus, the study of the figured numbers dates to Pythagoras himself. Eurytus' description of living things with pebbles proves that the early Pythagoreans used dots to construct figured numbers. There is also Aristotle's statement that the Pythagoreans considered boundaries and continuity to be to the various figures, just "as flesh and bones are to man and bronze and stone to the statue; they reduce all things to number and say the line is expressed by two."[1]

By placing the right number of dots in the proper positions, one point or dot was used to represent one; 2 dots placed apart represented 2 and also the straight segment joining the 2 points; 3 dots represented 3 and corresponded to the triangle, the first plane rectilinear figure; and 4 dots, one being outside the plane containing the other 3, represented 4 as well as a pyramid, the first rectilinear solid. Through similar operations, other polygonal and solid numbers were obtained. This assimilation of numbers and figures led the Pythagoreans to investigate their mutual properties and left a lasting mark on our mathematical language, as we still speak of the "square" or the "cube" of a number.

The varieties of polygonal and solid numbers are discussed by Nicomachus and Theon of Smyrna. These numbers represent the shapes of the various polygons and solids. Polygonal numbers are obtained by adding to one the successive terms of a series with a definite difference; the product of three terms is a solid number, the cube being a special case. The simple construction of the *triangular numbers,* formed by the sum of any successive terms of the series of natural numbers, scarcely requires an elaborate explanation. A triangular number with side n is generally represented in modern notation by the formula

$$1 + 2 + 3 + \ldots + n = n(n + 1)/2$$

But the Pythagoreans *showed* each number by a separate diagram; Figure 1 corresponds to the first four triangular numbers, 1, 3, 6, and 10, respectively.

The *square numbers* have a peculiar construction calling for special comment. If we consider the series of the successive odd numbers, and if we add the first to the second, their sum to the third, and so on, we obtain a new series of numbers forming successive squares (Figure 2). Moreover, if we have a number of dots forming and filling up a square, for example, 9 dots, the next higher square, or 16, can be formed by adding rows of dots around two sides of the original square (Figure 3), and so on. In modern notation, the expression of a square number is

$$1 + 3 + 5 + \ldots + (2n - 1) = n^2$$

Figure 1

The addition of the next odd number makes the next higher square $(n + 1)^2$, and so on.

1	= 1
1 + 3	= 4
1 + 3 + 5	= 9
1 + 3 + 5 + 7	= 16
1 + 3 + 5 + 7 + 9	= 25

Figure 2

The Pythagoreans would thus define an odd number as the difference of two square surfaces having for their sides two successive integers.

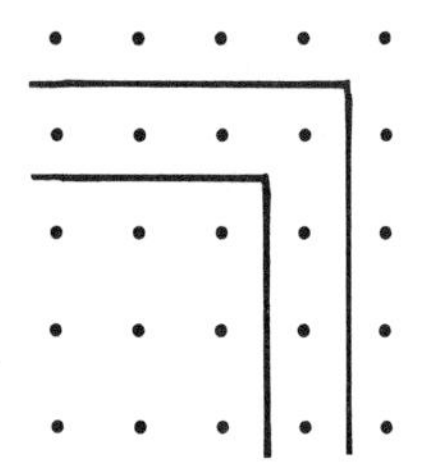

Figure 3

This geometrical representation of odd numbers probably suggested their being called *gnomons* by analogy with the primitive astronomical instrument measuring time, consisting of an upright stick casting a shadow on a surface. Owing to its shape, the gnomon was used to describe what remained of a square when a smaller square was removed from it. Euclid extended the geometrical meaning of gnomon by applying it to figures similarly related to parallelograms. Heron of Alexandria and Theon of Smyrna used the term in a more general sense, defining it as that which, added to a number or figure, makes a whole similar to what it is added to. In this sense, the gnomon can be applied to polygonal and solid numbers.

The generation of triangular and square numbers led naturally to the kind of numbers produced by adding the successive terms of the series of even numbers beginning with 2. Taking two dots and placing an even number of dots around them, gnomon-wise and successively, we obtain the *oblong numbers* with sides or factors differing by a unit. The diagrams in Figure 4 give the first four oblong numbers beginning with 2. The successive numbers of this kind are 2; $2 \cdot 3 = 6$; $3 \cdot 4 = 12$; $4 \cdot 5 = 20$; $5 \cdot 6 = 30$ . . . and generally $n(n + 1)$.

Any oblong number is twice a triangular number (Figure 5) and any square is made up of two

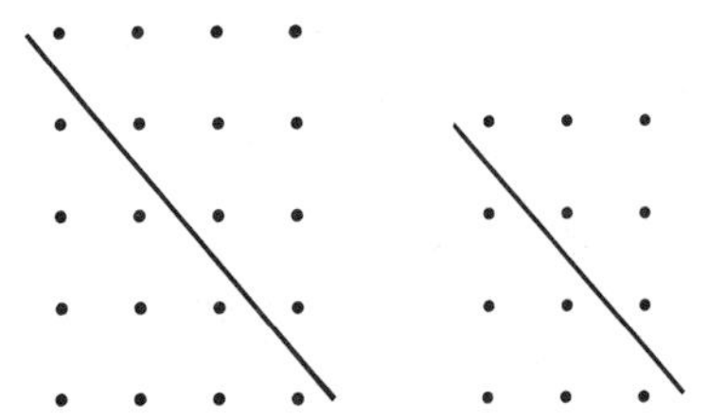

Figure 5

triangular numbers (Figure 6), the sides of the triangles differing by a unit. The oblong numbers are dissimilar, the ratio $n:(n + 1)$ being different for every value of n. By adding the successive odd numbers to one, we obtain always the form of the

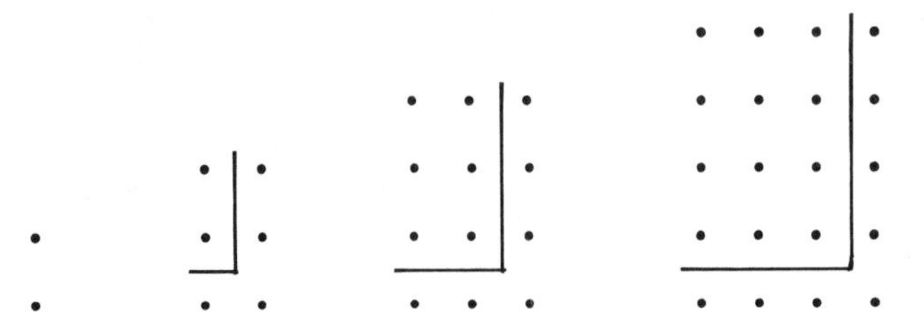

Figure 4

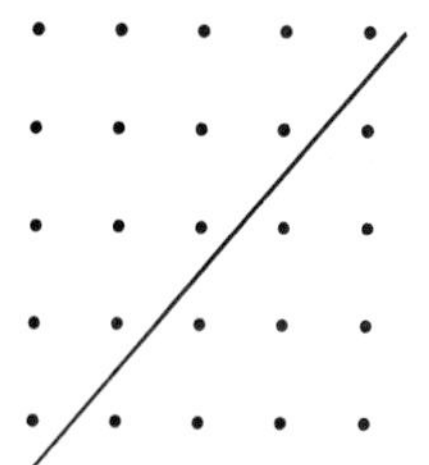

Figure 6

square; by adding the successive even numbers to 2, we get a series of oblong numbers all dissimilar in form.

These relations may explain some Pythagorean views reported by Aristotle. One is the identification of "odd" with "limited" and the inclusion of "square and oblong" in the Pythagorean scheme of the 10 pairs of opposites where odd, limited, and square are opposed to even, unlimited, and oblong, respectively. Another is the identification of the unlimited with the even, which provides things with the element of indeterminacy. Thus, when gnomons are placed around the unit and then around any other number, the resulting figure in the latter case is always different or undetermined, while in the former it is always the same. The figure referred to as being the same is, of course, the square formed by adding the odd numbers as gnomons around the unit.

The words "without the unit" are quite proper, provided their elliptic meaning is adequately understood. Aristotle surely refers to gnomons placed around the numbers beginning with the unit, and then around the numbers beginning without it, in other words, around any number other than the unit. In Figure 7 we start respectively with 3 dots placed in a row, then with 4, then with 5 dots; by placing gnomons around these linear groups of dots, we still obtain oblong numbers. Such an interpretation requires "without the unit" to mean "separately" from the one, or other than the unit. This rendering is confirmed by Iamblichus in a passage dealing with the unity of shape preserved in figured numbers beginning with one, and the diversity of shapes obtained with series beginning with numbers other than the one. Further, both Plato and Aristotle give the word "oblong" the wider meaning of any non-square number with 2 unequal factors. This interpretation is fixed by the term *prolate* in the writings of Nicomachus and Theon of Smyrna, possibly using much earlier sources.

We have no details about the Pythagorean investigations concerning solid numbers, although later commentators discussed them in detail. Nicomachus speaks of the "pyramid" as the first solid number; he argues that the base of the pyramid may be a triangular number, a square, or any polygonal number. But he only mentions the first triangular pyramids 1, 4, 10, 20, 35, 56, and 84, and he explains the formation of pyramids on square bases. He also classifies other solid numbers, speaking of cubes and of "scalene" solid numbers such as beams, columns, and tiles, as well as other combinations. The early Pythagoreans probably thought of the solid numbers in connection with the construction of regular solids, which they may have identified with the material elements of the universe.

The manipulation of figured numbers alone could not provide a comprehensive interpretation of the world, especially as the Pythagoreans had no means of generalizing the relations among the various figured numbers. For a closer assimilation of number and figure, they could propose to represent numbers with straight segments propor-

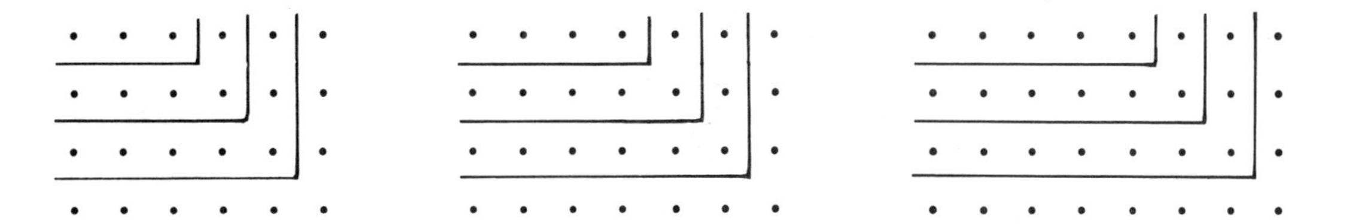

Figure 7

tional in length to their corresponding numbers. When two such segments formed a plane right angle, their product was the rectangle having them as adjacent sides; when three such segments formed a solid right angle, their product was the parallelepiped having them as adjacent sides. This method, used extensively by Euclid, is implied in Plato and in the Pythagorean use of the proportionals and of the application of areas. These theories represent the culmination of the earliest combination of arithmetic and geometry as urged by the Pythagorean doctrines.

The discovery of proportionals attributed to Pythagoras[2] is a natural extension of his number theory. If a point is a unit in position, then a line is made up of points, just as a necklace is made up of beads. Disregarding later disputes about the continuous and the discrete, homogeneous and identical points may be taken as the ultimate units of spatial measurement. Consequently, the ratio of 2 given segments is merely the ratio of the numbers of points in each. Moreover, because any magnitude involves a ratio between the number of units it contains and the unit itself, the comparison of 2 magnitudes implies 2 ratios, or 4 terms. When any 2 pairs of terms have equal ratios, these 4 terms are said to form a *proportion.*

Development of the theory of numerical proportion leads to the consideration of *means* which the Pythagoreans studied, probably for their musical experiments. They knew of 3 means: the arithmetic, the geometric, and the harmonic, which Archytas explained for the first time. In the arithmetic mean, the first of the 3 terms exceeds the second by the same amount as the second exceeds the third. In the geometric mean, the first term is to the second as the second is to the third. In the harmonic mean, the 3 terms are such that, by whatever part of itself the first exceeds the second, the second exceeds the third by the same part of the third. Their corresponding algebraic expressions are as follows:

Arithmetic $\dfrac{a-m}{m-b} = \dfrac{a}{a}$ or $2m = a + b$

and m is the arithmetic mean between a and b.

Geometric $\dfrac{a-m}{m-b} = \dfrac{a}{m}$ or $m^2 = ab$

and m is the geometric mean between a and b.

Harmonic $\dfrac{a-m}{m-b} = \dfrac{a}{b}$ or $\dfrac{2}{m} = \dfrac{1}{a} + \dfrac{1}{b}$

and m is the harmonic mean between a and b.

The name "harmonic mean" was adopted by Archytas in accordance with the views of Philolaus concerning geometrical harmony, instead of the older name "subcontrary," which may imply a previous arithmetical definition before its discovery in the intervals of the octave.[3] This name was applied to the cube because it has 12 edges, 8 angles, and 6 faces, while 8 is the harmonic mean between 12 and 6, as the following expression shows:

$$\frac{a-m}{m-b} = \frac{a}{b}, \quad \text{hence} \quad \frac{12-8}{8-6} = \frac{12}{6}$$

The relation of this proportion to musical harmony can be explained thus: "If 6 is made to correspond to the first note of the scale, so as to have exclusively integers in the relations between the tones of the scale, then 8 corresponds to the fourth instead of 4/3 and 12 corresponds to the octave instead of 2."[4]

The generalization of this proportion, called the "musical mean," is considered by Iamblichus after Nicomachus as the "most perfect proportion." It consists of 4 terms so combined that the two middle terms are the arithmetic and harmonic means between the extremes:

$$a : \frac{a+b}{2} = \frac{2ab}{a+b} : b$$

While the Babylonians are credited with its discovery, Pythagoras introduced it into Greek science and Plato used it in *Timaeus.* This numerical connection between geometrical and musical harmony was another link in the cosmic significance of the number theory.

The doctrine of means developed by later Pythagoreans consists of 10 types described by Nicomachus and Pappus, 7 of them elaborations of the 3 fundamental means already mentioned. No information is available about the developments of the early Pythagorean theory of proportion, except that it was based on considerations involving integers only. When the irrationals were discovered, the Pythagoreans did not build a more adequate theory of proportion. This was left to Eudoxus, who was probably influenced by Plato's views. But the Pythagoreans used their own restricted method of proportionals, especially in their theory of application of areas.

This method is thus outlined by Proclus:

> These ancient things, says Eudemus, are discoveries of the Pythagorean Muse: I mean the application of areas, their exceeding and their falling short. These men inspired later geometers to give their names to the so-called "conic" lines, calling one of these the parabola (application), another the hyperbola (exceeding), and another the ellipse (falling short). Those godlike men of old saw the things signified by these names in the construction of areas upon a given straight segment in a plane. For when you take a segment and lay the given area exactly alongside the whole of it, they say you apply that area. When you make the length of the area greater than the segment, it is said to "exceed"; and when you make it less, so that after drawing the area a portion of the segment extends beyond it, it is said to fall short. In the Sixth Book, Euclid speaks in this way both of exceeding and falling short.[5]

The general form of the problem involving application of areas is as follows: "Given 2 figures, to construct a third equal to the one and similar to the other." The ratio of one area to the other, or the ratio of the contents of the 2 figures, can be expressed as a ratio between straight segments, and such ratios can be manipulated in various ways. The application of areas is an important part of Greek mathematics; it is the foundation of the Euclidian theory of irrationals and of the Apollonian treatment of the conics. In performing the geometrical operations involved, the Greeks did the equivalent of the algebraical processes of addition, subtraction, multiplication, division, squaring, extracting roots, and solving mixed quadratic equations with real roots.

The method of application of areas has been systematized by Euclid, who used the Eudoxian theory of proportion in the *Elements* to prove such Pythagorean problems as the following:

I, 45—To construct in a given rectilinear angle a parallelogram equal to a given rectilinear figure.[6]

II, 5—If a straight line is cut into equal and unequal segments, the rectangle contained by the unequal segments of the whole and the square on the line between the points of section equal the square on the half.[7]

II, 11—To cut a given straight line so that the rectangle contained by the whole and one of the segments equals the square on the remaining segments.[8]

VI, 28—To apply on a given straight line a parallelogram equal to a given rectilinear figure and deficient by a parallelogrammic figure similar to the given one; the given rectilinear figure must not be greater than the parallelogram described on half the segment and similar to the defect.

VI, 29—To apply on a given straight line a parallelogram equal to a given rectilinear figure and exceeding by a parallelogrammic figure similar to the given one.

These examples show how the Pythagoreans could use geometry as a substitute for modern algebraic operations, and why a large part of their geometry may properly be called "geometrical algebra." When dealing with problems involving similar figures, their method of applying areas required the notion of proportionals. But when dealing with mere elementary problems of simple transformation of a given area into another of a different form, the Pythagoreans could use their theory of figured numbers.

Such methods may be linked with the remarkable theorem of the square of the hypotenuse, which illustrates all of these other methods. Tradition attributes it to Pythagoras, who may have obtained it either by generalizing particular Babylonian and Egyptian cases, by means of purely numerical considerations, by the application of areas, or by restricted use of proportion. According to the method of obtaining square numbers explained by Heath,[9] the sum of any successive terms of the odd numbers series is a square. It then suffices to pick out the odd numbers which are squares and to find an expression for all sets of such 3 numbers giving the hypotenuse relation. More specifically, the right-angled triangle with sides (3, 4, 5) illustrates the case of a square with side 5 being transformed into 2 others of sides 3 and 4 respectively, and together equivalent to the first. This fact may have led Pythagoras to state the more general theorem in terms of the theory of application of areas, the problem being to transform a given square into 2 squares together equivalent to the first; or, conversely, to transform 2 given squares into one square equivalent to the sum of their areas.

Several proofs by proportion can be given of this theorem. One of the simplest is as follows (Figure 8). Let ABC be a triangle right-angled at A, with AD perpendicular to BC. The similarity of the triangles DBA and DAC to the triangle ABC gives the relations $BA^2 = BD \cdot BC$ and $AC^2 = CD \cdot BC$; hence $BA^2 + AC^2 = BC^2$. In this proof, the square on BC is equal to the sum of 2 rectangles, which is precisely what Euclid proves in Book I by a different method. Although the Pythagorean theories of proportion and of numbers in general were applicable to commensurable quantities only, the restricted use of proportion in a special proof was possible; when the discovery of the irrationals made revision of the Pythagorean proof necessary, neither the Master nor his immediate disciples tried any substitute for it. Indeed, the invention of irrationals was so far-reaching that it demanded revision of the number theory itself, a momentous task the early Pythagoreans could not undertake with their restricted methods.

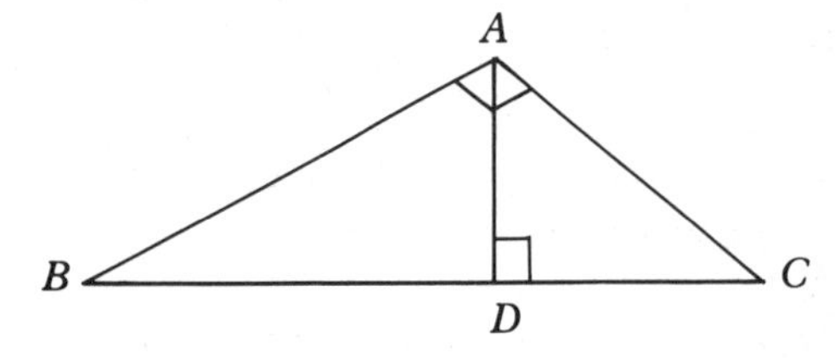

Figure 8

The Crisis of the Irrationals

There are strong reasons to believe that Pythagoras discovered the irrationals when considering the relation between the diagonal of the square and its side. His method of obtaining square-numbers may have prompted him to question why every number is not a square, when he had found some exception to his practical rule. But geometry offered the best examples of the existence of irrationals, as in the cases of the isosceles right-angled triangle and the numerical relation between the diagonal of the square and its side. The very names of *side-numbers* and *diagonal-numbers* seem to justify this view.

If the primitive treatment of the theorem of the right-angled triangle was arithmetical, the impossibility of finding a root for the square of the hypotenuse of an isosceles right-angled triangle would naturally yield the notion of $\sqrt{2}$, the first irrational. According to a scholium to Euclid, the Pythagoreans discovered the irrationals by observing numbers; for "though the unit is a common measure of all numbers, they were unable to find a common measure of all magnitudes."[10] This was because magnitudes are endlessly divisible, without leaving any part too small for further division.

After this discovery, the Pythagoreans probably investigated the properties of such magnitudes, but nothing definite seems to have been done before Theodorus of Cyrene and Theaetetus of Athens. The title of the lost work of Democritus, *On Irrational Lines and Solids,* suggests that surds were known before his time. The irrationality of $\sqrt{2}$ is alluded to in Plato's *Republic* as a well-known fact; in *Theaetetus* we are told that Theodorus was the first to prove the irrationality of $\sqrt{3}, \sqrt{5}, \ldots, \sqrt{17}$, which implies that $\sqrt{2}$ has been dealt with earlier.

The traditional proof of the irrational character of certain numbers is indicated by Aristotle as an example of *reductio ad absurdum:* "All who effect an argument *per impossibile* infer syllogistically what is false, and they prove the original conclusion hypothetically when something impossible results from the assumption of its contradictory; for example, the diagonal of the square is incommensurate with its side because odd numbers are equal to evens if it is assumed to be commensurate.[11] This is evidently the proof interpolated in the tenth book of Euclid's *Elements,* where the fraction m/n in its lowest terms is supposed to equal $\sqrt{2}$; then both m and n must be even because $m^2/n^2 = 2$ and $m^2 = 2n^2$; yet this cannot be, because of the condition of the fraction. With this contradiction, no fraction m/n can have 2 as its square.

Their fruitless effort to find the exact root of $\sqrt{2}$ led the Pythagoreans to prove its incommensurability, and to determine any number of successive approximations to its value by finding the integral solutions of an indeterminate equation (of the form $2x^2 - y^2 = -1$ in modern notation). The pairs of values of x and y were called *side-numbers* and *diameter-numbers* or *diagonal-numbers,* respectively; as these values increase, the ratio of y to x approximates $\sqrt{2}$ more closely. Theon of Symrna gives an interesting explanation of the formation of these numbers, whose names are justified by their particular function and purpose.

The Greeks treated the irrationals in general as a part of geometry rather than arithmetic, because of the difficulty of handling irrationals arithmetically and of the successful Pythagorean combination of geometry with number theory. For want of a notation, any irrational was represented by a rectilinear segment or a combination of lines. This is illustrated in the tenth book of Euclid's *Elements,* where simple and compound irrationals are dealt with geometrically. Yet the Pythagoreans could not develop the theory of irrationals on this basis, precisely because their geometry depended on their restricted number theory. Furthermore, as their arithmetic involved a theory of proportion applicable to commensurable magnitudes only, discovery of the irrationals must have dealt a severe blow to its whole structure. In fact, it involved the restriction of their method of proportion, pending the discovery of the generalized theory of proportion established by Eudoxus during Plato's time. The discovery of surds also shattered the geometrical methods of the Pythagoreans, as the proof of several geometrical theorems rested on their primitive theory of proportion.

When the invention of surds became known to members of the Brotherhood, a rift was opened between arithmetic and geometry; this led to the investigation of various problems involving irrational magnitudes. Pythagorean and other schools of mathematicians tried laboriously to find whether the rift could be bridged by purely mathematical methods. This might explain why such problems as the squaring of the circle, trisection of the angle, and duplication of the cube were popular after the death of Pythagoras. In spite of the interesting results obtained by some mathematicians, these circumstances may account for the setback suffered by mathematics at the end of the fifth century B.C.

It may be questioned whether the Pythagoreans knew the incommensurability of a circumference in relation to its diameter, although the circle played an important part in their cosmogonies. Tradition mentions Anaxagoras of Clazomenae (*ca.* 500–428 B.C.) as the first to deal openly with the problem of squaring the circle. But he may have heard of it from the early Pythagoreans, of whom he was a younger contemporary, and who could not have dealt openly with this question because of their oath of secrecy. To be sure, discovery of the irrationals must have been kept secret a long time, as it had far-reaching consequences for the practical and philosophical doctrines of the Brotherhood. The esoteric disciples must have heard with awe from the Master that number, which was and explained everything, could not account for some simple geometrical magnitudes related to their number theory.

This situation must have weakened the cosmic and moral applications of the number theory as a whole. The square, a most fundamental and beautiful type of number, was found to bear within itself an element of irrationality. By using a square number as a physical explanation of facts, reason was appealing, so to speak, to "unreasonable" elements. It may also have been observed that many rules of harmony apparently subject to number entailed irrationals. When the irrationality of π was established later, many must have thought that the laws of the heavens themselves could not be as true as Pythagoras said they were, for if the distances between the heavenly bodies were proportional to the lengths of the vibrating strings which produced the musical scale, the perfect circular paths of these bodies had no common number with the proportional distances between the center of the world and the various bodies moving in space. Consequently, it was impossible to maintain any longer that number was the essence of all existing things, or that all things were made of number.

This awkward crisis encouraged the Pythagorean esoterics to maintain their oath to withhold from the public the existence of irrationals. This accounts for the legend that the first Pythagorean who made it public, whether it was Hippasus or another, perished at sea for his impiety. According to Proclus, the unutterable and formless were to be concealed; those who uncovered and touched this image of life were instantly destroyed and shall remain forever the play of the eternal waves.

Meanwhile, destruction of the primitive mathematical balance between the cosmos and man called for new conceptions to satisfy man's yearning for truth. If nature contained elements beyond reason, then man himself should be studied in order to find out his limitations and their eventual remedy. This task was performed by the Socratic schools, although the serious interest of the Pythagoreans in the practical rules of life was originally responsible for introducing ethics and social theory into the range of philosophical inquiry. This interest was intensified with Xenophanes and Heraclitus, and it reached its highest mark with Democritus and the Sophists.

The Pythagorean experiment was certainly discussed at the time. If the existing fragments of the pre-Socratic thinkers were not so scanty, we could probably trace many more references to Pythagorean doctrines before the time of Plato. The founder of the Academy must have borrowed a good deal from them, although he scarcely mentions them in his Dialogues. We have to turn to Aristotle, who disagreed with both Platonism and Pythagoreanism, for the first serious criticism of these views. In the first book of his *Metaphysics*, Aristotle says the Pythagoreans treat of principles and elements stranger than those of the Ionian philosophers, for these principles are taken from non-sensible things, since the objects of mathematics are things without movement. Yet the Pythagoreans claim to discuss and investigate nature, for they generate the heavens and explain their parts and functions by observing the natural phenomena and referring them to principles and causes. This attitude was shared by the Ionian philosophers, for whom the real is all that is perceptible and contained by the so-called heavens.

For Aristotle, the causes and principles mentioned by the Pythagoreans may lead gradually to the higher levels of reality, but they are less suited to theories about nature. Elaborating this view, he criticizes the Pythagoreans for neglecting to make clear whether the limit and the indefinite, the odd and the even are the only principles assumed. As a result, they are unable to account for the existence of motion and the facts of generation and destruction. Consequently, they offer no explanation of the particular movements of the heavenly bodies and of the difference between light and heavy objects. Although Plato assimilated the elements with the regular solids, the Pythagoreans apparently said nothing about fire, earth, air, or water. Engrossed with the problem of motion, Aristotle fails to see how number and its attributes are causes of what exists and of what happens in nature, when the only acknowledged numbers are those out of which the world is composed. Similar difficulties may have led Zeno to formulate his famous arguments.

Referring to some aspects of Pythagorean mysticism, Aristotle cannot reconcile its implications with his logical vision of the world. He fails to see why opinion and opportunity are placed in one particular region, while injustice and mixture are above or below. He also disagrees with the alleged proof that each of these is number, and that a plurality of the extended bodies composed of numbers is already in those places because the attributes of numbers are attached to various places. In this case, one cannot determine whether or not the number identified with each of these abstractions is also the number exhibited in the material universe. Although Plato says it is different, for him both the bodies and their causes are numbers.

Notwithstanding the effectiveness of Aristotle's criticism, the Pythagorean doctrines and methods have suggested a basic approach to the problems of the world which has greatly influenced the development of philosophy and science. Commenting upon the Pythagorean doctrine that mathematical entities are the ultimate stuff of existence and experience, Whitehead asserts that, with this bald and crude statement, Pythagoras had hit upon

> a philosophical notion of considerable importance, a notion which has a long history, and which has moved the minds of men and has even entered into Christian theology. About a thousand years separate the Athanasian Creed from Pythagoras, and about two thousand four hundred years separate Pythagoras from Hegel. Yet for all these distances in time, the importance of definite numbers in the constitution of the Divine Nature, and the concept of the real world as exhibiting the evolution of an idea, can both be traced back to the train of thought set going by Pythagoras. . . . So today, when Einstein and his followers proclaim that physical facts such as gravitation are to be construed as exhibitions of local peculiarities of spatio-temporal properties, they are following the pure Pythagorean tradition.[12]

In short, the Pythagorean conception of science corresponds definitely to a fundamental attitude of the mind in its search for truth.

Notes

1. Aristotle, *Metaphysics*, 1036^b 10.
2. Proclus *Commentary on Euclid*, p. 65. There are two readings of the Greek text, one for proportionals and the other for irrationals. The latter is dismissed, as it was difficult to believe that Pythagoras developed a "theory of irrationals." But either reading may be true, as Pythagoras did discover both the proportionals and the irrationals.
3. In his *Commentary on Plato's Timeaus* (Oxford, 1928), A. E. Taylor shows how the Pythagoreans would have observed the 3 means in their musical studies.
4. Gaston Milhaud, *Les Philosophes Géomètres de la Grèce*, 2nd ed. (Paris, 1934), p. 93.
5. *Commentary on Euclid* i. p. 420.
6. Proclus observes that ancient geometers were led to investigate the squaring of the circle as a consequence of this problem.
7. This theorem yields the Pythagorean rule for finding integral square numbers. Thomas L. Heath obtains through it the geometrical solution of the equation $ax - x^2 = b^2$; but there is no direct evidence that the Pythagoreans or even Euclid used this proposition for such a solution. Cf. *The Thirteen Books of Euclid's Elements* (Cambridge, 1908), I, p. 384 (repub. in New York, 1956).
8. This problem gives a geometrical solution of the equation $x^2 + ax = a^2$.
9. Thomas L. Heath, *A Manual of Greek Mathematics* (Oxford, 1931), pp. 46–48.
10. Thomas L. Heath, trans., *The Thirteen Books of Euclid's Elements*, ed. Heiberg (Cambridge, 1908), I, p. 415 (repub. in New York, 1956).
11. Aristotle, *Analytica priora*, 41^a 23.
12. A. N. Whitehead, *Science and the Modern World* (New York, 1941), p. 36.

The Evolution of Mathematics in Ancient China

Frank Swetz

Frank Swetz received a grant from the East Asian Institute of Columbia University, which made much of his research possible. He is currently a professor of mathematics at Capitol Campus, Pennsylvania State University. He has written extensively on the history of the exact sciences in China.

That Western historians routinely and systematically ignore or belittle Chinese cultural achievements is very nearly a cliché. It should surprise no one that this bias extends to the history of mathematics. There are a number of reasons for this, the chief among them being language. The problem is not so much that modern Chinese is difficult to read, which it is, but that historical Chinese documents use arcane and outdated forms of Chinese characters, whose meaning is often obscure even to Chinese scholars. In some academic fields when this problem occurs, other types of evidence—such as archaeological finds or accounts in other languages—may be used to determine the meaning of a difficult passage. But in the case of mathematics often too little can be translated. Who is willing to take the time to become both a trained mathematician and a scholar of ancient Chinese? Where is the demand for such a combination and what would be the reward?

Yet another difficulty might be called the gunpowder syndrome. There is no question that the Chinese invented gunpowder hundreds of years before it was known in the West. And they used it principally for fireworks. Not until Western civilization acquired it was gunpowder applied to the practical task of killing people. The Chinese were the first to make a number of other truly significant technological achievements that, as with gunpowder, they did not systematically exploit; they simply didn't care. They had a different set of values that gave technology, including mathematics, a decidedly secondary role. The Chinese may in fact have known a great deal more about mathematics than is revealed in this excellent article by Frank Swetz.

A POPULAR SURVEY book on the development of mathematics has its text prefaced by the following remarks:

> Only a few ancient civilizations, Egypt, Babylonia, India and China, possessed what may be called the rudiments of mathematics. The history of mathematics and indeed the history of western civilization begins with what occurred in the first of these civilizations. The role of India will emerge later, whereas that of China may be ignored because it was not extensive and moreover has no influence on the subsequent development of mathematics.[1]

Source: Frank Swetz, The Evolution of Mathematics in Ancient China, *Mathematics Magazine* 52(1979): 10–19. Reprinted by permission of *Mathematics Magazine*. (Some references omitted.)

Even most contemporary works on the history of mathematics reinforce this impression, either by neglecting or depreciating Chinese contributions to the development of mathematics. Whether by ignorance or design, such omissions limit the perspective one might obtain concerning both the evolution of mathematical ideas and the place of mathematics in early societies. In remedying this situation, western historians of mathematics may well take heed of Whittier's admonition:

We lack but open eye and ear
To find the Orient's marvels here.[2]

Language barriers may limit this quest for information; however, a search of English language sources will reveal that there are many "marvels" in Chinese mathematics to be considered.

Legend and Fact

The origins of mathematical activity in early China are clouded by mysticism and legend. Mythological Emperor Yü is credited with receiving a divine gift from a Lo river tortoise. The gift in the form of a diagram called the *Lo shu* is believed to contain the principles of Chinese mathematics, and pictures of Yü's reception of the *Lo shu* have adorned Chinese mathematics books for centuries. This fantasy in itself provides some valuable impressions about early Chinese science and mathematics. Yü was the patron of hydraulic engineers; his mission was to control the flood-prone waters of China and provide a safe setting in which a water-dependent civilization could flourish. The users of science and mathematics in China were initially involved with hydraulic engineering projects, the construction of dikes, canals, etc., and with the mundane tasks of logistically supporting such projects. A close inspection of the contents of the *Lo shu* reveals a number configuration (Figure 1) which would be known later in the West as a magic square. For Chinese soothsayers and geomancers from the Warring State period of Chinese history (403–221 B.C.) onward, this square, comprised of numbers, possessed real magical qualities because in it they saw a plan of universal harmony based on a cosmology predicated on the dualistic theory of the *Yin* and the *Yang*.[3]

When stripped of ritualistic significance, the principles used in constructing this first known magic square are quite simple and can best be described by use of diagrams as shown in Figure 2. The construction and manipulation of magic squares became an art in China even before the

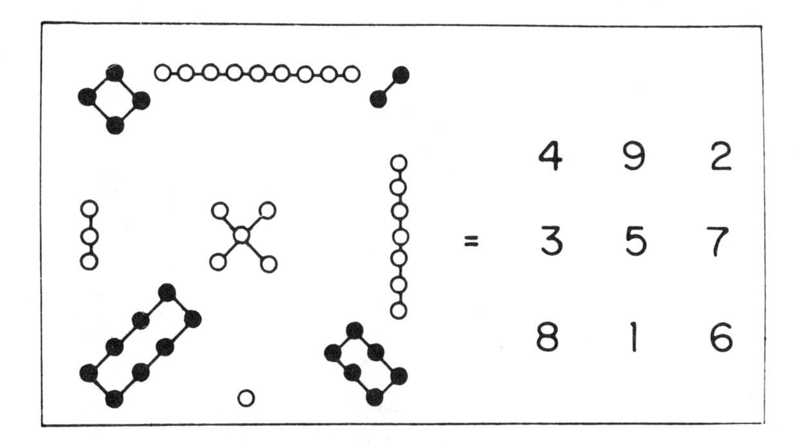

Figure 1

7	4	1
8	5	2
9	6	3

Construct a natural square.

		1		
	4		2	
7		5		3
	8		6	
		9		

Distort it into a diamond.

		9		
	4		2	
3		5		7
	8		6	
		1		

Exchange corner elements.

4	9	2
3	5	7
8	1	6

Compress back into a square.

Figure 2

concept was known in the West. Variations of the *Lo shu* technique were used in constructing magic squares of higher order with perhaps the most impressive square being that of order nine; see Figure 3.

While the *Lo shu* provides some intriguing insights into early mathematical thinking, its significance in terms of potential scientific or technological achievement is negligible. Historically, the first true evidence of mathematical activity can be found in numeration symbols on oracle bones dated from the Shang dynasty (14th century B.C.). Their numerical inscriptions contain both tally and code symbols, are clearly decimal in their conception, and employ a positional value system. The Shang numerals for the numbers one through nine were:

一 二 三 亖 ㄨ ∧ +)(ㄅ

By the time of the Han Dynasty (2nd century B.C.–4th century A.D.), the system had evolved into a codified notation that lent itself to computational algorithms carried out with a counting board and set of rods. The numerals and their computing-rod configurations are

1	2	3	4	5	6	7	8	9
𝍩	𝍪	𝍫	𝍬	𝍭	𝍮	𝍯	𝍰	𝍱

for coefficients of 10^{2n-2} $n = 1,2. . .$

1	2	3	4	5	6	7	8	9
𝍠	𝍡	𝍢	𝍣	𝍤	𝍥	𝍦	𝍧	𝍨

for coefficients of 10^{2n-1} $n = 1,2. . . .$

Thus in this system 4716 would be represented as 𝍣𝍯𝍠𝍮. (Occasionally the symbol × was used as an alternative to 𝍭.)

Counting boards were divided into columns designating positional groupings by 10. The resulting facility with which the ancient computers could carry out algorithms attests to their full understanding of decimal numeration and computation. As an example, consider the counting board

1	10	19	28	37	46	55	64	73
2	11							.
3	12							.
4	13							.
5	14							.
6	15							.
7	16							.
8	17							.
9	18							81

(a)

55	28	1
64	37	10
73	46	19

(b)

28	73	10
19	37	55
64	1	46

(c)

31	76	13	36	81	18	29	74	11
22	40	58	27	45	63	20	38	56
67	④	49	72	⑨	54	65	②	47
30	75	12	32	77	14	34	79	16
21	39	57	23	41	59	25	43	61
66	③	48	68	⑤	50	70	⑦	52
35	80	17	28	73	10	33	78	15
26	44	62	19	37	55	24	42	60
71	⑧	53	64	①	46	69	⑥	51

(d)

Start with a natural square (a) then fold each row into a square (b) of order 3 (example using row 1) and apply the *Lo Shu* technique (c). The nine resulting magic squares of order 3 (d) are then positionally ordered according to the correspondence of the central element in their bottom rows with the numbers of the *Lo shu,* i.e., 4, 9, 2; 3, 5, 7; 8, 1, 6.

Figure 3

Counting board

		2	4	6	(multiplier)
					(product)
		3	5	7	(multiplicand)

Accompanying rod computations

$2 \times 3 = 6$
$2 \times 5 = \underline{10}$
70

		2	4	6
7	1	4		
3	5	7		

$2 \times 7 = \underline{\ 14}$
714
$4 \times 3 = \underline{12\ }$
834

			4	6
8	5	6	8	
	3	5	7	

$4 \times 5 = \underline{\ 20}$
854
$4 \times 7 = \underline{\ 28}$
8568
$6 \times 3 = \underline{18\ }$
8748

				6	
8	7	8	2	2	(answer)
		3	5	7	

$6 \times 5 = \underline{\ 30}$
8778
$6 \times 7 = \underline{\ 42}$
87822

Figure 4

method of multiplying 2 three-digit numbers, as illustrated in Figure 4. The continual indexing of partial products to the right as one multiplies by smaller powers of ten testifies to a thorough understanding of decimal notation. In light of such evidence, it would seem that the Chinese were the first society to understand and efficiently utilize a decimal numeration system. If one views a popular schematic of the evolution of our modern system of numeration (Figure 5) and places the Chinese system in the appropriate chronological position, an interesting hypothesis arises, namely that the numeration system commonly used in the modern world had its origins 34 centuries ago in Shang China!

The Systematization of Early Chinese Mathematics

The oldest extant Chinese text containing formal mathematical theories is the *Arithmetic Classic of the Gnomon and the Circular Paths of Heaven,* [*Chou pei suan ching*]. Its contents date before the

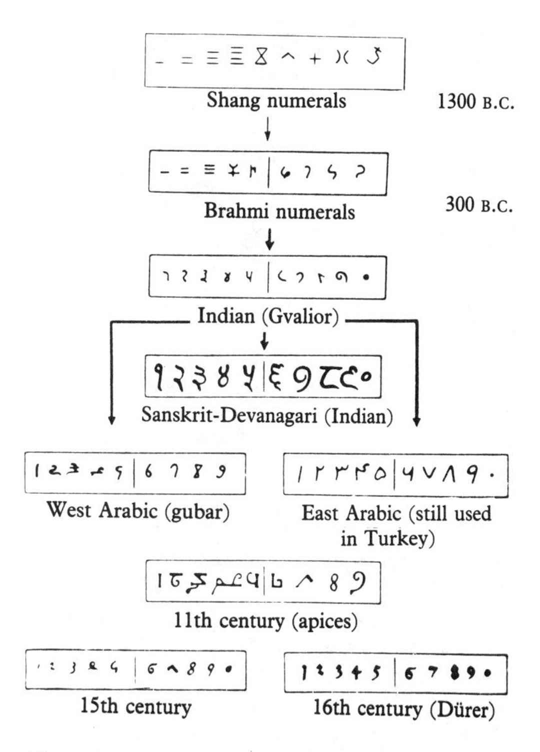

Figure 5

third century B.C. and reveal that mathematicians of the time could perform basic operations with fractions according to modern principles employing the concept of common denominator. They were knowledgeable in the principles of an empirical geometry and made use of the "Pythagorean theorem." A diagram (see Figure 6) in the *Chou pei* presents the oldest known demonstration of the validity of this theorem. This diagram, called the *hsuan-thu* in Chinese, illustrates the arithmetic-geometric methodology that predominates in early Chinese mathematical thinking and shows how arithmetic and geometry could be merged to develop algebraic processes and procedures. If the oblique square of the *hsuan-thu* is dissected and the pieces rearranged so that two of the four congruent right triangles are joined with the remaining two to form two rectangles, then the resulting figure comprised of two rectangles and one small square have the same area as their parent square. Further, since the new configuration can also be viewed as being comprised of two squares whose sides are the legs of the right triangle, this figure demonstrates that the sum of the squares of the legs of a right triangle is equal to the square of the hypotenuse. The process involved in this intuitive, geometric approach to obtain algebraic results was called *chi-chü* or "the piling up of squares."

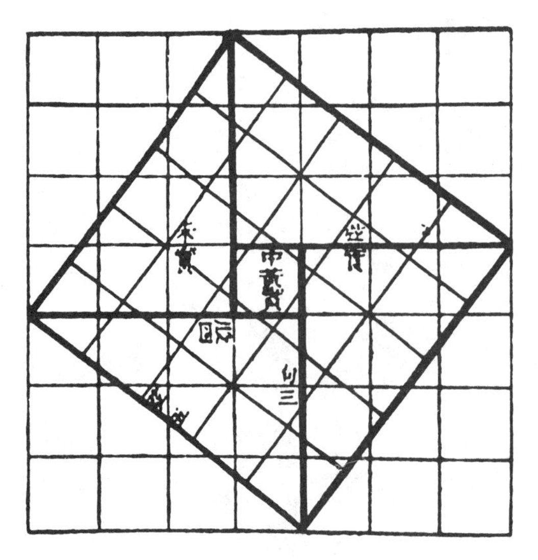

Figure 6

The next historical text known to us is also a Han work of about the third century B.C. It is the *Nine Chapters on the Mathematical Art*, [*Chiu chang suan shu*], and its influence on oriental mathematics may be likened to that of Euclid's *Elements* on western mathematical thought. The *Chiu chang's* chapters bear such titles as surveying of land, consultations on engineering works, and impartial taxation, and confirm the impression that the Chinese mathematics of this period centered on the engineering and bureaucratic needs of the state. Two hundred and forty-six problem situations are considered, revealing in their contents the fact that the Chinese had accumulated a variety of formulas for determining the areas and volumes of basic geometric shapes. Linear equations in one unknown were solved by a rule of false position. Systems of equations in two or three unknowns were solved simultaneously by computing board techniques that are strikingly similar to modern matrix methods. While algebraists of the ancient world such as Diophantus or Brahmagupta used various criteria to distinguish between the variables in a linear equation,[4] the Chinese relied on the organizational proficiency of their counting board to assist them in this chore. Using a counting board to work a system of equations allowed the Chinese to easily distinguish between different variables.

Consider the following problem from the *Chiu chang* and the counting board approach to its solution.

> Of three classes of cereal plants, 3 bundles of the first; 2 of the second and 1 of the third will produce 39 *tou* of corn after threshing; 2 bundles of the first; 3 of the second and 1 of the third will produce 34 *tou;* while 1 of the first, 2 of the second and 3 of the third will produce 26 *tou*. Find the measure of corn contained in one bundle of each class.
>
> (1 *tou* = 10.3 liters)

This problem would be set up on the counting board as:

1	2	3	1st class grain
2	3	2	2nd class grain
3	1	1	3rd class grain
26	34	39	Number of *tou*

Using familiar notation this matrix of numbers is equivalent to the set of equations

$$3x + 2y + z = 39$$
$$2x + 3y + z = 34$$
$$x + 2y + 3z = 26$$

which are reduced in their tabular form by appropriate multiplications and subtractions to

$$\begin{aligned} 3x + 2y + z &= 30 \\ 36y &= 153 \\ 36z &= 99 \end{aligned} \quad \text{and} \quad \begin{aligned} 36x &= 333 \\ 36y &= 153 \\ 36z &= 99. \end{aligned}$$

Thus $x = 333/36$, $y = 153/36$ and $z = 99/36$.

A companion problem from the *Chiu chang* involves payment for livestock and results in the system of simultaneous equations:

$$-2x + 5y - 13z = 1000$$
$$3x - 9y + 3z = 0$$
$$-5x + 6y + 8x = -600.$$

Rules provided for the solution treat the addition and subtraction of negative numbers in a modern fashion; however, procedures for the multiplication and division of negative numbers are not found in a Chinese work until the Sung dynasty (+1299). Negative numbers were represented in the computing scheme by the use of red rods, while black computing rods represented positive numbers. Zero was indicated by a blank space on the counting board. This evidence qualifies the Chinese as being the first society known to use negative numbers in mathematical calculations.

The *Chou pei* contains an accurate process of extracting square roots of numbers. The ancient Chinese did not consider root extraction a separate process of mathematics but rather merely a form of division.[5] Let us examine the algorithm for division and its square root variant. The division algorithm is illustrated in Figure 7 for the

$166536 \div 648$

Counting board layout		Accompanying rod computations		Explanations
2	(quotient)	166500		200 is chosen
		−120000	= (200 × 600)	as the first
166536	(dividend)	46500		partial
		−8000	= (200 × 40)	quotient
648	(divisor)	38500		
		−1600	= (200 × 8)	
		36900		
25		36930		50 is chosen
		−30000	= (50 × 600)	as the second
36936		6930		partial
		−2000	= (50 × 40)	quotient
648		4930		
		400	= (50 × 8)	
		4530		
257		4536		7 is chosen
		−4200	= (7 × 600)	as the third
4536		336		partial
		−280	= (7 × 40)	quotient
648		56		
		−56	= (7 × 8)	
		0		process is finished

Figure 7

problem 166536 ÷ 648. The Chinese technique of root extraction depends on the algebraic proposition

$$\begin{aligned}(a + b + c)^2 &= a^2 + 2ab + b^2 \\ &\quad + 2(a + b)c + c^2 \\ &= a^2 + (2a + b)b \\ &\quad + (2[a + b] + c)c\end{aligned}$$

which is geometrically substantiated by the diagram given in Figure 8. This proposition is incorporated directly into a form of division where

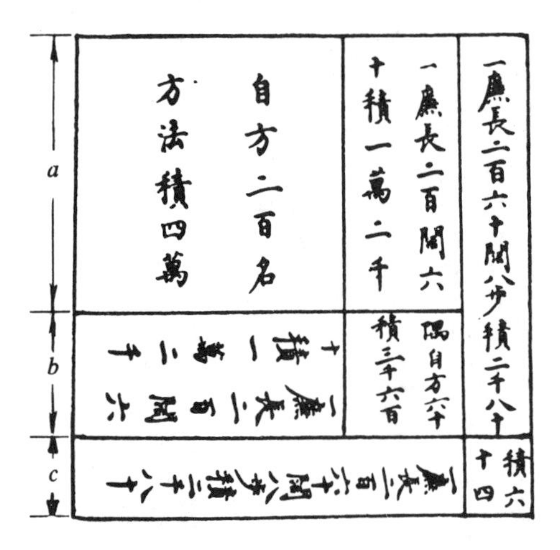

Figure 8 A geometric "proof" of the algebraic proposition (above) which justifies the calculations (Figure 9) leading to $\sqrt{55225} = 235$. The 1 in the upper box represents an indexing rod that determines the decimal value of the divisors used. At the beginning of the process, it is moved to the left in jumps of two decimal places until it establishes the largest power of ten that can be divided into the designated number. After each successful division, the rod is indexed two positional places to the right.

$\sqrt{N} = a + b + c$. The counting board process for extracting the square root of 55225 is briefly outlined in Figure 9. Root extraction was not limited to three digit results, for the Chinese were able to continue the process to several decimal places as needed. Decimal fractions were known and used in China as far back as the 5th century B.C. Where a root was to be extracted to several decimal places, the computers achieved greater

Algebraic Significance	Numerical entries on board
N 1	55225 1
a $N - a^2$ $a \times 10000$ 10000	2 15225 20000 10000
$a + b$ $N - a^2$ $(2a + b)b \times 100$ $(2a + b) \times 100$ 100	23 15225 12900 4300 100
$a + b$ $N - [a^2 + (2a + b)b]$ $(2a + b) \times 100$ 100	23 2325 4300 100
$a + b + c$ $N - (a + b)^2 - [2a(a + b) + c]c$ $2(a + b) + c$	235 0 465

Figure 9

accuracy by use of the formula $\sqrt[n]{m} = \sqrt[n]{m10^{kn}} / 10^k$. Cube root extraction was conceived on a similar geometric-algebraic basis and performed with equal facility.

Historians of mathematics often devote special consideration to the results obtained by ancient societies in determining a numerical value for π as they believe that the degree of accuracy achieved supplies a comparative measure for gauging the level of mathematical skill present in the society. On the basis of such comparisons, the ancient Chinese were far superior to their contemporaries in computational mathematical ability. Aided by a number system that included the decimalization of fractions and the possession of an accurate root extraction process the Chinese had obtained by the first century a value of π of 3.15147. The scholar Liu Hui in a third century commentary on the *Chiu chang* employed a "cutting of the circle method"—determining the area of a circle with known radius by polygonal approximations—to determine π as 3.141024. A successor, Tsu Chung-chih, refined the method in the fifth century to derive the value of π as 355/113 or 3.1415929. This accuracy was not to be arrived at in Europe until the 16th century.

Trends in Chinese Algebraic Thought

While the Chinese computational ability was indeed impressive for the times, their greatest accomplishments and contributions to the history of mathematics lay in algebra. During the Han period, the square and cube root extraction processes were being built upon to obtain methods for solving quadratic and other higher order numerical equations. The strategy for extending the square root process to solve quadratic equations was based on the following line of reasoning. If $x^2 = 289$, 10 would be chosen as a first entry approximation to the root, then

$$289 - (10)^2 = 189.$$

Let the second entry of the root be represented by y; thus, $x = 10 + y$ or $(10 + y)^2 = 289$ which, if expanded, gives the quadratic equation $y^2 + 20y - 189 = 0$. By proceeding to find the second entry of the square root of 289, 7, we obtain the positive root for the quadratic $y^2 + 20y - 189 = 0$.

By the time of Sung Dynasty in the 13th century, mathematicians were applying their craft to solve such challenging problems as:

> This is a round town of which we do not know the circumference or diameter. There are four gates (in the wall). Three *li* from the northern (gate) is a high tree. When we go outside of the southern gate and turn east, we must walk 9 *li* before we see the tree. Find the circumference and the diameter of the town.
>
> (1 *li* = .644 kilometers)

If the diameter of the town is allowed to be represented by x^2, the distance of the tree from the northern gate, a, and the distance walked eastward, b, the following equation results:

$$x^{10} + 5ax^8 + 8a^2x^6 - 4a(b^2 - a^2)x^4 - 16a^2b^2x^2 - 16a^3b^2 = 0.$$

For the particular case cited above, the equation becomes

$$x^{10} + 15x^8 + 72x^6 - 864x^4 - 11{,}664x^2 - 34{,}992 = 0.$$

Sung algebraists found the diameter of the town to be 9 *li*.

The earliest recorded instance of work with indeterminate equations in China can be found in a problem situation of the *Chiu chang* where a system of four equations in five unknowns results. A particular solution is supplied. A problem in the third century *Mathematical Classic of Sun Tzu,* [*Sun Tzu suan ching,*] concerns linear congruence and supplies a truer example of indeterminate analysis.

> We have things of which we do not know the number; if we count by threes, the remainder is 2; if we count by fives, the remainder is 3; if we count by sevens, the remainder is 2. How many things are there?

In modern form, the problem would be represented as:

$$N \equiv 2 \pmod 3 \equiv 3 \pmod 5 \equiv 2 \pmod 7.$$

Sun's solution is given by the expression

$$70 \times 2 + 21 \times 3 + 15 \times 2 - 105 \times 2 = 23$$

which when analyzed gives us the first application of the Chinese Remainder Theorem.

If $m_1, \ldots, m_k$ are relatively prime in pairs, there exist integers x for which simultaneously $x \equiv a_1 \pmod{m_1}, \ldots, x \equiv a_k \pmod{m_k}$. All such integers x are congruent modulo $m = m_1m_2 \ldots m_k$. The existence of the Chinese Remainder Theorem was communicated to the west by Alexander Wylie, an English translator and mathematician in the employ of the nineteenth century Chinese court. Wylie recorded his findings in a series of articles, "Jottings on the Science of the Chinese; Arithmetic" which appeared in the *North China Herald* (Aug.–Nov.) 1852. The validity of the theorem was questioned until it was recognized as a variant of a formula developed by Gauss.

Perhaps the most famous Chinese problem in

indeterminate analysis, in the sense of its transmission to other societies, was the problem of the "hundred fowls" (ca. 468).

> A cock is worth 5 *ch'ien,* a hen 3 *ch'ien,* and 3 chicks 1 *ch'ien.* With 100 *ch'ien* we buy 100 fowls. How many cocks, hens, and chicks are there? (*ch'ien,* a small copper coin)

The development of algebra reached its peak during the later part of the Sung and the early part of the following Yuan dynasty (13th and 14th centuries). Work with indeterminate equations and higher order numerical equations was perfected. Solutions of systems of equations were found by using methods that approximate an application of determinants, but it wasn't until 1683 that the Japanese Seki Kowa, building upon Chinese theories, developed a true concept of determinants.

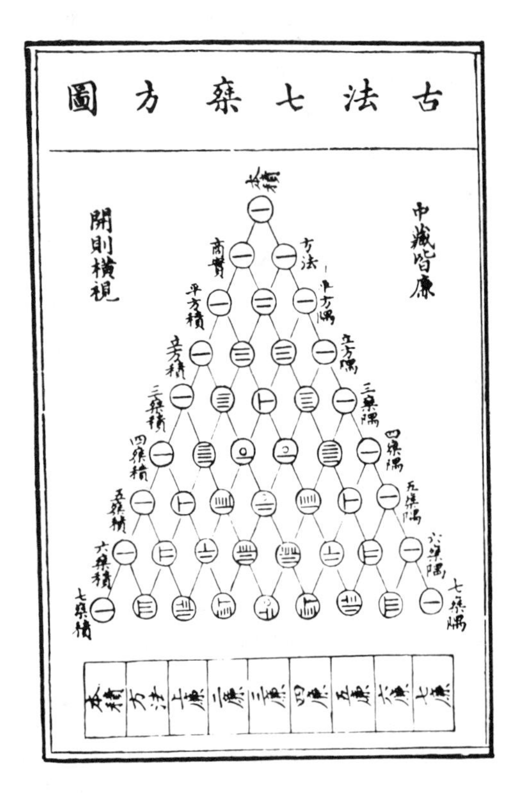

Figure 10

Work with higher numerical equations is facilitated by a knowledge of the binomial theorem. The testimony of the *Chiu chang* indicates that its early authors were familiar with the binomial expansion $(a + b)^3$, but Chinese knowledge of this theorem is truly confirmed by a diagram (Figure 10) appearing in the 13th century text *Detailed Analysis of the Mathematical Rules in the Nine Chapters.* [*Hsiang chieh chiu chang suan fa.*]. It seems that "Pascal's Triangle" was known in China long before Pascal was even born.

While mathematical activity continued in the post-Sung period, its contributions were minor as compared with those that had come before. By the time of the Ming emperors in the 17th century, western mathematical influence was finding its way into China and the period of indigenous mathematical accomplishment had come to an end.

Conclusions

Thus, if comparisons must be made among the societies of the pre-Christian world, the quality of China's mathematical accomplishments stands in contention with those of Greece and Babylonia, and during the period designated in the West as pre-Renaissance, the sequence and scope of mathematical concepts and techniques originating in China far exceeds that of any other contemporary society. The impact of this knowledge on the subsequent development of western mathematical thought is an issue that should not be ignored and can only be resolved by further research. In part, such research will have to explore the strength and vitality of Arabic-Hindu avenues of transmission of Chinese knowledge westward. The fact that western mathematical traditions are ostensibly based on the logico-deductive foundations of early Greek thought should not detract from considering the merits of the inductively-conceived mathematics of the Chinese. After all, deductive systemization is a luxury afforded only after inductive and empirical experimentation has established a foundation from which theoretical considerations can proceed. Mathematics, in its primary state, is a tool for societal survival; once that survival is assured, the discipline can then become more of an intellectual and aesthetic pursuit. Unfortunately, this second stage of mathematical development never occurred in China. This phenomenon—the fact that mathematics in China, although developed to a high art, was never elevated further to the status of an abstract deductive science—is yet another fascinating aspect of Chinese mathematics waiting to be explained.

Notes

1. Morris Kline, *Mathematics: A Cultural Approach* (Reading, Mass.: Addison-Wesley Publishing Co., 1962), p. 12.
2. John Greenleaf Whittier, "The Chapel of the Hermits."
3. Under this system, the universe is ruled by Heaven through means of a process called the *Tao* ("the Universal way"). Heaven acting through the Tao expresses itself in the interaction of two primal forces, the *Yin* and the *Yang*. The *Yang*, or male force, was a source of heat, light and dynamic vitality and was associated with the sun; in contrast, the *Yin*, or female force, flourished in darkness, cold and quiet inactivity and was associated with the moon. In conjunction, these two forces influenced all things and were present individually or together in all physical objects and situations. In the case of numbers, odd numbers were *Yang* and even, *Yin*. For a harmonious state of being to exist, *Yin-Yang* forces had to be balanced.
4. Diophantus (275 A.D.) spoke of unknowns of the first number, second number, etc., whereas Brahmagupta (628 A.D.) used different colors in written computations to distinguish between variables.
5. For a discussion of the Chinese ability at root extraction, see Wang Ling and Joseph Needham, "Horner's Method in Chinese Mathematics: Its Origins in the Root Extraction Procedures of the Han Dynasty", *T'oung Pao* (1955), 43: 345-88; Lam Lay Yong, "The Geometrical Basis of the Ancient Chinese Square-Root Method", *Isis* (Fall, 1970), pp. 92–101.

The Muslim Contribution to Mathematics

Ali Abdullah Al-Daffa'

Ali Abdullah Al-Daffa' is currently professor of mathematics at the University of Petroleum and Minerals in Saudi Arabia. He has written extensively on the history of mathematics in Islam.

Europe was in the Dark Ages from 400 A.D. to 1500 A.D. Too often we make the mistake of assuming that the rest of the world was in the Dark Ages then as well. The Muslim civilization originated in the Prophet Muhammad's Hegira in 622 A.D. Within a century of this time the Muslim empire stretched from Spain to India. This culture encouraged the development of the arts and sciences while its common religion and language provided a skeleton upon which its civilization was formed from parts of the old Byzantine, Egyptian, Roman, and other empires. The rich mathematical works of the Hindus and the Greeks fused in new ways under the religious philosophy of the Muslims.

The European Renaissance of 1400 to 1600 A.D. was dependent on the Greek documents that the Muslims transmitted. But the Muslims did much more than act as caretakers of the Greek heritage. Probably the greatest contributions of the Muslim mathematicians were in the field of algebra. Al-Daffa' surveys some of the contributions of Al-Khwarizmi, the most important mathematician of the medieval period. The industrial and scientific revolutions of the West would have been utterly impossible without algebra. The current Western civilization owes much to the Muslim contributions made during our Dark Ages.

IT WOULD BE an injustice to pioneers in mathematics to stress modern mathematical ideas with little reference to those who initiated the first and possibly the most difficult steps. Nearly everything useful that was discovered in mathematics before the seventeenth century has either been so greatly simplified that it is now part of every regular school course, or it has long since been absorbed as a detail in some work of greater generality.

The Muslims translated numerous Greek works in mathematics as they did in other fields of science. At the same time, they turned to the East and gathered all that was available in India in the way of science, and particularly in mathematics. In the field of algebra the Muslims soon made original contributions which proved to be the greatest of their distinctive achievements in mathematics.

Definition of Algebra

Algebra is that "branch of mathematical analysis which reasons about quantities using letters to symbolize them."[1] Algebra is defined in the *Mathematics Dictionary* as "a generalization of arithmetic; e.g., the arithmetic facts that $2 + 2 + 2 = 3 \times 2$, $4 + 4 + 4 = 3 \times 4$, etc., are all special cases of the (general) algebraic statement

Source: Excerpts from Ali Abdullah Al-Daffa', *The Muslim Contribution to Mathematics,* 1977, 49–63. Reprinted by permission of Humanities Press Inc., Atlantic Highlands, N.J., and Croom Helm Ltd. Publishers, Kent, England. (Some references omitted.)

that $x + x + x = 3x$, where x is any number."[2] The Muslim scholar ibn Khaldun defined algebra as a "subdivision" of arithmetic. This is a craft in which it is possible to discover the unknown from the known data if there exists a relationship between them.

In the ninth century the Muslim mathematician Al-Khwarizmi wrote his classical work on algebra, *Al-Jabr wa-al-Muqabala*. In this title the word *Al-Jabr* means transposing a quantity from one side of an equation to another, and *Muqabala* signified the simplification of the resulting expressions. Figuratively, *al-jabr* means restoring the balance of an equation by transposing terms. Because of the double title, the explanation given contains a comment upon the second word, *al-muqabala,* as well as the first one. According to David Eugene Smith:

> In the 16th century it is found in English as *algiebar* and *almachabel,* and in various other forms but was finally shortened to *algebra*. The words mean restoration and opposition, and one of the clearest explanations of their use is given by Beha Eddin (1600 A.D.) in his *Kholasat Al-Hisab* (essence of arithmetic): The member which is affected by a minus sign will be increased and the same added to the other member, this being algebra; the homogeneous and equal terms will then be cancelled, this being *Al-Muqabala.*[3]

That is, given

$$x^2 + 5x + 4 = 4 - 2x + 5x^3,$$

Al-Jabr gives

$$x^2 + 7x + 4 = 4 + 5x^3,$$

Al-Muqabala gives

$$x^2 + 7x = 5x^3.$$

Therefore, the best translation for *Hisab Al-Jabr Wa-al-Muqabala* is "the science of equations."

The Origin of the Term "Algebra"

Al-Khwarizmi's text of algebra, entitled *Al-Jabr Wa-Al-Muqabala* (the science of cancellation and reduction) was written in 820 A.D. A Latin translation of this text became known in Europe under the title *Al-Jabr*. Thus "the Arabic word for reduction, *al-Jabr,* became the word *algebra.*"[4]

Al-Khwarizmi

From the eighth to the thirteenth centuries, the center of scientific activity was Arabia. Scientific activity was centered in the Muslim world, especially at the court of the Caliph Al-Ma'mum. It was there that Al-Khwarizmi (825 A.D.) influenced mathematical thought more than any other medieval writer by finding a system of analysis for solving equations of first- and second-degree in one unknown by both algebraic and geometric means.

The first half of the ninth century is characterized by Sarton in his *Introduction to the History of Science* as "the time of Al-Khwarizmi," because he was "the greatest mathematician of the time, and if one takes all circumstances into account, one of the greatest of all times."[5] E. Wiedmann has said: "His works, which are in part important and original, reveal in Al-Khwarizmi a personality of strong scientific genius."[6]

David Eugene Smith and Louis Charles Karpinski characterized Al-Khwarizmi as:

> . . . the great master of the golden age of Baghdad, one of the first of the Muslim writers to collect the mathematical classics of both the East and the West, preserving them and finally passing them on to the awakening Europe. This man was . . . a man of great learning and one to whom the world is much indebted for its present knowledge of algebra and of arithmetic.[7]

It was Mohammad Khan who stated:

> In the foremost rank of mathematicians of all times stands Al-Khwarizmi. He composed the oldest works on arithmetic and algebra. They were the principal source of mathematical knowledge for centuries to come both in the East and the West. The work on arithmetic first introduced the Hindu numbers to Europe, as the very name algorism signifies; and the work on algebra not only gave the name to this important branch of mathematics in the

> European world, but contained in addition to the usual analytical solution of linear and quadratic equations (without, of course, the conception of imaginary quantities) graphical solution of typical quadratic equations.[8]

The mathematics that the Muslims inherited from the Greeks made the division of an estate among the children extremely complicated, if not impossible. It was the search for a more accurate, comprehensive, and flexible method that led Al-Khwarizmi to the innovation of algebra. While engaged in astronomical work at Baghdad and Constantinople, he found time to write the algebra which brought him fame. His book, *Al-Kitab al-Mukhtasar fi hisab al-jabr wa-al-Muqabala,* is devoted to finding solutions to practical problems which the Muslims encountered in daily life.

> In evolving his algebra, Al-Khwarizmi transformed the number from its earlier arithmetical character as a finite magnitude into an element of relation and of infinite possibilities. It can be said that the step from arithmetic to algebra is in essence a step from "being" to "becoming" or from the static universe of the Greek to the dynamic ever-living, God-permeated one of the Muslims.[9]

Al-Khwarizmi emphasized that he wrote his algebra book to serve the practical needs of the people concerning matters of inheritance, legacies, partition, lawsuits, and commerce. He dealt with the topic which in Arabic is known as *'ilm al-fara'id* (the science of the legal shares of the natural heirs). Gandz stated in *The Source of Al-Khwarizmi's Algebra:*

> Al-Khwarizmi's algebra is regarded as the foundation and cornerstone of the sciences. In a sense, Al-Khwarizmi is more entitled to be called "the father of algebra" than Diophantus because Al-Khwarizmi is the first to teach algebra in an elementary form and for its own sake. Diophantus is primarily concerned with the theory of numbers.[10]

In the twelfth century the algebra of Al-Khwarizmi was translated into Latin by Gerhard of Cremona and Robert of Chester. It was used by Western scholars until the sixteenth century. Of the translation by Robert of Chester, Sarton judicially remarked: "The importance of this particular translation can hardly be exaggerated. It may be said to mark the beginning of European algebra."[11]

After Al-Khwarizmi, there were many other Muslims who studied and taught algebra, but they made few new discoveries. They were content to know what he had written in his great book.

Roots

The term "root" had its origin in the Arab language. "Latin works translated from the Arabic have *radix* for a common term, while those inherited from the Roman civilization have *latus.*" *Radix* (root) is the Arabic *jadhr,* while *latus* referred to the side of a geometric square.[12]

The Arabic word for root was used by Al-Khwarizmi to denote the first-degree term of a quadratic equation. Al-Khwarizmi wrote, "the following is an example of squares equal to roots: a square is equal to 5 roots. The root of the square then is 5, and 25 forms its square, which of course equals 5 of its roots."[13]

Square Root. The method of extracting the square root employed by the Muslims resembled their method of division. For example, to find the square root of 107584, vertical lines are drawn and numerals are partitioned into periods of two digits. See details in Figure 1. The nearest root of 10 is 3, which is placed both below and above, and its square, 9, is subtracted from 10. The 3 is now doubled and the result is written in the next column. Six is contained twice in 17, the remainder with first figure of the next period. The 2 is set down both above and below, and being multiplied by 6 gives 12, which subtracted from 17, leaves 5. The square of 2, or 4, is now subtracted from 55. The difference 51, together with the succeeding figure, or 518, is divided by the double of 32, or 64, giving 8 for the quotient. Then 8 times 64, or 512, is subtracted from 518 with a difference of 6. This digit together with the succeeding fig-

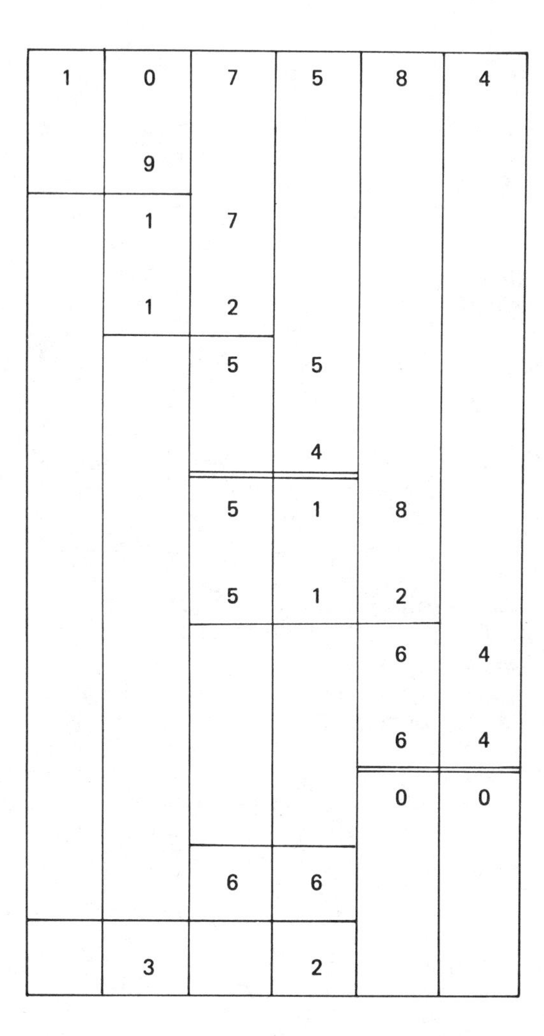

Figure 1 An illustration of the Arabian method for extracting square root.

ure forms 64 which is exhausted by subtracting from it the square of 8. Therefore, the square root of 107584 is 328. It has been said that this method was adopted from the Muslims by the Hindus.

Al-Karkhi, another Muslim mathematician, employed a method of approximation to find the square root of numbers using the formula:

$$\sqrt{a} = w + \frac{a - w^2}{2w + 1}.$$

The approximate root for $\sqrt{17} = 4 = w$. Therefore

$$\sqrt{17} = 4 + \frac{17 - 16}{2(4) + 1} = 4\frac{1}{9} = 4.111$$

and this value checks fairly well with 4.123106, the value precise to six figures.

Linear and Quadratic Equations

The Egyptians solved equations of the first degree more than four thousand years ago. That is, they found that the solution of the equation $ax + b = 0$ is $x = -b/a$. The graph of the equation is represented in geometry by a straight line. The quadratic equation, however, $ax^2 + bx + c = 0$ was solved by the Muslims with the formula:

$$x = \frac{-b \pm \sqrt{(b^2 - 4ac)}}{2a}$$

The various conic sections, such as the circle, the ellipse, the parabola, and the hyperbola are the geometric representations of quadratic equations in two variables which were studied by the Muslims.

In his work on linear and quadratic equations, Al-Khwarizmi used special technical terms for the various multiples or powers of the unknown. The unknown is referred to as a "root" and the unknown squared is called the "power." With this vocabulary, Al-Khwarizmi would describe general linear equations as "roots equal to numbers." In present-day notation it would appear as $bx = c$. Instances of linear equations are: one root equals three, $x = 3$; four roots equal twenty, $4x = 20$; and one-half a root equals ten $(1/2)x = 10$.

Al-Khwarizmi separated general quadratic equations into five cases for purposes of finding the solution to a given equation. The five cases he considered were: (1) squares equal to roots, $ax^2 = bx$; (2) squares equal to numbers, $ax^2 = b$; (3) squares and roots equal to numbers, $ax^2 + bx = c$; (4) squares and numbers equal to roots, $ax^2 + c = bx$; and (5) squares equal to roots and numbers, $ax^2 = bx + c$. In all applications, Al-Khwarizmi considered a, b, c positive integers with $a = 1$. He was concerned with only positive real roots, but he recognized the existence of a second root which was not conceived of previously. Examples are given of cases (3), (4), and

(5) above to illustrate the methods of Al-Khwarizmi.

Case (3): Square and roots equal to numbers, $x^2 + 10x = 39$. Construct the square ABCD with side AB = x. Extend AD to E and AB to F such that DE = BF = (1/2)10 = 5, then complete the square AFKE. By extending DC to G and BC to H, the area of square AFKE may be expressed as $x^2 + 10x + 25$. However, the equation to be solved is $x^2 + 10x = 39$. Therefore, 25 must be added to each member of this equation to yield $x^2 + 10x + 25 = 39 + 25 = 64$, which is the required area. In other words, $x^2 + 10x + 25$ is a perfect square $(x+5)^2$, and this is equal to another perfect square, 64. Hence, the dimensions of the area $(x + 5)^2$ must be 8 by 8. However, since AF = $x + 5 = 8$, this means $x = 3$.

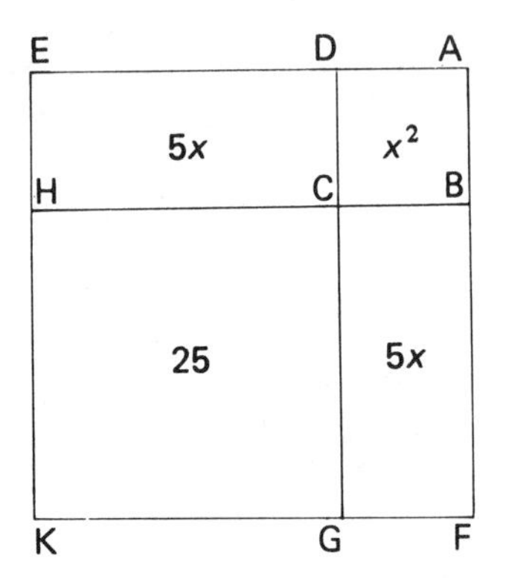

Figure 2 Paradigm of Al-Khwarizmi for quadratic equation $x^2 + 10x = 39$.

Case (4): Squares and numbers equal to roots, $x^2 + 21 = 10x$, $x<b/2$, *where a is coefficient of* x. Construct a rectangle ABCD with side AB = x and BC = 10. The area of rectangle ABCD = $10x = x^2 + 21$. On side BC mark off a point E such that BE = BA, then complete the square ABEF. It follows that the area of rectangle CDFE = 21. Let H be the midpoint of BC. Extend side CD to N such that CN = CH = 5 and complete the square HCNM, whose area is 25. From I, the midpoint of AD, construct the point S such that IS = IF = $5 - x$ and complete the square MISW, whose area is $(5 - x)^2$. Since DS = x, the area of rectangle DSWN = $x(5 - x)$ = the area of rectangle FEHI. Therefore, the area of rectangle CDIH plus the area of rectangle

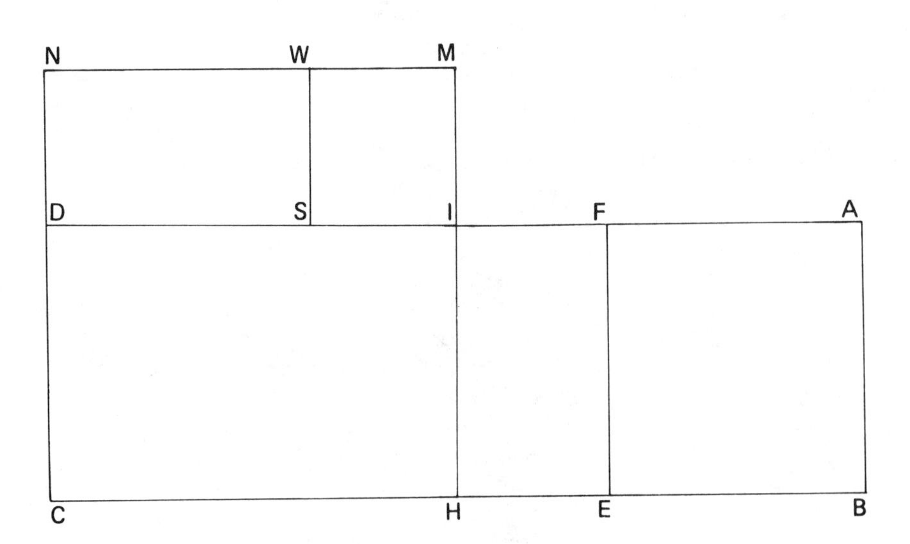

Figure 3 Paradigm of Al-Khwarizmi for quadratic equation $x^2 + 21 = 10x$.

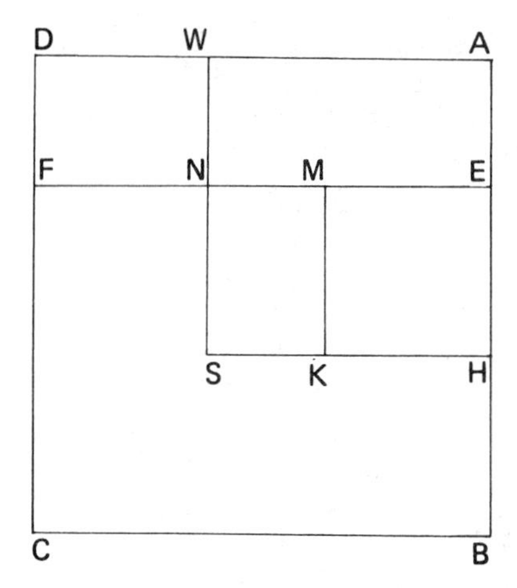

Figure 4 Paradigm of Al-Khwarizmi for quadratic equation $x^2 = 3x + 4$.

DSWN = 21. Thus the area of square HCNM = the area of rectangle CDFE + the area of square WSIM = $21 + (5 - x)^2 = 25$. Then $(5 - x)^2 = 4$ or $x = 3$.

Case (5): Square equal to roots and numbers, $x^2 = 3x + 4$. Construct square ABCD with sides = x. Select a point E on side AB such that BE = 3 and complete the rectangle BEFC. The area of rectangle BEFC = $3x$ and the area of rectangle AEFD = 4. Let H bisect segment EB, and construct square EHKM with area = 9/4. By extending HK to S such that KS = AE = DF and constructing SW perpendicular to DA, rectangles MKSN and DWNF have equal areas. The equality of these areas follows from DW = HB = HE = KM. The area of square AHSW is (SENW + MKSN) + EHKM or 4 + 9/4. The side AH = 5/2 and the side AB = AH + HB or 5/2 + 3/2, therefore $x = 4$.

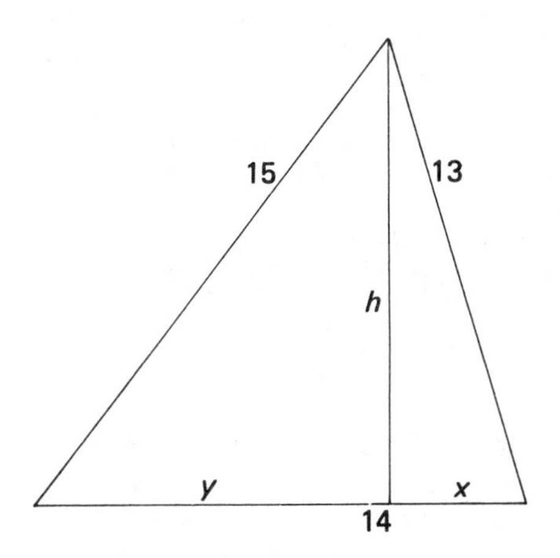

Figure 5 Paradigm for Al-Khwarizmi's triangle.

Al-Khwarizmi gave an algebraic method for finding the altitude and the two segments of the base formed by the foot of the altitude, x and y, of the triangle when the three sides (13, 14, and 15) are given, as in Figure 5. The square of the height, h^2, is equal to $13^2 - x^2 = 15^2 - y^2 = 15 - (14 - x)^2$. Hence $169 - x^2 = 255 - 196 + 28x - x^2$. And upon simplification, $x = 5$; and it follows that $h^2 = 169 - 25$ and that $h = 12$.

The change from the Greek conception of a static universe to a new dynamic one was initiated by Al-Khwarizmi who was the herald of modern algebra, and the first mathematician to make algebra an exact science. After dealing with equations of second degree, Al-Khwarizmi discussed algebraic multiplication and division.

Miscellaneous

Following the period of Al-Khwarizmi's works came those of Thabit ibn Qurra (836–901 A.D.), mathematician and linguist. His chief contribution to mathematics was in his translations of Euclid, Archimedes, Apollonius, and Ptolemy. Fragments of some original writing in the area of algebraic geometry have also been preserved. This particular branch of algebra received considerable attention from Muslim mathematicians. According to Karl Fink:

> Al-Khwarizmi calls a known quantity a number, the unknown quantity *jidr* (root) and its square *mal* (power). In Al-Karkhi we find the expression *kab* (cube) for the third power, and there are formed from these expressions mal mal = x^4, mal kab = x^5, kab kab = x^6, mal mal kab = x^7, etc.[14]

In his *History of Mathematics,* David Eugene Smith stated:

> . . . Al-Haitham of Basra, who wrote on algebra, astronomy, geometry, gnomic, and optics attempted the solution of the cubic equation by the aid of conics. . . .[15]

The Muslims discovered the theorem that for integers the sum of two cubes can never be a cube. This theorem was later rediscovered by P. Fermat, a French physician, and claimed by him. Creditable work in number theory and algebra was done by Al-Karkhi of Baghdad, who lived at the beginning of the eleventh century. His treatise on algebra is sometimes considered the greatest algebraic work of Muslim mathematicians; it shows the influence of Diophantus. For the solution of quadratic equations, he gave both arithmetical and geometrical proofs.

Al-Karkhi's work contains basic algebraic theory with application to equations and especially to problems to be solved for positive rational numbers. For instance, to find two numbers the sum of whose cubes is a square number yields the algebraic expression: $x^3 + y^3 = z^2$. To solve the equation in rational numbers, let:

$$y = mx,\ z = nx;$$
$$x^3 + m^3x^3 = n^2x^2;$$
$$x^3(1 + m^3) = n^2x^2.$$

By cancellation of x^2, therefore,

$$x = \frac{n^2}{1 + m^3},$$

where m and n are arbitrary positive rational numbers. As a special solution, Al-Karkhi gave the following values: $x = 1, y = 2, z = 3$. The same method is clearly applicable to many more general rational problems having the form

$$ax^n + by^n = cz^{n-1}.$$

One of the oldest methods for approximating the real root of an equation $ax + b = 0$ is often called the rule of double false position. The Muslims called the rule *hisab al-Khataayn;* it is found in the works of Al-Khwarizmi. This rule seems to have come from India, but it was the Muslims who made it known to European scholars. In order to explain the rule, let g_1 and g_2 be any guessing values of x, and let f_1 and f_2 be the errors. Therefore, if the guesses were right, then $ag_1 + b = 0$; $ag_2 + b = 0$. However, if the guesses were wrong, then

$$ag_1 + b = f_1 \qquad (1)$$
$$ag_2 + b = f_2 \qquad (2)$$
$$a(g_1 - g_2) = f_1 - f_2 \qquad (3) \quad \text{by subtraction of (2) from (1)}$$

From (1)

$$ag_1g_2 + bg_2 = f_1g_2$$

and from (2) by subtraction

$$ag_1g_2 + bg_1 = f_2g_1.$$

Therefore

$$b(g_2 - g_1) = f_1g_2 - f_2g_1. \qquad (4)$$

Dividing (4) by (3)

$$\frac{b(g_2 - g_1)}{a(g_1 - g_2)} = \frac{f_1g_2 - f_2g_1}{f_1 - f_2}$$

or

$$\frac{-b}{a} = \frac{f_1g_2 - f_2g_1}{f_1 - f_2}$$

But, since

$$\frac{-b}{a} = x,$$

therefore,

$$x = \frac{f_1g_2 - f_2g_1}{f_1 - f_2}.$$

Suppose, for example, that $2x - 5 = 0$; guessing value for x: $g_1 = 5$, $g_2 = 1$.

Then, $2 \cdot 5 - 5 = 5 = f_1$ and $2 \cdot 1 - 5 = -3 = f_2$.

But

$$x = \frac{f_1g_2 - f_2g_1}{f_1 - f_2} = \frac{5 \cdot 1 - (-3) \cdot 5}{5 - (-3)} = \frac{20}{8} = 2.5$$

According to Howard Eves, this method was used by the Muslims and can be illustrated geometrically by letting x_1 and x_2 be two numbers lying close to and on each side of a solution x of the equation $f(x) = 0$. The intersection with the

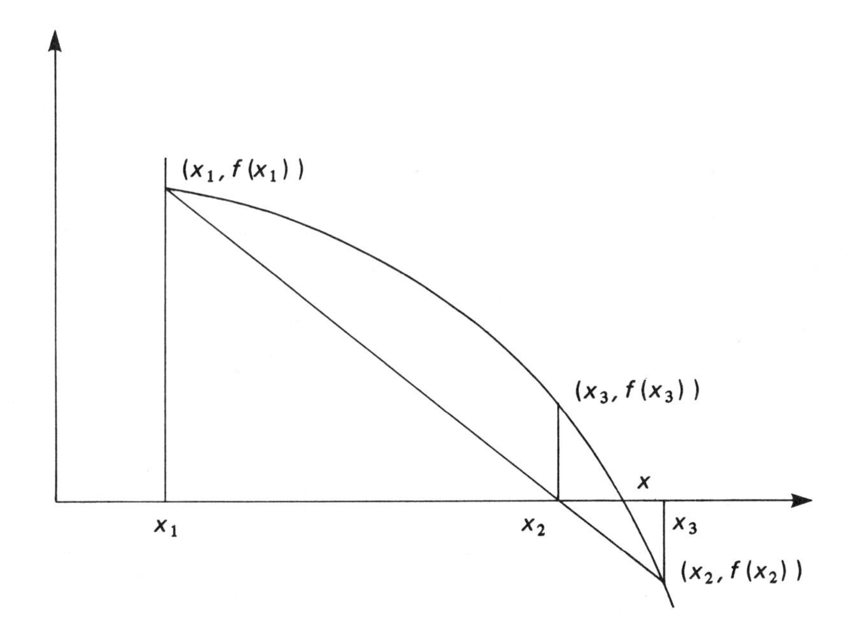

Figure 6 Rule of double false position of Al-Khwarizmi.

x-axis of the chord joining the points $(x_1, f(x_1))$, $(x_2, f(x_2))$ gives a better approximation to the solution:

$$x_3 = \frac{x_2 f(x_1) - x_1 f(x_2)}{f(x_1) - f(x_2)}.$$

The process can now be applied with appropriate pairs x_1, x_3 or x_3, x_2 depending on circumstances. This is the numerical method of "False Position" (*Regular Falsi*) which is used in numerical analysis today.

Summary

The Muslims not only created algebra, which was to become the indispensable instrument of scientific analysis, but they laid the foundations for methods in modern experimental research by the use of mathematical models. Since Mohammed ibn Musa Abu Djefar Al-Khwarizmi was the founder of the Muslim school of mathematics, the subsequent Muslim and early medieval works on algebra were largely founded on his algebraic treatise. Al-Khwarizmi's work plays an important role in the history of mathematics, for it is one of the main sources through which Arabic numerals and Muslim algebra came to Europe.

The contribution of Muslim mathematicians to the field of algebra includes methods for finding the solution to linear and quadratic equations. Solutions to these equations were also given by geometric methods. Al-Karkhi contributed rational solutions to certain special equations of degree higher than two and a method for approximating the solution to linear equations. These are but a few of the more outstanding developments in algebra that resulted directly from the efforts of Muslim mathematicians.

Notes

1. Isaac Funk, Calvin Thomas, and Frank H. Vizetelly (Supervisors), *New Standard Dictionary of the English Language* (New York, Funk and Wagnalls Company, 1940), p. 70.
2. Glenn James and Robert C. James (eds.), *Mathematics Dictionary* (New York, D. Van Nostrand Company, 1963), p. 17.
3. David Eugene Smith, *History of Mathematics*

(New York, Ginn and Company, 1925), Vol. II, p. 388.
4. Solomon Gandz, "The Origin of the Term Algebra," *The American Mathematical Monthly,* XXXIII (May 1926), 437.
5. George Sarton, *Introduction to the History of Science* (Baltimore, The Williams and Wilkins Company, 1927), Vol. I, p. 563.
6. M. Th. Houtsma, T. W. Arnold, R. Basset, and R. Hartmann (eds.), *The Encyclopaedia of Islam* (London, Luzac and Company, 1913), Vol. I, p. 912.
7. David Eugene Smith and Louis Charles Karpinski, *The Hindu-Arabic Numbers* (Boston and London, Ginn and Company, 1911), pp. 4–5.
8. Mohammad Abdur Rahman Khan, *A Brief Survey of Moslem Contribution to Science and Culture* (Lahore, Sh. Umar Daraz at the Imperial Printing Works, 1946), pp. 11–12.
9. Rom Landau, *Arab Contribution to Civilization* (San Francisco, The American Acadamy of Asian Studies, 1958), p. 33.
10. Solomon Gandz, "The Source of Al-Khwarizmi's Algebra," *Osiris* (Bruges, Belgium, The Saint Catherine Press Ltd., 1936), Vol. I, p. 264.
11. George Sarton, *Introduction to the History of Science* (Baltimore, The Williams and Wilkins Company, 1953), Vol. II, Part I, p. 176.
12. Solomon Gandz, "Arabic Numerals," *American Mathematical Monthly,* XXXIII (January 1926), 261.
13. Philip S. Jones, ' "Large" Roman Numerals,' *The Mathematics Teacher,* CXVII (March 1954), 196.
14. Karl Fink, *A Brief History of Mathematics* (The Open Court Publishing Company, 1900), p. 75.
15. David Eugene Smith, *History of Mathematics* (New York, Dover Publications, 1958), Vol. I, pp. 175–6.

The Japanese Mathematics

Yoshio Mikami

Yoshio Mikami was educated at the Imperial University in Tokyo when it was under the direct sponsorship of the Emperor Meiji. He was, in the early part of this century, the foremost interpreter of the history of the exact sciences in Japan for Western audiences.

Yoshio Mikami's article may read in parts like a parody, but it is the real McCoy. Phrases such as "while the controversy was furiously raging between Fujita and Aida, there was living a worthy person enjoying deep-going studies in quiet" and "Wada lived in poverty, and being very loose in character and given to a drinking habit, it is said, he sometimes sold his solutions to make a little money for more drink" make it easy to write parodies of the Japanese art of understatement.

Mikami's article was originally written in 1913, merely forty-five years after the Restoration of 1868, the event that marked Japan's entry into the Western world. This was the Japan that had just finished humiliating Russia. This was the Japan of which Mikami could say without the slightest embarassment, "We don't know, however, of any material influence the Occidental mathematics exercised upon learning in the Land of the Rising Sun."

An intensely proud nationalist, Mikami explains, ". . . the improvement of mathematical studies in Japan after that remarkable event may at the first glance be wondered at as having been exceedingly rapid." The reader who is aware of the remarkable and rapid growth of the Japanese computer industry might wonder why technology in some countries seems to take off and in others seems to stagnate.

Mikami's explanation of the Japanese mathematics is hardly exciting and appears primitive. But the writings of a mathematician one generation removed from the Westernization of Japan give us a unique account of the mathematics of a culture before and after an important turning point. It is as significant as would be the account of an Egyptian scribe forty-five years after the Greek conquest of Egypt and the introduction of the Pythagorean mathematical philosophy.

Finally, notice the title "The Japanese Mathematics." This is not a slight mistranslation; it is part of Mikami's attitude, as evidenced in such phrases as "the circular theory as improved by Ajima is strikingly Japanese in character" and "the Japanese mathematics was destitute of geometry." Mikami thinks of mathematics in nationalistic terms that we may find strange. His article helps us understand how the Newton-Leibniz controversy and the subsequent development of "Continental" mathematics could really have occurred. Perhaps a parallel in modern America is the concern for a "Christian economics" or a "Christian biology."

Source: Excerpts from Yoshio Mikami, *The Development of Mathematics in China and Japan*, 2nd edition, 1974, pp. 156–165, 168–178. Reprinted by permission of Chelsea Publishing Company. (References omitted.)

A General View of the Japanese Mathematics

Of the mathematical knowledge possessed by the Japanese in the oldest times we know practically nothing. After the introduction of Buddhism through Korea in the 6th century, something of the Chinese mathematics was brought to Japan and studied for some time by native scholars. But their knowledge must have been exceedingly limited, for they did not ever rise to the point of carrying on their work independently. Blind adherence to Chinese masters was all that they could do. Even such a menial state did not continue very long, for with the loosening of political regulations learning too shared the same fate, gradually degenerating until at last practically nothing remained.

It was after a reintroduction of the Chinese mathematics that the Japanese became really observant of mathematical studies. After the long duration of the dark ages the 17th century at last saw the science taken up into favor for a second time, when there appeared various scholars, in whose hands the Japanese mathematics properly so called was destined to dawn and flourish for more than two centuries that followed.

The first Japanese mathematician noted in history was Mōri Kambei who lived at the beginning of the 17th century. He is sometimes referred to as having traveled to China, but the story is not certainly authentic.

The Chinese abacus *suan-pan,* which the Japanese call *soroban,* has extensively been used in Japan since Mōri's days. As is usually believed, it was first brought by him from China, but its introduction seems to be more ancient. In any case, however, Mōri was a dexterous hand in the art of manipulating the instrument. He taught the *soroban* arithmetic at Kyōto, when a multitude of pupils crowded around him. In popularizing the abacus Mōri was certainly no little instrumental. Mōri published, it is said, a treatise on the use of the *soroban,* but it has since been lost.

Of Mōri's numerous pupils there were three, who distinguished themselves specially. These three were Yoshida Kōyū, Imamura Chishō and Takahara Kisshu. The *Sanshi,* meaning "Three Honorable Mathematicians or Scholars," was the appellation respectfully given them by their contemporaries.

Yoshida's *Jinkōki* of 1627 was a successful publication. After its appearance the *soroban* arithmetic experienced a rapid progress. The book was destined to earn such a popularity that it went through various editions even in the life-time of its illustrious author. Sometimes booksellers made reprints of it without the author's knowledge or permission. In after ages there were published tens of treatises with the same title, to which some prefixes were added.

Yoshida's work was followed by various other publications, which were, like it, mostly based on the contents of the Chinese Ch'êng Tai-wei's *Suan-fa T'ung-tsung* of 1593. This Chinese master was reprinted in Japan in 1675.

The first half of the 17th century was a period for the Japanese mathematics, when the arithmetic practiced on the *soroban* was becoming popularized in every part of the country. But now another stimulation to the progress of the science appeared. It was the introduction of another Chinese treatise, Chu Shih-chieh's *Suan-hsiao Chi-mêng* of 1299. . . . The said treatise had been long missing in China until it was restored by the discovery of a Korean edition of 1660 and was reprinted by Lo Shih-lin in 1839. Even the learned Yüan Yüan was not able to find the book at the time of his historical studies. This work, although missing in China, had the happy destiny of being found in Japan, perhaps introduced from Korea, where it was once reprinted in 1660. But the Japanese knew it from before that date; for the first Japanese edition was given in 1658. It was widely studied by Japanese mathematicians of the 17th century and supplied them with the rudiments of the instrumental algebra that was carried on by means of the calculating pieces or *sangi,* as they were now called in Japan. This was the so-called *t'ien-yuen-shu* or the Japanese *tengen jutsu,* which literally means "the celestial element method."

The *sangi* or calculating pieces were no new makeshifts. Their use in China we have already referred to. When the Chinese mathematics was introduced for the first time from Korea to Japan, these pieces were also undoubtedly brought over, and they have remained in use ever since. But with the popularizing of the *soroban* the calculating pieces were fast giving way, for they are far more cumbrous and non-practical, when compared to the new sort of abacus. If the algebra of the celestial element method had not come to be studied, these pieces would certainly have altogether sunk into oblivion ere long. It was not however the case. While the *soroban* was suitable for the practice of daily-life, the calculating pieces served best for the discussion of complicated problems, being applicable to the solution of numerical equations of any degree. Thus it came about that the two sorts of abaci, the *soroban* and the *sangi*, had to flourish side by side. Both were practiced during the whole era of the ascendency of the old Japanese school of mathematics.

Mathematical studies had been in continual progress since the appearance of Yoshida's *Jinkōki* in 1627, but the introduction of the celestial element method contributed to considerably improve the scholars' knowledge, and it was not long afterwards that the Japanese began to enter upon independent work.

While the Chinese mathematicians of the 13th and 14th centuries studied the algebra of the celestial element, they all contented themselves with a single solution of an equation, whatever its degree might be; they did not come to discover the existence of more than one root in an equation. But the Japanese soon found that the roots of an equation are not necessarily restricted to a single one. Satō Seikō (1666) adopted various solutions to a problem that arise from the multiplicity of roots. Thereupon Sawaguchi Kazuyuki (1670) attacked the looseness of such a procedure, though he had admitted the existence of more than one solution under certain circumstances. He ascribed the origin of such solutions to the inappropriateness of the quantities constituting the data of the problems. He thus preferred to change some of the constants occurring in these data so as to yield a single solution.

Soon after the publication of the works of these scholars there appeared a great genius who is called the Japanese Newton. Seki Kōwa was his name. Seki was born in the same year in which Galileo died and Newton was born, namely in 1642. If Seki did not surpass Newton in his achievements, yet he was no inferior of the two. The same uplift Newton gave the mathematics of his country, Seki was also able to render for Japan. The Japanese mathematics properly so called had awaited his genius for its sound establishment. No doubt was Seki the father of Japanese mathematics.

It is said, Seki wrote a great number of manuscript works in his life, but what now remain of the transcripts of these manuscripts are not very numerous. It is a great pity, we must confess, that nothing is known of works in his own hand-writing. But the few written works that are still preserved are sufficient to immortalize his memory. Who can deny his having been a mathematician of the first rank?

Seki wrote very much but printed little. He did not publish any of his works, except the single book entitled *Hatsubi Sampō,* that appeared in 1674. Its appearance was however the occasion of bringing him the fame he deserved.

In a later edition of the *Jinkōki* Yoshida published some mathematical questions at the end of the book and proposed to solve them for his successors. Isomura solved these problems in his *Ketsugishō* of 1660 and published some other questions at the end of his work. These were in turn solved in some succeeding publications that were printed in a few years. In this way there arose the usage of proposing new questions and solving them in successive publications, a usage that continued for a century and a half. It was known as the usage of the *idai shōtō*—the successive proposals and solutions. It had stimulated no little the progress of mathematical studies. The works of Satō (1666) and of Sawaguchi (1670), already referred to, were, with others, publications in this line. Sawaguchi proposed fifteen new questions, and it was Seki's work that answered them. One of Seki's solutions had led to an equation of the 1458th degree. It may be better imagined than described how tremendous such an equation should have been. Seki does not, it is true, give the equation itself, but mentions it expressly. A score of years afterwards Miyagi Seikō wrote the *Wakan Sampō,* printed in 1695, in which he considered the same problem, when he gave the detailed analysis leading to the equation of the 1458th degree mentioned by Seki. The consideration of equations of wonderfully great degrees was not rare among the Japanese mathematicians of the 17th and 18th centuries. Sometimes equations whose degrees were 3000 or 4000 were taken into account. However great the degree might be, the Japanese solved them invariably by applying the celestial element method. In solving these tremendous equations the usual paper or wooden boards would be too insignificant for the arrangement of the *sangi* pieces, and therefore, it is said, a certain mathematician prepared a *sangi*-board ruling over the matts of his study or sitting room, whereupon he arranged the small pieces to solve equations. Perseverance and hard study were a part of the spirit that characterized Japanese mathematicians of old times.

The mathematical method Seki had used in writing his published work was that he called the *yendan jutsu,* which literally means "the method of explanations or of analysis." It was an extension of the celestial element algebra of the Chinese and served as the first corner-stone of the gigantic building he accomplished in the course of his famous life. If he did not illustrate the nature of the method in the printed work, only giving the results of calculations in the form of equations, yet the elegance with which he had expressed his results did not fail to advance his reputation among his contemporaries. A detailed analysis of the work was attempted a dozen years afterwards by his pupil Takebe Kenkō, and then the method became generally known to mathematical circles outside of his school.

After the establishment of the *yendan* algebra, Seki pursued his studies further and was able to arrive at the proud system of Japanese mathematics that was known by the name of *tenzan jutsu.* This name was not one that Seki himself had used. He called it the *kigen-seihō,* which was perhaps intended to mean "a method or art by which the buried origin or relations of things may be made clear." The very use of such a term is enough to show how proud he had felt at his success. Seki's *kigen-seihō* was an algebraical system in the writing style.

The name *tenzan jutsu* was adopted by Matsunaga half a century after Seki's invention of the subject, fulfilling the desire of the master he was serving, Naitō Masaki, feudal lord of Nobeoka in Kyūshū, who was himself a mathematician. The word *tenzan* consists of the two ideograms *ten* that means "taking away" and *zan* that means "taking in." It appears therefore that the constitution of the term highly resembles the Arabian origin of the term *algebra,* and thus it has been a very appropriate terminology designating the nature of what is practically nothing but algebra. On the introduction of the Occidental science of algebra, however, the old term *tenzan* gave way to another new one, *daisū gaku.* It was a mere adoption of the name used by the Chinese. It means literally "the science in which letters are employed in place of numbers."

Seki's *yendan* algebra was early made known to the public but the *tenzan* system remained secret. It was not until 1769 that the method was treated in a printed work, when Lord Arima, daimyō of Kurume in Kyūshū, published the *Shūki Sampō* under the feigned name of Toyota. The usage of keeping inventions in secrecy must have considerably delayed the progress of science, for those to whom were imparted the secret subjects were not, and could not be, always the best minds of their times. Unfortunately the spirit of invested interests ruled the conduct of scholars.

The two subjects of *yendan* and *tenzan,* both established by Seki, equally relate to algebraical considerations, and of these the one was early published while the other had the fate of remaining in secret for nearly a century. The old mathematicians had evidently put a stress on the distinction between these two sciences, or how could such a distinct way of treatment happen? But for us who have not been trained in the old Japanese school of learning this distinction appears very obscure and groundless, at least at the first glance. If we go, however, deep into the study of the Japanese science, it becomes clear, partly at least. The *yendan* was intended to mean explanation or analysis of problems. In solving problems we must first consider the relations that exist between the various quantities given in the data, and calculations and devices must be duly used to get the desired solution, or before arriving at the equation from which a solution or solutions may be obtained. All these considerations belong to what was called the *yendan jutsu* or "explanation process." It was therefore invariably applicable, whether these steps were carried on arithmetically, or geometrically, or in the algebraical treatment in writing or in the celestial element method as arranged by calculating pieces. The word *yendan* was replaced in subsequent ages by another, *kaigi*, which definitely means "explanation." But in a limited and more exact sense of the term, it seems to have represented, particularly in the first years of its establishment, the construction of two or more relations, each of which is obtained by the common and simple application of the celestial element method, and then deducing the final equation in the celestial element from these relations. Thus, as T. Endō rightly understands it, the *yendan* process was the repeated application of the usual algebraical treatment of the celestial element method. The final goal of the *yendan* algebra lay in constructing the numerical equation to be solved by means of the calculating pieces. The *yendan* algebra was carried out in writing.

In the algebra of the *tenzan* method the considerations were all carried out in writing. Where it differs essentially from the *yendan* algebra was in the fact that here the analysis leading to the final results was also considered as constituting a part of available knowledge and the whole was to be done in writing, although to the final equations was sometimes applied the celestial element method for solution. The subjects studied under the common heading of the *tenzan* algebra covered a wide scope, including the theory of equations, the solution of indeterminate equations, algebraical treatment of geometrical relations, and so forth.

The algebraical notations that were employed in the *yendan* and *tenzan* algebras were the same as adopted by the Chinese mathematicians of the 13th and 14th centuries. They are indeed cumbersome and not so convenient as those we are now employing. When we consider the circumstances that reigned in the second half of the 17th century, when Seki Kōwa was briskly engaged in the establishment of his algebraical science, and reflect on the symbols borrowed by Seki and his contemporaries from the Chinese masters, we feel driven to the conclusion that the Japanese did not borrow anything from the Western World in the course of the work. The *tenzan jutsu* or the Japanese algebra must have been brought to light genuinely by the Japanese mind.

Seki's predecessors were already convinced of the multiplicity of roots in an equation, as we have said, but Seki went considerably further in the theory of equations. He knew that the roots of an equation are not restricted to positive ones, that an equation will in general have as many roots as the number of its degree, that there are equations in which this maximum number of roots does not occur, that in such a case the number of roots that exist will be less than the number of the degree by an even number, that there are equations that have no roots, positive or negative, that equations with only even powers of the unknown may be simplified by taking the square as new unknown, that equations may be conveniently transformed by taking new unknowns, etc., etc. Seki also knew that the data in problems may sometimes lead to insufficiency or to impossibility, that in such cases the limits of some constants may be so determined that the problems become sufficient or possible, that the solutions obtained from the equations constructed from the data of problems do sometimes not satisfy their requirements.

Seki gives a rule, which is the same as employing the derived functions. But it seems, he did not apply the rule to the study of maxima and minima, as believed by some scholars. Such an application was left to his successors.

Before Seki's time the solution of equations as effected by the calculating pieces was to be calculated digit after digit, no abbreviation of labor being ever tried. It was Seki who attempted such an abbreviation for the first time in China or Japan, so far as we know. That the notion of determinants is employed by Seki is very noteworthy. . . .

The *shōsa* method, the *jōichi* method, etc., which are usually believed to be Seki's inventions, were certainly learned by him from the Chinese mathematics. The *shōsa*, or *ch'ao-cha* in Chinese, is a method by which the undetermined multipliers $\alpha, \beta, \gamma, \ldots$ are to be determined from the relation $S = \alpha x + \beta x^2 + \gamma x^3 + \ldots$, on the condition that the values of S for some known values of x are known. T. Hayashi is of the opinion that the *shōsa* method is identical with the European method of finite differences, which may be very correct. Seki would have learned it most probably from Kuo Shou-ching's writings. The *jōichi jutsu* is the same as Ch'in Chiu-shao's solution of the indeterminate relation $p\alpha - q\beta = 1$. Seki may have known of Ch'in's work.

It is undeniable that the Japanese mathematics of the 17th century had been influenced by the Chinese science. But not many Chinese works are known as studied in Japan. Ch'êng Tai-wei's *Suan-fa T'ung-tsung* of 1593, Chu Shih-chieh's *Suan-hsiao Chi-mêng*, Kuo Shou-ching's calendrical works, were almost the whole that were brought to Japan, so far as is known to us. At the beginning of the 18th century Mei Wen-ting's *Li-suan Ch'üan-shu* and the *Su-li Ching-yün* of 1713, etc., were imported. Seki's pupil Takebe studied Mei's work and wrote remarks upon it, while Takebe's pupil Nakane Genkei translated some parts of it. Trigonometry was introduced into Japan from the Occidental sources through these works. Logarithms were also obtained by the Japanese from China. We don't know, however, of any material influence the Occidental mathematics exercised upon learning in the Land of the Rising Sun.

The highest development of the Japanese mathematics must of course be looked upon as the invention of the *yenri*—literally "circle-principle" or "circular theory." It was the measurement of the circle carried out by means of analytical considerations. The method has gone through various phases of development, but the first instances of the theory as handed down to us are those recorded by Takebe, Tanzan (or Awayama) and others. The rudiments of Takebe's manuscript *Yenri Tetsujutsu* or *Yenri Kohaijutsu* are as follows:

Let in a circular segment be inscribed regular polygonal lines of 2, 2^2, 2^3, . . . sides in succession. The sagitta to one side of the 2-gonal line will be obtained from the relation: $sd + 4dx - 4x^2 = 0$, where d is the diameter and s the sagitta or altitude of the given arc. One root of this equation may be expanded in a binomial series by means of a mathematical method called *tetsujutsu*. The same method of expansion may be applied to the sagitta of one side of the 2^2-gonal line, using the expanded form for the sagitta of the 2-gonal line. It may be repeated any number of times. But since this procedure is impracticable, Takebe only finds the expansions of the sagittae of the 2, 2^2, 2^3 and 2^4-gonal lines to the first 10 or 6 terms in this way, and then comparing the coefficients he establishes the relations that underlie them through an imperfect induction and forms a table for the sagittae of the polygonal lines up to that of 2^{10} sides. The product of the sagittae of the 2^i-gonal line multiplied by the diameter is the square of one side of the 2^{i+1}-gonal line. If we multiply the result by $(1/2 \times 2^{i+1})^2$, we get the square of half the length of the inscribed 2^{i+1}-gonal line. A table of this quantity is also constructed, from which the value of $\lim_{n\to\infty} \{1/2 \times (\text{length of the } 2^n\text{-gonal line})\}^2$ is deduced by resorting to an imperfect induction. The result is the series

$$\frac{1}{4}(\text{arc})^2 = sd\left\{1 + \frac{2^2}{3\cdot4}\left(\frac{s}{d}\right) + \frac{2^2\cdot4^2}{3\cdot4\cdot5\cdot6}\left(\frac{s}{d}\right)^2 + \frac{2^2\cdot4^2\cdot6^2}{3\cdot4\cdot5\cdot6\cdot7\cdot8}\left(\frac{s}{d}\right)^3 + \cdots\right\}.$$

It will be noticed that this series is the same as the one of the so-called nine formula of Tu Tê-mei or the French Pierre Jartoux. . . .

The measurement of the circle the analysis of which we have just described is usually ascribed to Seki, but this does not seem to be very authentic. According to an old manuscript (Tanzan Shōkei's *Yenri Hakki* of 1728) this analysis or the resulting series was obtained by Takebe in 1722, after he had persevered in his studies for tens of years.

In any case, it is true, Seki contributed much to the subject of the measurement of the circle. He was of course crowned with various results worthy to be told in history. Ōtaka's *Kwatsuyō Sampō*, compiled in 1709 and published in 1712, was a work wherein were embodied some of the results of Seki's writings. In the fourth book of it is given something of the circle-measurement. The formula Seki gives for the determination of the square of a circular arc is of the 7th degree in the diameter and the sagitta. The formula, it seems, was deduced from the numerical values of certain known arcs. . . .

In calculating the numerical value of an arc of a circle, Seki of course uses numerical devices, starting from the inscribed 2-gonal line and successively doubling the number of sides. This is the same method used in the quadrature of the circle. Such a kind of measurement of the circle had existed from before his time, but here one thing remained for him to accomplish. It was the rectifying of the result thus numerically calculated. If a, b, c be three successive values of the perimeters of the inscribed polygons or polygonal lines, the rectified result will be $b + (b - a) \times (c - b)/[(b - a) - (c - b)]$, a rectification which, we learn from later commentaries, has been tried by the comparison of the differences of the successive values to an infinite geometric series.

This way of rectification may be applied any number of times, which was not however done by Seki himself. It was his pupil Takebe who applied the method of repeated rectifications. Takebe's value of π correct to 41 figures was calculated in this way. Calculating in a similar way a great number of figures of the value of the square of an arc of a circle, he also deduced by imperfect induction the series for the square of an arc. By dint of similar reasonings he gave two further series for the same quantity. These results are embodied in his *Fukyū Tetsujutsu* of 1722.

Takebe's studies about the circle-principle were followed by those of Matsunaga, whose *Hōyen Sankyō* of 1739 contains eight series for the circle, its square, a circular arc, its square, an altitude of a circular segment, and its chord. One of these series is the same as Takebe's series and three others are found in Mei Ku-ch'êng's "Pearls dropped in the Red River." Matsunaga gave in this manuscript his value of π correct to 50 figures. He gives no account of how he derived the series he has recorded.

It will be worthy of notice that all the Japanese circle-squarers have resorted to only inscribed polygons beginning with the square. Neither the circumscribed polygons nor the hexagon are employed, perhaps with the single exception of Takuma Genzayemon's using circumscribed polygons besides those that are inscribed. Even he does not use the hexagon, though some subsequent writers used it. It is easy to conjecture that Takuma received the idea of resorting to circumscribed figures from the Chinese *Su-li Ching-yün* of 1713.

When we examine the manuscripts concerning the circular theory, we never fail to be struck with the frequent resort to imperfect induction, which was, at any rate, one of the characteristic features of the Japanese mathematics. The old Japanese seem to have considered mathematics as a branch of natural science; mathematical rules or methods devised or used by them were all treated as a kind of art. They never thought of demonstration, they never once understood what an exact demonstration should be. Geometrical relations were freely studied in Japan, but all these studies were invariably carried on algebraically, never being considered from a geometrical point of view. If by geometry is meant a demonstrative system, the Japanese never had the science of geometry. The Chinese translation of Euclid, printed early in the 17th century, was no doubt imported into Japan, but no trace remains of the Japanese having taken notice of the book.

One of the most noteworthy achievements belonging to the 18th century is certainly Nakane Genjun's solution of an equation. He takes

$$x_r = x_{r-2} + \frac{f(x_{r-2})(x_{r-1} - x_{r-2})}{f(x_{r-2}) - f(x_{r-1})}$$
$$(r = 3, 4, \ldots),$$

for the successive approximations to one root of the equation $f(x) = 0$, the arbitrary quantities x_1 and x_2 being so selected at first that $f(x_1)$ and $f(x_2)$ are of different signs.

The results of the studies of Sakabe and his pupil Kawai upon the solution of equations . . . were obtained in the opening of the 19th century. Subsequent scholars have devoted attention to further studies about the same subject.

In connection with the circular theory we have mentioned the expansion of one root of a quadratic equation whose coefficients are denoted by letters. The expanded series is binomial and this reminds us of Newton's studies. But it does not seem that the Japanese borrowed the result from the English savant. The *tetsujutsu* by which the expansion was effected is nothing but an extension to the case of a literal equation of the method of solving equations of the celestial element method. If the latter was applicable to equations of any degree, the *tetsujutsu* expansion of literal equations was restricted to the case of the quadratic, at least in the beginning of its practice. The scheme for the expansion was at first very complicated, but by and by simplifications were tried in the hands of various scholars. In later years the method was extended to equations whose degrees exceeded the second. It was even extended to the case of an equation of the degree infinity. Thus by the *tetsujutsu* expansion the Japanese effected the reversion of series. Hasegawa Kan's *Sampō Shinsho*, printed in 1830, gives an instance of the subject. Wada Nei also solved equations of infinite degree. . . .

In the course of the development of mathematics in Japan there appeared some transcendental equations in connection with the solution of some problems. These equations were solved by means of successive approximations. Kurushima's studies on the *yenri* theory may be mentioned in this connection. . . . Wada, Saitō, and others who flourished in the first half of the 19th century also studied the subject. The method of solution devised or practiced by these mathematicians was called the *kanrui jutsu* or literally "the method of recurrence."

We have mentioned the tremendous equations whose degrees were counted by thousands. In the 17th and in the first half of the next century scholars took pleasure in the great value of the degrees of the equations they had obtained. Consequently problems were considered as exalted and supreme, if they would lead to equations of very high degrees. But they soon became aware of the true nature of the subject, when they endeavored to reduce the degree as much as might be possible. Simplicity was then sought for. The "Gion Temple Problem" may afford a deserving example. The problem is one that relates to a figure where a circle and a square are inscribed in a circular segment with its altitude between them. It was first suspended at the Gion Temple in Kyōto by Tsuda Yenkyū, pupil to Nishimura Yenri. Tsuda's solution had consisted in an equation of the 1024th degree, which Nakata reduced to one of the 46th degree. Afterwards Ajima Chokuyen solved the problem in an equation of only 10th degree. His manuscript bears the date of 1773. The seeking after simplicity caused naturally a change in the nature of problems considered. What would involve needless complications were by and by neglected.

The Japanese did not content themselves with merely arriving at the formulae of calculations; they always preferred to state them in the form of rules expressed in words. The length of such a statement or the number of words employed are naturally independent of the simplicity or complexity of the formula embodied therein. But as they were too eager after simplicity or neatness of expression, they were at one time driven to such an extreme as to consider a rule better than another, simply because a less number of words was sufficient for its statement. This inclination was especially conspicuous in the case of Aida Ammei, whose genius enriched largely the material of the Japanese mathematics. This defect has some relation to the neglect of logical or demonstrative culture of the science.

The Japanese mathematics was destitute of geometry. Certainly the Japanese was possessed of various geometrical theorems, some of which were even, it is said, ahead of their establishment in the Western World. But the mere enunciation of these propositions, which are mostly expressed in formulae, does not constitute the science of geometry. Some of the relations were established by algebraical considerations, but sometimes intuition was the sole guide determining such matters. It resulted therefore that correct propositions intermingled with false results, which even the best mathematicians were not able to detect.

At the end of the 18th century there broke out a controversy between Aida and Fujita Sadasuke, which raged for some years and various controversial works were published on both sides. If we want to properly understand the matter we must look into the environment in which both parties lived. In Japan mathematics was not cultivated by the government; it was entirely the science of the people. But there were various rival schools, the teachings of which were restricted to a few adherents. Of these schools Fujita belonged to that of Seki which was the most powerful of all. Almost all the leading mathematicians of the 18th and 19th centuries belonged to it. Fujita was a leader of the school. If he was not a man of creative genius, his reputation was a very honorable one. His

Seiyō Sampō of 1779 became a standard of mathematical instruction. In a word he was the most renowned mathematician of his days. Aida who was a gifted man endowed with productive talent, was on the contrary not yet at the height of his fame. Besides, being an almost self-made man, he had had little opportunity of receiving instruction in the secret subjects of Seki's school. He therefore went to Fujita with the hope of being entrusted with the treasured gems. But he was as proud and immodest as he was talented, while Fujita was narrow-minded and despotic in disposition, as men in power, which their virtues do not deserve, are frequently apt to be. The first interview therefore resulted in the incurring of displeasure on both sides. Aida's respect towards the renowned man rapidly subsided, for he saw nothing in him that was excellent. Thus a difference arose between the two contemporaries.

In the course of the controversy sometimes problems were solved and criticisms were attempted. But as the mathematical knowledge of those times lacked any scientific basis, there was nothing that could be taken for a standard of reference, and so it came that frequently the accusation and the defense both failed in precision. Mathematics being cultivated as art, not as science, the old Japanese in pursuing their studies of the subject must have lost much time and labor in useless entanglement. When therefore we study the history of the Japanese mathematics and especially about the solution of problems in connection with the Aida-Fujita controversy, we acknowledge the truth of Ernst Mach's economical theory of science. The Japanese are in general of a practical turn of mind, and the development of mathematics, if cultivated independently of practical applications, was carried on in the main from a practical point of view, the term practical being understood to mean not-scientific. This usage led to unavoidable complications.

In spite of the long controversy, Fujita enjoyed a continuance of fame, until he passed away peacefully in 1807. Aida gradually rose in renown, and the school he had founded, the Saijō-Ryū or "the Superior School," thrived. His writings numbered over a thousand, most of which, however, have gone tracelessly. What now remains of his writings abounds with the results of original research. . . .

While the controversy was furiously raging between Fujita and Aida, there was living a worthy person enjoying deep-going studies in quiet. If he was a fellow pupil of Fujita and his friend, both being pupils to Yamaji, yet he did not take any part in the contention. His leisure hours were all devoted to the progress of mathematics. He did not strive after fortune, he did not yearn after renown; the culture of mathematics was his sole aim of life, it was his best consolation too. If he did not acquire the fame he deserved, the science owed a considerable improvement to his hand. This devoted savant was Ajima Chokuyen by name.

Of Ajima's simplification of the Gion Temple problem already mention has been made. . . . His study of the problem where three circles are inscribed in a triangle is also very famous. In this last subject Ajima was anticipated by Ban Seiyei, but his result was by far the more general. These investigations certainly precede that of the famous Italian Malfatti. It must however be mentioned that the Japanese did not consider it from the point of geometrical construction, the magnitude relations only being the goal of their studies. If Aida's method could be compared to that of Euler, Ajima must fill the position of Lagrange. Scientific generality, if not scientific precision, was largely gained in the Japanese mathematics through the rich ideas of Ajima.

The most splendid of Ajima's successes was perhaps his improvement of the circular theory. While his predecessors had been invariably accustomed to divide the arc into equal parts, he preferred to cut the chord instead of the arc itself by parallel ordinates drawn at equal intervals. The equal division of the chord had formerly been attempted by the thoughtful Takebe but he was not able to deduce anything profitable from it, and the success of this attempt remained for Ajima's genius. This was indeed a small improvement but the result was a considerable gain in simplicity and now the circular theory was rendered applicable to a wide scope of problems, which extended over the whole range of calculations about the lengths and areas of curves, volumes and surfaces of solids, centers of gravity, etc., subjects that were left untouched by his predecessors as unmanageable.

In Ajima's circular theory the chord of a circular segment will be divided into a number of equal parts, say n, parallel ordinates being drawn through the points of division. Thus the arc of the segment will also be divided into n parts. Instead of taking the element of length, Ajima first considers the area comprised between each pair of successive ordinates, which are assumed to terminate both ways at the circumference. The length of any one of them may be easily calculated, and the quadratic surd being expanded by the *tetsujutsu* method, its value will be obtained in a binomial series. Multiplying it by the length of one part of the chord the product is taken as the element of area. If we sum up such areas and go to the limit the number of divisions is increased without limit, the result is the area of the part of the circle which is bounded by the two extreme ordinates and the two opposite arcs lying between them. From this area the length of the arc sought for may be derived at once. Ajima's successors, among whom we may specially mention Wada Nei, adopted his construction but took the element of length directly, whence they effected the operation of "folding" or summing up and going to the limit.

Ajima's way of the circular theory is not essentially different from the operation of integration carried out between two prescribed limits, whereby the differentials are considered as always constant. It is therefore a particular case of definite integration. Ajima's time was a century subsequent to the days of Newton and Leibnitz, when the differential and integral calculus had been established. Ajima may or may not have borrowed his conception from his European predecessors. We have however no way whatever of definitely conjecturing the influence of the Occidental mode of integration. The equal division of the chord is a slight step from the achievements of the preceding ages and the rest of the analysis does not considerably differ from what had been previously done by Takebe and his followers. Besides in the measurement of the sphere a similar procedure had been long practiced. . . . If the process could be considered a particular case of integration, its deviation from the Occidental science seems much greater than its difference from the previous phases of the circular theory. Would it not be most likely for a man of his genius and his eager devotion to be able to independently strike the slight step he had added? Indeed it was a mere slight step; the tremendous effect that necessarily followed was certainly nothing but chance. The circular theory as improved by Ajima is strikingly Japanese in character.

After the improvement of the circular theory Ajima applied himself to extend the result to the calculation of the volume of a solid which is obtained by piercing a circular cylinder by another one, the axes of the two intersecting at right angles. Here drawing parallel planes at equal intervals and at right angles to the axis of the pierced cylinder, the sections of the solid in question made by these planes are all parts of circles such as just considered. Their areas being found as in the above and multiplied by the equal portion of the diameter of the piercing cylinder which is divided by the parallel planes, we obtain the elements of volume, from which "folding" gives the desired expression for the required volume. If we denote the diameters of the pierced and piercing cylinders by d and k, respectively, the expression obtained assumes the form

$$k^2 d\frac{\pi}{4}\left\{1 - \frac{1}{8}\left(\frac{k^2}{d^2}\right) - \frac{1}{8\cdot 8}\left(\frac{k^2}{d^2}\right)^2 - \frac{1\cdot 5}{8\cdot 8\cdot 16}\left(\frac{k^2}{d^2}\right)^3 - \frac{1\cdot 5\cdot 7}{8\cdot 8\cdot 16\cdot 16}\left(\frac{k^2}{d^2}\right)^4 - \cdots\right\}.$$

Some parts of the analysis of the problem as studied by Ajima had been left imperfect and complicated, which were simplified subsequently by Wada, Shiraishi, and other mathematicians. This analysis has sometimes been compared to double integration, but it was carried out, as we have seen, by first finding the area of a portion of a circle and then constructing the element of volume from it and effecting the operation of folding. Thus the procedure was really equivalent to a double application of the operation of integrating, but nevertheless it is not comparable properly to any conception of a double integral.

After Ajima's time there appeared a host of talented mathematicians almost simultaneously and carried on their studies in happy emulation. Among these scholars we may mention Wada Nei, Kawai Kyūtoku, Shiraishi Chōchū and numerous others. The analysis of the single problem of the pierced cylinder that had been studied by Ajima was extended by these men to the consideration of other problems of allied nature. The ellipse was rectified. Not only the volume of an ellipsoid, and especially of a spheroid, but its surface also were studied. . . . The curve in space arising from the intersection of the cylinders in Ajima's problem were investigated. The area of the portion of the cylindrical surface bounded by this curve did not escape the attention of scholars. One of the cylinders was sometimes replaced by an elliptic cylinder or a prism. The case where the piercing cylinder touches a generating line of the pierced instead of having their axes intersecting with each other was taken up. The intersection of a sphere and a cylinder or prism was considered. The category of the subjects that now constituted the favorite theme of Japanese mathematics is not easy to be exhausted in such a brief sketch as we are here trying. Suffice it to say that all these formed the subject matter of the *yenri* or circular theory. But sometimes they were counted under the new branch of mathematics called *yenri katsujutsu* or simply *katsujutsu*, a term that was intended to comprise the study of problems that required the double or multiple application of the *tetsujutsu* expansion. This agrees in general with the consideration of problems in double and multiple integrations. But the Japanese *katsujutsu* is not equivalent to the latter. The rectification of the ellipse required a double application of the *tetsujutsu* expansion and so belonged to the *katsujutsu*, but it does not relate to a double integration. The Japanese did not happen to have the conception of a double or multiple integral, at least, in the earlier days of the development of the *katsujutsu* calculus. If at all, it must have appeared in connection with the calculation of the skew surface and similar problems in the hands of Saitō, Kobayashi, Hōdōji, Hagiwara and others, but as to the existence of this conception the results of further studies are needed.

Of the numerous expounders of the *katsujutsu* theory, who appeared subsequent to Ajima's time, the most noted was undoubtedly Wada Nei. The circular theory received an improvement at his hands starting from the form Ajima had left. Some historical writers would therefore take it as worthy in the same degree as Ajima's achievements. He had been exceedingly successful in the solution of difficult problems and so his contemporary mathematicians of the first rank had to go to him to school. Wada lived in poverty, and being very loose in character and given to a drinking habit, it is said, he sometimes sold his solutions to make a little money for more drink. Thus among the writings of his contemporaries there might be results learned from him in this way. But it is a matter of course that the whole of these had not come from such a source. It is a great pity that the manuscripts embodying Wada's wide resources were burnt in a fire that occurred in 1836.

In the calculations of the *yenri* calculus or of the *katsujutsu* the summation of some quantities and the transit to the limit is required. But it was very tedious to go over these operations every time the calculations are effected. It was therefore desirable to save labor by means of a table giving the results of preliminary calculations. So Wada Nei constructed various tables. There are some tables calculated by Aida, but we don't know their dates. Wada's successors, Saitō, Hagiwara, etc., also attempted to complete the task.

In Japan the operation of integrating was called to *tatamu* or "to fold," and so the tables used in integrating were called *jō-hyō*—"folding tables" or "tables used in folding." The two words *tatamu* and *jō* are of the same meaning, being different ways of reading of one and same ideogram. The completion of these tables proved a convenience to the Japanese *yenri* scholars. They afforded the same advantage the logarithmic tables had been to the work of practical calculators in Europe.

Here we have accidentally spoken of logarithms. The first logarithmic tables brought to Japan were certainly those given in the Chinese work, *Su-li Ching-yün* of 1713. But it appears, the Japanese did not take any interest in the subject for a long time. It was at the end of the 18th century that they began to study it. Even in the 19th century logarithmic tables were rarely used.

The Occidental trigonometry had a better destiny than logarithms in Japan. There were various native scholars who studied it. Napier's rods or tallies were also studied, being certainly introduced through China. About the influence of the Occidental mathematics during the 18th century we know almost nothing any further than these trifles. If we cannot conclude on that account of no further influence received from foreign sources, yet it does not appear that there had been any material source of knowledge transmitted from Europe that contributed greatly to the progress of the mathematical studies of the Japanese. In the latter part of the 18th century there arose various Occidentalists and various Occidental sciences were studied. But in the domain of mathematics their studies seem to have been exceedingly insignificant.

As the Japanese mathematicians of those days used to maintain, the European science excelled in astronomical subjects but the Japanese was the better of the two about mathematics. From the end of the 18th and towards the next century certainly were brought numerous treatises on astronomy and mathematics from Holland to Japan, but it seems there was not any book treating of mathematics of advanced nature. If the contents of elementary works were compared to the results of native higher mathematics, the latter were necessarily ahead. The opinion of old mathematics about the European mathematics was not necessarily the outcome of their prejudiced pride that shuts away anything foreign to them. The mathematical knowledge that they could receive from abroad must have been of a very meager kind. This was the reason why the mathematics of the Japanese could proceed straight on in the same course it had trod for one or two centuries in the past, never once deviating, until at last the political changes of the Restoration of 1868 brought Japan under the influence of the all-swallowing inundations of the universal civilization.

If we are led to such a conclusion as this, yet nothing restrains us from gratefully acknowledging the influence received from without upon some points of the science. At the middle of the 19th century there arose vigorously studies about the center of gravity. A germ of the theory had indeed been seen among the writings of Seki in connection with his method of finding the volume of a solid generated by revolving a circular arc. But it had long remained undeveloped, and it was after a contact with the European science that it was taken up in earnest. In Hashimoto's treatise published in 1830 there are two problems of finding the center of gravity, which were solved geometrically. This way of solution was certainly borrowed from the Dutch science. In subsequent years the problem was connected with the method of the circular theory.

The cycloid and epicycloid and the like were a theme favored in Japan. These curves were all designated by the name *tenkyo kiseki*, literally meaning "loci described by rolling." The first instance of the subject was perhaps the board suspended by Wada's pupil at the Atago Temple in Yedo. The conception of a roulette, it is usually believed, was learned from a Dutch source. The treatment is however decidedly Japanese in character.

Wada's investigation of the "pointed circles" or "ovals of different species" was beyond doubt original with him. These curves, which are closed curves of the 4th degree, and the solids generated by their revolution received much attention from subsequent writers. . . .

The catenary or "hanging string" as it was called in old Japan was another curve that was deeply studied. Its study happened to arise some time later than the problem of the center of gravity. The very complexity of treatment strongly reminds us of its unindebtedness to any foreign source of knowledge. . . .

There are of course still numerous other performances effected by the Japanese mind, which are worthy of notice, but our brief sketch is not sufficient to make a full reference to them. We shall therefore content ourselves with merely mentioning one or two more instances. The writers of the 18th and 19th centuries had here and there tried some problems by the geometrical way, however imperfect they might have been. Hasegawa's *hengyō-jutsu* may be mentioned as an example, though it is not geometrical merely. It was indeed a defective one, as the author's contemporaries were already aware of, sometimes leading to gross error. This was a plan for considering geometrical problems in their special cases and extending the results to the general case. Though erroneous if considered from a general point of view, yet it was not a method utterly futile, as some scholars were inclined to think. When managed with care, it may have helped us in guiding our thought in the attacks upon problems. The idea of generalizing the results obtained in special cases to the case in the general form was itself profitable. If the old Japanese had remained a little longer isolated from the rest of the world, they would certainly have improved this plan to a blameless theory or method. It is said that Saitō Gigi was possessed of an exact method that resembled Hasegawa's method. It was exact and not defective, and was so highly valued by its author that he never dared show it to others. His manuscript *Kyōdai Benshiki* was lost without being acknowledged in the mathematical circle. Hōdōji's *Kanshinkō Sampen* was a development on the basis of a similar idea. He treated the straight line as a special case of a circle or the plane as that of a sphere. In a word he based his theory on the treatment of the limiting case. This point awaits the results of further studies.

The Japanese considered the ellipse as a section of a circular cylinder. Though they were aware of the fact that the ellipse may be obtained from the section of a circular cone, it was seldom studied as a section of a cone. This fact was perhaps the cause of their never conceiving of other sections of the cone. The parabola and the hyperbola were not studied in Japan. In fact, Isomura's *Ketsugishō* of 1660 contains a diagram of a parabola, and some astronomical works of the 19th century were illustrated by the same. But the mathematicians never had the pleasure of studying about them. The hyperbola was perhaps never known to the Japanese. This state of things continued for the whole period of the old Japanese school of mathematics.

As we have said, the Japanese mathematics did not exist as a science but as art. On that account everything studied in Japan had borne the character of speciality, lacking in generality. But the Japanese were by no means wanting in the scientific spirit, they were on the contrary endowed strongly with the zealous yearning after truth or knowledge, which prevailed throughout the whole history of the Japanese mathematics. Consequently scientific precision was gained step by step, a considerable degree of generality being won by Ajima. In vivid contrast to what had been treasured in former days, the final results only, later writers came to acknowledge the value and significance of analysis or detailed accounts of calculations. Thus it happened that various attempts were made at general rules and general methods. Some of these, it may be, were false and defective, but the attempts themselves were not sterile, they were of great importance. The attitude of the old Japanese was perfectly progressive, which we cannot but admire. Before, therefore, the Japanese came in full contact with the civilization of the world at the time of the Restoration, it must be admitted, they had been genuinely prepared to receive anything that was to enter into his grasp. The progress of civilization in general and particularly the improvement of mathematical studies in Japan after that remarkable event may at first glance be wondered at as having been exceedingly rapid. But when we reflect on the intrinsic circumstances that had gradually developed with the constant culture of the restless spirit of the Japanese, the rapid progress there will be no surprise at all.

Some of the mathematicians in the first part of the 19th century were able to read Dutch works, though their knowledge of the language was of an exceedingly limited kind. A certain number of Dutch astronomical works were possessed by the Astronomical Board of the Shogunate, but we know practically nothing of what were the mathematical treatises brought from Holland to Japan in those days. Nor are we able to find traces of the Dutch influence upon the writings belonging to this epoch. No quotations, no references are found. The relation of the Dutch science and the mathematics cultivated in Japan still remains unexplored.

It is almost the whole of the Dutch influence, of which we know, that some of the writings of Kawai, Shiraishi, Ichino and others contain some deformed Roman characters as symbols. Reflecting on the incorrectness with which the names of the authors are spelled, their knowledge learned from Dutch works appears to have been very limited if any. We have no knowledge of any Occidentalist, who was at the same time a mathematician.

The first material introduction of the European mathematics into Japan was certainly Alexander

Wylie's translation of Loomis' *Calculus*, which was brought to Japan soon after its publication. Just about these times Dutch sciences were taught by Dutch men to the students of navigation. The political changes of 1868 then took place, which were followed by the opening of the country to receive the importations of every kind of Occidental civilization. Mathematics was now studied after the Occidental knowledge. The mathematicians of the old Japanese school struggled for some time longer for the existence of their science; but with the lapse of years their influence was destined to give way and entirely disappeared in a short time. . . .

The Music of the Spheres

Jacob Bronowski

Jacob Bronowski was born in Poland and educated in England. He received his Ph.D. in mathematics from Cambridge and held scientific positions in both the university and government. He was the author of a number of expository works on mathematics and science and is probably best known for his book *The Ascent of Man*, which became the basis for a very popular series of television specials on science. Bronowski died in 1974.

This essay by Jacob Bronowski is a chapter from his book The Ascent of Man. *It is much more than just an introduction to the history of mathematics. On the skeleton of historical fact Bronowski creates a body of plausible inference to suggest how mathematical ideas grew, spread, and gave birth to yet other ideas. Bronowski suggests that ideas are linked with the civilizations that created them and yet the ease with which these ideas cross language, time, and culture suggests that ideas transcend their spawning civilization.*

An important theme in this essay is the interaction of mathematical ideas and nearly every other aspect of the cultures in which they exist. This idea has special relevance for our own time. Many aspects of modern mathematics are divorced from the physical universe in which they exist, which creates at least two problems. First, an increasingly large proportion of the general population has not even the vaguest understanding of modern mathematics. Second, the monumental towers of abstraction that are the home of modern mathematics are no longer subject to the discipline of physical reality. That these two conditions could continue indefinitely seems unlikely, and in this there is a moral for the professional mathematician. At some level, mathematics must be relevant to the society that supports its creation. That is, there must be an audience for the current theme and variations on the music of the spheres.

MATHEMATICS IS IN many ways the most elaborated and sophisticated of the sciences—or so it seems to me, as a mathematician. So I find both a special pleasure and constraint in describing the progress of mathematics, because it has been part of so much human speculation: a ladder for mystical as well as rational thought in the intellectual ascent of man. However, there are some concepts that any account of mathematics should include: the logical idea of proof, the empirical idea of exact laws of nature (of space particularly), the emergence of the concept of operations, and the movement in mathematics from a static to a dynamic description of nature. They form the theme of this essay.

Even very primitive peoples have a number system; they may not count much beyond four, but they know that two of any thing plus two of the same thing makes four, not just sometimes but always. From that fundamental step, many cultures have built their own number systems, usually as a written language with similar conventions. The Babylonians, the Mayans, and the people of India, for example, invented essentially the

Source: Excerpts from Jacob Bronowski, "The Music of the Spheres," *The Ascent of Man*, Little, Brown and Company, 1973, pp. 155–187.

same way of writing large numbers as a sequence of digits that we use, although they lived far apart in space and in time.

So there is no place and no moment in history where I could stand and say "Arithmetic begins here, now." People have been counting, as they have been talking, in every culture. Arithmetic, like language, begins in legend. But mathematics in our sense, reasoning with numbers, is another matter. And it is to look for the origin of that, at the hinge of legend and history, that I went sailing to the island of Samos.

In legendary times Samos was a center of the Greek worship of Hera, the Queen of Heaven, the lawful (and jealous) wife of Zeus. What remains of her temple, the Heraion, dates from the sixth century before Christ. At that time there was born on Samos, about 580 B.C., the first genius and the founder of Greek mathematics, Pythagoras. During his lifetime the island was taken over by the tyrant, Polycrates. There is a tradition that before Pythagoras fled, he taught for a while in hiding in a small white cave in the mountains which is still shown to the credulous.

Samos is a magical island. The air is full of sea and trees and music. Other Greek islands will do as a setting for *The Tempest*, but for me this is Prospero's island, the shore where the scholar turned magician. Perhaps Pythagoras was a kind of magician to his followers, because he taught them that nature is commanded by numbers. There is a harmony in nature, he said, a unity in her variety, and it has a language: numbers are the language of nature.

Pythagoras found a basic relation between musical harmony and mathematics. The story of his discovery survives only in garbled form, like a folk tale. But what he discovered was precise. A single stretched string vibrating as a whole produces a ground note. The notes that sound harmonious with it are produced by dividing the string into an exact number of parts: into exactly two parts, into exactly three parts, into exactly four parts, and so on. If the still point on the string, the node, does not come at one of these exact points, the sound is discordant.

As we shift the node along the string, we recognize the notes that are harmonious when we reach the prescribed points. Begin with the whole string: this is the ground note. Move the node to the midpoint: this is the octave above it. Move the node to a point one third of the way along: this is the fifth above that. Move it to a point one fourth along: this is the fourth, another octave above. And if you move the node to a point one fifth of the way along, this (which Pythagoras did not reach) is the major third above that.

Pythagoras had found that the chords which sound pleasing to the ear—the western ear—correspond to exact divisions of the string by whole numbers. To the Pythagoreans that discovery had a mystic force. The agreement between nature and number was so cogent that it persuaded them that not only the sounds of nature, but all her characteristic dimensions, must be simple numbers that express harmonies. For example, Pythagoras or his followers believed that we should be able to calculate the orbits of the heavenly bodies (which the Greeks pictured as carried round the earth on crystal spheres) by relating them to the musical intervals. They felt that all the regularities in nature are musical; the movements of the heavens were, for them, the music of the spheres.

These ideas gave Pythagoras the status of a seer in philosophy, almost a religious leader, whose followers formed a secret and perhaps revolutionary sect. It is likely that many of the later followers of Pythagoras were slaves; they believed in the transmigration of souls, which may have been their way of hoping for a happier life after death.

I have been speaking of the language of numbers, that is arithmetic, but my last example was the heavenly spheres, which are geometrical shapes. The transition is not accidental. Nature presents us with shapes: a wave, a crystal, the human body, and it is we who have to sense and find the numerical relations in them. Pythagoras was a pioneer in linking geometry with numbers, and since it is also my choice among the branches of mathematics, it is fitting to watch what he did.

Pythagoras had proved that the world of sound is governed by exact numbers. He went on to prove that the same thing is true of the world of vision. That is an extraordinary achievement. I look about me; here I am, in this marvelous, colored landscape of Greece, among the wild natural forms, the Orphic dells, the sea. Where under this beautiful chaos can there lie a simple, numerical structure?

The question forces us back to the most primitive constants in our perception of natural laws. To answer well, it is clear that we must begin from universals of experience. There are two experiences on which our visual world is based: that gravity is vertical, and that the horizon stands at right angles to it. And it is that conjunction, those cross-wires in the visual field, which fixes the nature of the right angle; so that if I were to turn this right angle of experience (the direction of "down" and the direction of "sideways") four times, back I come to the cross of gravity and the horizon. The right angle is defined by this fourfold operation, and is distinguished by it from any other arbitrary angle.

In the world of vision, then, in the vertical picture plane that our eyes present to us, a right angle is defined by its fourfold rotation back on itself. The same definition holds also in the horizontal world of experience, in which in fact we move. Consider that world, the world of the flat earth and the map and the points of the compass. Here I am looking across the straits from Samos to Asia Minor, due south. I take a triangular tile as a pointer and I set it pointing there, south. (I have made the pointer in the shape of a right-angled triangle, because I shall want to put its four rotations side by side.) If I turn that triangular tile through a right angle, it points due west. If I now turn it through a second right angle, it points due north. And if I now turn it through a third right angle, it points due east. Finally, the fourth and last turn will take it due south again, pointing to Asia Minor, in the direction in which it began.

Not only the natural world as we experience it, but the world as we construct it is built on that relation. It has been so since the time that the Babylonians built the Hanging Gardens, and earlier, since the time that the Egyptians built the pyramids. These cultures already knew in a practical sense that there is a builder's set square in which the numerical relations dictate and make the right angle. The Babylonians knew many, perhaps hundreds of formulae for this by 2000 B.C. The Indians and the Egyptians knew some. The Egyptians, it seems, almost always used a set square with the sides of the triangle made of three, four, and five units. It was not until 550 B.C. or thereabouts that Pythagoras raised this knowledge out of the world of empirical fact into the world of what we should now call proof. That is, he asked the question, "How do such numbers that make up these builder's triangles flow from the fact that a right angle is what you turn four times to point the same way?"

His proof, we think, ran something like this. (It is not the proof that stands in the school books.) The four leading points—south, west, north, east—of the triangles that form the cross of the compass are the corners of a square. I slide the four triangles so that the long side of each ends at the leading point of a neighbor. Now I have constructed a square on the longest side of the right-angled triangles—on the hypotenuse. Just so that we should know what is part of the enclosed area and what is not, I will fill in the small inner square area that has now been uncovered with an additional tile. (I use tiles because many tile patterns, in Rome, in the Orient, from now on derive from this kind of wedding of mathematical relation to thought about nature.)

Now we have a square on the hypotenuse, and we can of course relate that by calculation to the squares on the two shorter sides. But that would miss the natural structure and inwardness of the figure. We do not need any calculation. A small game, such as children and mathematicians play, will reveal more about calculation. Transpose two triangles to new positions, thus. Move the triangle that pointed south so that its longest side lies along the longest side of the triangle that pointed north. And move the triangle that pointed east so that its longest side lies along the longest side of the triangle that pointed west.

Now we have constructed an L-shaped figure with the same area (of course, because it is made of the same pieces) whose sides we can see at once in terms of the smaller sides of the right-angled triangle. Let me make the composition of the L-shaped figure visible: put a divider down that separates the end of the L from the upright part. Then it is clear that the end is a square on the shorter side of the triangle; and the upright part

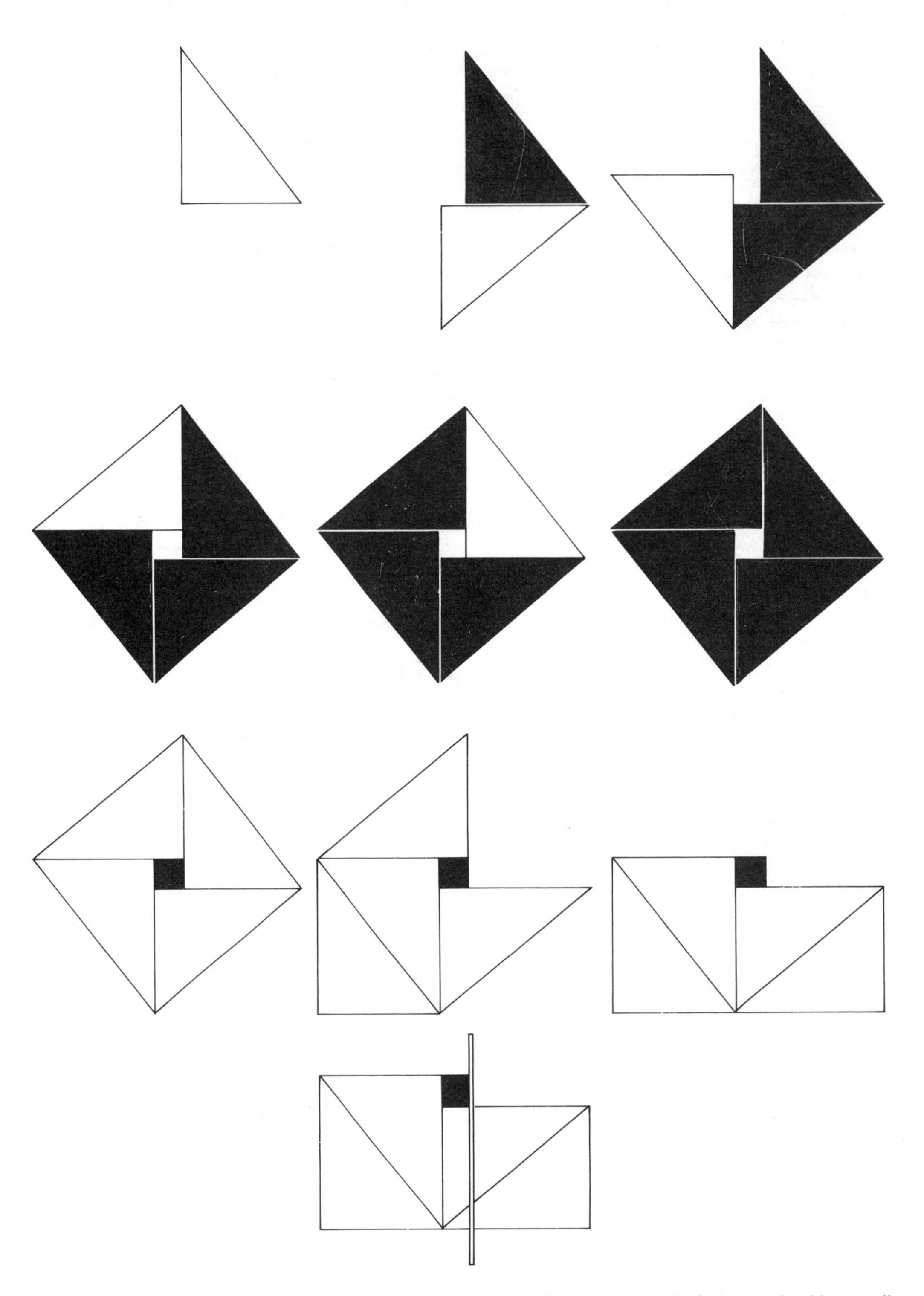

Pythagoras raised this knowledge out of the world of empirical fact into the world of what we should now call proof. *The Pythagorean proof, described in the text, that in a right-angled triangle the square on the hypotenuse is equal to the sum of the squares on the other two sides.*

of the L is a square on the longer of the two sides enclosing the right angle.

Pythagoras had thus proved a general theorem: not just for the 3:4:5 triangle of Egypt, or any Babylonian triangle, but for every triangle that contains a right angle. He had proved that the square on the longest side or hypotenuse is equal to the square on one of the other two sides plus the square on the other if, and only if, the angle they contain is a right angle. For instance, the sides 3:4:5 compose a right-angled triangle because

$$\begin{aligned} 5^2 &= 5 \cdot 5 = 25 \\ &= 16 + 9 = 4 \cdot 4 + 3 \cdot 3 \\ &= 4^2 + 3^2. \end{aligned}$$

And the same is true of the sides of triangles found by the Babylonians, whether simple as 8:15:17, or forbidding as 3367:3456:4825, which leave no doubt that they were good at arithmetic.

To this day, the theorem of Pythagoras remains the most important single theorem in the whole of mathematics. That seems a bold and extraordinary thing to say, yet it is not extravagant; because what Pythagoras established is a fundamental characterization of the space in which we move, and it is the first time that it is translated into numbers. And the exact fit of the numbers describes the exact laws that bind the universe. In fact, the numbers that compose right-angled triangles have been proposed as messages which we might send out to planets in other star systems as a test for the existence of rational life there.

The point is that the theorem of Pythagoras in the form in which I have proved it is an elucidation of the symmetry of plane space; the right angle is the element of symmetry that divides the plane four ways. If plane space had a different kind of symmetry, the theorem would not be true; some other relation between the sides of special triangles would be true. And space is just as crucial a part of nature as matter is, even if (like the air) it is invisible; that is what the science of geometry is about. Symmetry is not merely a descriptive nicety; like other thoughts in Pythagoras, it penetrates to the harmony in nature.

When Pythagoras had proved the great theorem, he offered a hundred oxen to the Muses in thanks for the inspiration. It is a gesture of pride and humility together, such as every scientist feels to this day when the numbers dovetail and say, "This is a part of, a key to, the structure of nature herself."

Pythagoras was a philosopher, and something of a religious figure to his followers as well. The fact is there was in him something of that Asiatic influence which flows all through Greek culture and which we commonly overlook. We tend to think of Greece as part of the west; but Samos, the edge of classical Greece, stands one mile from the coast of Asia Minor. From there much of the thought that inspired Greece first flowed; and, unexpectedly, it flowed back to Asia in the centuries after, before ever it reached Western Europe.

Knowledge makes prodigious journeys, and what seems to us a leap in time often turns out to be a long progression from place to place, from one city to another. The caravans carry with their merchandise the methods of trade of their countries—the weights and measures, the methods of reckoning—and techniques and ideas went where they went, through Asia and North Africa. As one example among many, the mathematics of Pythagoras has not come to us directly. It fired the imagination of the Greeks, but the place where it was formed into an orderly system was the Nile city, Alexandria. The man who made the system, and made it famous, was Euclid, who probably took it to Alexandria around 300 B.C.

Euclid evidently belonged to the Pythagorean tradition. When a listener asked him what was the practical use of some theorem, Euclid is reported to have said contemptuously to his slave, "He wants to profit from learning—give him a penny." The reproof was probably adapted from a motto of the Pythagorean brotherhood, which translates roughly as "A diagram and a step, not a diagram and a penny"—"a step" being a step in knowledge or what I have called the Ascent of Man.

The impact of Euclid as a model of mathematical reasoning was immense and lasting. His book *Elements of Geometry* was translated and copied more than any other book except the Bible right into modern times. I was first taught mathematics by a man who still quoted the theorems of geometry by the numbers that Euclid had given them; and that was not uncommon even fifty years ago, and was the standard mode of reference in the past. When John Aubrey about 1680 wrote an account of how Thomas Hobbes in middle age had suddenly fallen "in love with geometry" and so with philosophy, he explained that it began when Hobbes happened to see "in a gentleman's library, Euclid's *Elements* lay open, and 'twas the 47 *Element libri* I." Proposition 47 in Book I of Euclid's *Elements* is the famous theorem of Pythagoras.

The other science practiced in Alexandria in the centuries around the birth of Christ was astronomy. Again, we can catch the drift of history in the undertow of legend: when the Bible says that three wise men followed a star to Bethlehem, there sounds in the story the echo of an age when wise men are star-gazers. The secret of the heavens that wise men looked for in antiquity was read by a Greek called Claudius Ptolemy, working in Alexandria about A.D. 150. His work came to Europe in Arabic texts, for the original Greek manuscript editions were largely lost, some in the pillage of the great library of Alexandria by Christian zealots in A.D. 389, others in the wars and invasions that swept the Eastern Mediterranean throughout the Dark Ages.

The model of the heavens that Ptolemy constructed is wonderfully complex, but it begins from a simple analogy. The moon revolves round the earth, obviously; and it seemed just as obvious to Ptolemy that the sun and the planets do the same. (The ancients thought of the moon and the sun as planets.) The Greeks had believed that the perfect form of motion is a circle, and so Ptolemy made the planets run on circles, or on circles running in their turn on other circles. To us, that scheme of cycles and epicycles seems both simple-minded and artificial. Yet in fact the system was a beautiful and a workable invention, and an article of faith for Arabs and Christians right through the Middle Ages. It lasted for fourteen hundred years, which is a great deal longer than any more recent scientific theory can be expected to survive without radical change.

It is pertinent to reflect here why astronomy was developed so early and so elaborately, and in effect became the archetype for the physical sciences. In themselves, the stars must be quite the most improbable natural objects to rouse human curiosity. The human body ought to have been a much better candidate for early systematic interest. Then why did astronomy advance as a first science ahead of medicine? Why did medicine itself turn to the stars for omens, to predict the favorable and the adverse influences competing for the life of the patient—surely the appeal to astrology is an abdication of medicine as a science? In my view, a major reason is that the observed motions of the stars turned out to be calculable, and from an early time (perhaps 3000 B.C. in Babylon) lent themselves to mathematics. The pre-eminence of astronomy rests on the peculiarity that it can be treated mathematically; and the progress of physics, and most recently of biology, has hinged equally on finding formulations of their laws that can be displayed as mathematical models.

Every so often, the spread of ideas demands a new impulse. The coming of Islam six hundred years after Christ was the new, powerful impulse. It started as a local event, uncertain in its outcome; but once Mahomet conquered Mecca in A.D. 630, it took the southern world by storm. In a hundred years, Islam captured Alexandria, established a fabulous city of learning in Baghdad, and thrust its frontier to the east beyond Isfahan in Persia. By A.D. 730 the Moslem empire reached from Spain and Southern France to the borders of China and India: an empire of spectacular strength and grace, while Europe lapsed in the Dark Ages.

In this proselytizing religion, the science of the conquered nations was gathered with a kleptomaniac zest. At the same time, there was a liberation of simple, local skills that had been despised. For instance, the first domed mosques were built with no more sophisticated apparatus than the ancient builders' set square—that is still used. The Masjid-i-Jomi (the Friday Mosque) in Isfahan is one of the statuesque monuments of early Islam. In centers like these, the knowledge of Greece and of the east was treasured, absorbed and diversified.

Mahomet had been firm that Islam was not to be a religion of miracles; it became in intellectual content a pattern of contemplation and analysis. Mohammedan writers depersonalized and formalized the godhead: the mysticism of Islam is not blood and wine, flesh and bread, but an unearthly ecstasy.

> Allah is the light of the heavens and the earth. His light may be compared to a niche that enshrines a lamp, the lamp within a crystal of star-like brilliance, light upon light. In temples which Allah has sanctioned to be built for the remembrance of his name do men praise him morning and evening, men whom neither trade nor profit can divert from remembering him.

One of the Greek inventions that Islam elaborated and spread was the astrolabe. As an observational device, it is primitive; it only measures the elevation of the sun or a star, and that crudely. But by coupling that single observation with one or more star maps, the astrolabe also carried out an elaborate scheme of computations that could determine latitude, sunrise and sunset, the time for prayer and the direction of Mecca for the traveler. And over the star map, the astrolabe was embellished with astrological and religious details, of course, for mystic comfort.

For a long time the astrolabe was the pocket watch and the slide rule of the world. When the poet Geoffrey Chaucer in 1391 wrote a primer to teach his son how to use the astrolabe, he copied it from an Arab astronomer of the eighth century.

Calculation was an endless delight to Moorish scholars. They loved problems, they enjoyed finding ingenious methods to solve them, and sometimes they turned their methods into mechanical devices. A more elaborate ready-reckoner than the astrolabe is the astrological or astronomical computer, something like an automatic calendar, made in the Caliphate of Baghdad in the thirteenth century. The calculations it makes are not deep, an alignment of dials for prognostication, yet it is a testimony to the mechanical skill of those who made it seven hundred years ago, and to their passion for playing with numbers.

The most important single innovation that the eager, inquisitive, and tolerant Arab scholars brought from afar was in writing numbers. The European notation for numbers then was still the clumsy Roman style, in which the number is put together from its parts by simple addition: for example, 1825 is written as MDCCCXXV, because it is the sum of M = 1000, D = 500, C + C + C = 100 + 100 + 100, XX = 10 + 10, and V = 5. Islam replaced that by the modern decimal notation that we still call "Arabic." . . . To write 1825, the four symbols would simply be written as they stand, in order, running straight on as a single number; because it is the place in which each symbol stands that announces whether it stands for thousands, or hundreds, or tens, or units.

However, a system that describes magnitude by place must provide for the possibility of empty places. The Arabic notation requires the invention of a zero. . . . The words *zero* and *cipher* are Arab words; so are *algebra, almanac, zenith,* and a dozen others in mathematics and astronomy. The Arabs brought the decimal system from India about A.D. 750, but it did not take hold in Europe for another five hundred years after that.

It may be the size of the Moorish Empire that made it a kind of bazaar of knowledge, whose scholars included heretic Nestorian Christians in the east and infidel Jews in the west. It may be a quality in Islam as a religion, which, though it strove to convert people, did not despise their knowledge. In the east the Persian city of Isfahan is its monument. In the west there survives an equally remarkable outpost, the Alhambra in southern Spain.

Seen from the outside, the Alhambra is a square, brutal fortress that does not hint at Arab forms. Inside, it is not a fortress but a palace, and a palace designed deliberately to prefigure on earth the bliss of heaven. The Alhambra is a late construction. It has the lassitude of an empire past its peak, unadventurous and, it thought, safe. The religion of meditation has become sensuous and self-satisfied. It sounds with the music of water, whose sinuous line runs through all Arab melodies, though they are based fair and square on the Pythagorean scale. Each court in turn is the echo and the memory of a dream, through which the Sultan floated (for he did not walk, he was carried). The Alhambra is most nearly the description of Paradise from the Koran.

> Blessed is the reward of those who labor patiently and put their trust in Allah. Those that embrace the true faith and do good works shall be forever lodged in the mansions of Paradise, where rivers will roll at their feet . . . and honored shall they be in the gardens of delight, upon couches face to face. A cup shall be borne round among them from a fountain, limpid, delicious to those who drink. . . . Their spouses on soft green cushions and on beautiful carpets shall recline.

The Alhambra is the last and most exquisite monument of Arab civilization in Europe. The last Moorish king reigned here until 1492, when Queen Isabella of Spain was already backing the adventure of Columbus. It is a honeycomb of courts and chambers, and the Sala de las Camas is the most secret place in the palace. Here the girls from the harem came after the bath and reclined, naked. Blind musicians played in the gallery, the eunuchs padded about. And the Sultan watched from above, and sent an apple down to signal to the girl of his choice that she would spend the night with him.

In a western civilization, this room would be filled with marvelous drawings of the female form, erotic pictures. Not so here. The representation of the human body was forbidden to Mohammedans. Indeed, even the study of anatomy at all was forbidden, and that was a major handicap to Moslem science. So here we find colored but extraordinarily simple geometric designs. The artist and the mathematician in Arab civilization have become one. And I mean that quite literally. These patterns represent a high point of the Arab exploration of the subtleties and symmetries of space itself: the flat, two-dimensional space of what we now call the Euclidean plane, which Pythagoras first characterized.

In the wealth of patterns, I begin with a very straightforward one. It repeats a two-leaved motif of dark horizontal leaves, and another of light vertical leaves. The obvious symmetries are translations (that is, parallel shifts of the pattern) and either horizontal or vertical reflections. But note one more delicate point. The Arabs were fond of designs in which the dark and the light units of the pattern are identical. And so, if for a moment you ignore the colors [the illustration below appeared in color in the original publication], you can see that you could turn a dark leaf once through a right angle into the position of a neighboring light leaf. Then, always rotating round the same point of junction, you can turn it into the next position, and (again round the same point) into the next, and finally back on itself. And the rotation spins the whole pattern correctly; every leaf in the pattern arrives at the position of another leaf, however far from the center of rotation they lie.

Reflection in a horizontal line is a twofold symmetry of the colored pattern, and so is reflection in a vertical. But if we ignore the colors, we see that there is a fourfold symmetry. It is provided by the operation of rotating through a right angle, repeated four times, by which I earlier proved the theorem of Pythagoras; and thereby the uncolored pattern becomes in its symmetry like the Pythagorean square. . . .

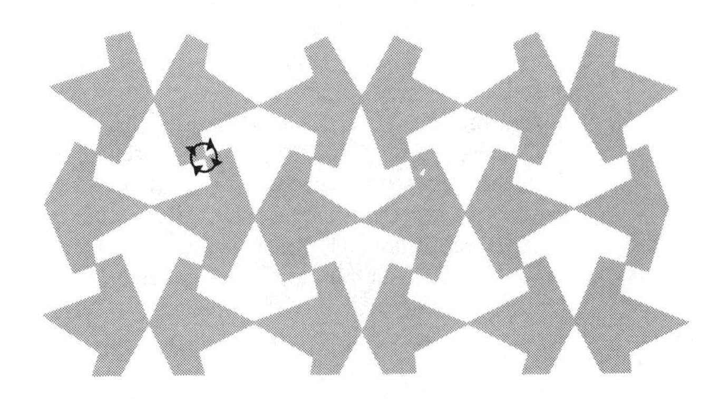

At this point, the non-mathematician is entitled to ask, "So what? Is that what mathematics is about? Did Arab professors, do modern mathematicians, spend their time with that kind of elegant game?" To which the unexpected answer is—Well, it is not a game. It brings us face to face with something which is hard to remember, and that is that we live in a special kind of space—three-dimensional, flat—and the properties of that space are unbreakable. In asking what operations will turn a pattern into itself, we are discovering the invisible laws that govern our space. There are only certain kinds of symmetries which our space can support, not only in man-made patterns, but in the regularities which nature herself imposes on her fundamental, atomic structures.

The structures that enshrine, as it were, the natural patterns of space are the crystals. And when you look at one untouched by human hand—say, iceland spar—there is a shock of surprise in realizing that it is not self-evident why its faces should be regular. It is not self-evident why they should even be flat planes. This is how crystals come; we are used to their being regular and symmetrical; but why? They were not made that way by man but by nature. That flat face is the way in which the atoms had to come together—and that one, and that one. The flatness, the regularity has been forced by space on matter with the same finality as space gave the Moorish patterns their symmetries that I analyzed.

Take a beautiful cube of pyrite. Or to me the most exquisite crystal of all, fluorite, an octahedron. (It is also the natural shape of the diamond crystal.) Their symmetries are imposed on them by the nature of the space we live in—the three dimensions, the flatness within which we live. And no assembly of atoms can break that crucial law of nature. Like the units that compose a pattern, the atoms in a crystal are stacked in all directions. So a crystal, like a pattern, must have a shape that could extend or repeat itself in all directions indefinitely. That is why the faces of a crystal can only have certain shapes; they could not have anything but the symmetries in the patterns. For example, the only rotations that are possible go twice or four times for a full turn, or three times or six times—not more. And not five times. You cannot make an assembly of atoms to make triangles which fit into space regularly five at a time.

Thinking about these forms of pattern, exhausting in practice the possibilities of the symmetries of space (at least in two dimensions), was the great achievement of Arab mathematics. And it has a wonderful finality, a thousand years old. The king, the naked women, the eunuchs and the blind musicians made a marvelous formal pattern in which the exploration of what exists was perfect, but which, alas, was not looking for any change. There is nothing new in mathematics, because there is nothing new in human thought, until the ascent of man moved forward to a different dynamic.

Christianity began to surge back in northern Spain about A.D. 1000 from footholds like the village of Santillana in a coastal strip which the Moors never conquered. It is a religion of the earth there, expressed in the simple images of the village—the ox, the ass, the Lamb of God. The animal images would be unthinkable in Moslem worship. And not only the animal form is allowed; the Son of God is a child, His mother is a woman and is the object of personal worship. When the Virgin is carried in procession, we are in a different universe of vision: not of abstract patterns, but of abounding and irrepressible life.

When Christianity came to win back Spain, the excitement of the struggle was on the frontier. Here Moors and Christians, and Jews too, mingled and made an extraordinary culture of different faiths. In 1085 the center of this mixed culture was fixed for a time in the city of Toledo. Toledo was the intellectual port of entry into Christian Europe of all the classics that the Arabs had brought together from Greece, from the Middle East, from Asia.

We think of Italy as the birthplace of the Renaissance. But the conception was in Spain in the twelfth century, and it is symbolized and expressed by the famous school of translators at Toledo, where the ancient texts were turned from Greek (which Europe had forgotten) through Arabic and Hebrew into Latin. In Toledo, amid other intellectual advances, an early set of astronomical tables was drawn up, as an encyclopedia of star positions. It is characteristic of the city and the time that the tables are Christian, but the numerals are Arabic, and are by now recognizably modern.

The most famous of the translators and the most brilliant was Gerard of Cremona, who had come from Italy specifically to find a copy of Ptolemy's book of astronomy, the *Almagest*, and who stayed on in Toledo to translate Archimedes, Hippocrates, Galen, Euclid—the classics of Greek science.

And yet, to me personally, the most remarkable and, in the long run, the most influential man who was translated was not a Greek. That is because I am interested in the perception of objects in space. And that was a subject about which the Greeks were totally wrong. It was understood for the first time about the year A.D. 1000 by an eccentric mathematician whom we call Alhazen, who was the one really original scientific mind that Arab culture produced. The Greeks had thought that light goes from the eyes to the object. Alhazen first recognized that we see an object because each point of it directs and reflects a ray into the eye. The Greek view could not explain how an object, my hand say, seems to change size when it moves. In Alhazen's account it is clear that the cone of rays that comes from the outline and shape of my hand grows narrower as I move my hand away from you. As I move it towards you, the cone of rays that enters your eye becomes larger and subtends a larger angle. And that, and only that, accounts for the difference in size. It is so simple a notion that it is astonishing that scientists paid almost no attention to it (Roger Bacon is an exception) for six hundred years. But artists attended to it long before that, and in a practical way. The concept of the cone of rays from object to the eye becomes the foundation of perspective. And perspective is the new idea which now revivifies mathematics.

The excitement of perspective passed into art in north Italy, in Florence and Venice, in the fifteenth century. A manuscript of Alhazen's *Optics* in translation in the Vatican Library in Rome is annotated by Lorenzo Ghiberti, who made the famous bronze perspectives for the doors of the Baptistry in Florence. He was not the first pioneer of perspective—that may have been Filippo Brunelleschi—and there were enough of them to form an identifiable school of the Perspectivi. It was a school of thought, for its aim was not simply to make the figures lifelike, but to create the sense of their movement in space.

The movement is evident as soon as we contrast a work by the Perspectivi with an earlier one. Carpaccio's painting of St. Ursula leaving a vaguely Venetian port was painted in 1495. The obvious effect is to give to visual space a third dimension, just as the ear about this time hears another depth and dimension in the new harmonies in European music. But the ultimate effect is not so much depth as movement. Like the new music, the picture and its inhabitants are mobile. Above all, we feel that the painter's eye is on the move.

Analyzing the changing movement of an object, as I can do on the computer, was quite foreign to Greek and to Islamic minds. They looked always for what was unchanging and static, a timeless world of perfect order. The most perfect shape to them was the circle. Motion must run smoothly and uniformly in circles; that was the harmony of the spheres.

This is why the Ptolemaic system was built up of circles, along which time ran uniformly and imperturbably. But movements in the real world are not uniform. They change direction and speed at every instant, and they cannot be analyzed until a mathematics is invented in which time is a variable. That is a theoretical problem in the heavens, but it is practical and immediate on earth—in the flight of a projectile, in the spurting growth of a plant, in the single splash of a drop of liquid that goes through abrupt changes of shape and direction. The Renaissance did not have the technical equipment to stop the picture frame instant by instant. But the Renaissance had the intellectual equipment: the inner eye of the painter, and the logic of the mathematician.

In this way Johannes Kepler after the year 1600 became convinced that the motion of a planet is not circular and not uniform. It is an ellipse along which the planet runs at varying speeds. That means that the old mathematics of static patterns will no longer suffice, nor the mathematics of uniform motion. You need a new mathematics to define and operate with instantaneous motion.

The mathematics of instantaneous motion was invented by two superb minds of the late seventeenth century—Isaac Newton and Gottfried Wilhelm Leibniz. It is now so familiar to us that we think of time as a natural element in a description of nature; but that was not always so. It was they who brought in the idea of a tangent, the idea of acceleration, the idea of slope, the idea of infinitesimal, of differential. There is a word that has been forgotton but that is really the best name for that flux of time that Newton stopped like a shutter: *Fluxions* was Newton's name for what is usually called (after Leibniz) the differential calculus. To think of it merely as a more advanced technique is to miss its real content. In it, mathematics becomes a dynamic mode of thought, and that is a major mental step in the ascent of man. The technical concept that makes it work is, oddly enough, the concept of an infinitesimal step; and the intellectual breakthrough came in giving a rigorous meaning to that. But we may leave the technical concept to the professionals, and be content to call it the mathematics of change.

The laws of nature had always been made of numbers since Pythagoras said that was the language of nature. But now the language of nature had to include numbers which described time. The laws of nature become laws of motion, and nature herself becomes not a series of static frames but a moving process.

Mathematics in Civilization

H. L. Resnikoff and R. O. Wells, Jr.

Howard L. Resnikoff received his Ph.D. in mathematics from Berkeley in 1963. He went to Rice University and was chairman there from 1975–78. He has been director of the NSF division of Information Science and Technology since 1978. An NSF fellow and a member of the Institute of Advanced Study, he has received a U.S. Senior Scientist Award.

Raymond O'Neil Wells, Jr., received his Ph.D. in mathematics in 1965 from NYU. He has been a professor of mathematics at Rice University since 1974. An editor of the Transactions of the American Mathematical Society from 1979–83 and a Guggenheim fellow, he also received a U.S. Senior Scientist Award.

This selection was taken from the introduction of a college textbook written by two active research mathematicians. The purpose of the book is to give general audiences some insight into the nature of mathematics. Resnikoff and Wells seek to paint a historical and social backdrop against which the developing ideas of mathematics may be brought on stage. While this device is common in such books, notice that the mathematical instincts of the authors can hardly be suppressed for as much as one page. No sooner is the title written than we are involved in a quantifiable discussion of how the density of notable mathematicians in the general population seems constant in time. Next, Resnikoff and Wells examine the manner in which this statistic impinges on the historical development of their subject. It is a fascinating approach and quite probably the last one that a professional historian would adopt. However, it is precisely because they are mathematicians that these authors bring such valuable insights to this subject.

YOUNG PEOPLE ENTERING college today should attempt to extract two things, at least, from their college years. First is a certain competence in an area or specialized field by means of which they can hope to find a place for themselves in society. Second, they should become aware of and knowledgeable in regard to the many other aspects of society that will engage their attention and to some extent determine the course and quality of the rest of their lives. It is in response to the latter purpose that in recent years college courses have been designed to convey an understanding of the nature and role of fields other than the area of a student's primary specialization. This book is intended for students and others who desire to understand the role that mathematics plays in science and society. It does not teach mathematical technique, nor will it prepare the reader to use mathematics as a tool in any serious way. It is sparing in its demands on the reader's mathematical knowledge; competence in the arithmetic of fractions and decimal expansions, elementary plane geometry, and the rudiments of algebraic manipulation and trigonometry are the only mathematical prerequisites.

One way to introduce mathematics to the uninitiated is to sample attractive topics of current or recent interest to mathematicians and by studying them in some detail attempt to teach the student what mathematics is about and how mathematicians think. Learning about mathematics this way is time consuming and difficult, much as it would be to acquire a knowledge of music by trying to learn to play a number of different musical instruments, but not very well. Just as it is possible to appreciate music and to understand its role in civilization in a more than superficial way without being able to play even one instrument nor even to understand the technicalities of how one is played, so also is it possible to attain a serious comprehension of the nature of mathematical achievements and the impact they have on civilization.

The structure and purpose of this book can be illuminated in terms of another analogy. Mathematics is a growing subject that can be likened to a tree: think of the height of a place on the tree as a measurement of *time*, with early mathematics located near the roots and the most recent advances flowering at the tips of the highest limbs. Those books that sample topics drawn from modern mathematics can be said to exhibit a horizontal section of the mathematical tree near its top, whereas a study of the historical development of one topic corresponds to tracing a vertical path that starts somewhere near the roots and continues up through its limbs. In this book we have concentrated on two major topics and traced their paths up through the tree from antiquity to modern times. These topics are fundamental and pervasive; we think they lie close to the essential nature of mathematics. Moreover, although each preserves its own identity, they have become inextricably intertwined throughout the centuries. By concentrating on them and following their development we hope to provide the reader with a perspective of the process of mathematical development and its symbiotic interaction with the corresponding development of civilization that is impossible to obtain from a study of a sectional selection of recent mathematical topics.

This evolutionary standpoint has another, pedagogical, advantage. The student who lacks technical proficiency in areas of current mathematical interest nevertheless has accumulated a store of mathematical knowledge that is elementary by current standards but once represented the research frontier. By tracing our way from the past to the present, although not always in strict chronological order, and by limiting ourselves to basic mathematical problems of ancient lineage we hope to be able to build on the reader's available technical knowledge to propel him to an understanding of the modern, more sophisticated, forms of these problems and therewith to an understanding of the value and implications of mathematical progress.

We have, as we have said, selected two paths. The first might be termed

the ability to compute.

It is directly related to the growth of technology and to the ability to organize increasingly complex forms of society. The second path can be called the

geometrical nature of space,

that is, the geometrical nature of the physical world in which we find ourselves. This path deals with the evolution of conceptions of the physical universe and their relation to abstract forms of geometry invented and studied by mathematicians through the centuries. In its practical applications this path is a determinant of our ability to control physical reality. It also lies at the foundation of some of the most profound philosophical speculations. The paths are interdependent because the ability to compute underlies the advance from simple and simplistic geometrical considerations to complex ones that can accurately describe portions of reality. On the other hand, the complexity of some geometrical models challenges the available computational capabilities and encourages their further development.

The purpose of this book is, as already asserted, to study the role of mathematics in the development and maintenance of civilization. It is part of our thesis that, although the techniques and personalities of mathematics are not and indeed should not be of much interest to the nonmathematician, the purposes and consequences of mathematics are of serious concern for the growth and health of society and therefore are a proper and necessary part of the workaday intellectual baggage that must be carried about by every educated and effective participant in civilized life.

Mathematics occupies a peculiar and unique role in that it is neither a science nor an art but partakes of both disciplines. Art provides the motivation for most pure mathematicians but science (this term understood in its broadest sense) reaps the harvest. That there are important differences between mathematics and science is not simply a matter of personal opinion; they show up in quantitative as well as qualitative ways, some of which are considered below.

Nevertheless, mathematics is usually considered to be a science, and in this guise it participates in the general increase of federal support for research activities. The competition for the taxpayers' dollars having now become quite keen, a serious inquiry into the role played by mathematics in society is well justified. Is the emphasis on mathematics in contemporary American society sufficient, too great, or just right? We have tried to provide you, the reader, with some tools that you can use to answer this difficult question yourself. After all, it is, or shortly will be, your tax dollar that will help to determine the future.

One way to evaluate the importance to society of an activity that it undertakes and supports is to estimate the fraction of its human and other resources this activity consumes. If this fraction is now large but was small in the past, we can reasonably assert that the activity has recently become more important; if the fraction has remained sensibly constant throughout long historical periods, then we should conclude that the activity has always had about the same relative importance as it now has.

Let us examine the size of the mathematics establishment today and compare it with the situation in the past as far back as we can. It will turn out that "memorable" mathematicians have always—at least for the last 2000 years—accounted for a virtually fixed fraction of the population: about one memorable mathematician for every 4 million people. First consider the current situation. The 1966 edition of the *World Directory of Mathematicians* lists about 11,000 persons. In 1965 the population of the world was about 3.3 billion—about 300,000 people for each mathematician. Not all mathematicians produce mathematics that will be memorable in years to come; perhaps only one in 50 will, which means that about 220 or so will be likely to pass into history as memorable. This estimate is, of course, just an opinion, but there is some independent evidence that suggests it cannot be far wrong.

To understand this evidence and to be able to compare the present size of the mathematics establishment with the past let us turn to the history books. Dirk J. Struik, himself a notable mathematician, has written on the history of science and mathematics. His *A Concise History of Mathematics*, now in its third edition, is a learned and sophisticated work that has been well received. Let us agree, for the sake of argument, that a mathematician is memorable if he is listed (with his date of birth) in Struik's index. This "defines" what we mean by a "memorable mathematician," and although it is a definition that is open to question with regard to any particular mathematician it certainly will reflect in general what is intuitively meant when it is said that any historical figure is memorable—he occurs in the history books. When counting the number of memorable mathematicians, it makes sense to accumulate all those who were born before a given date, since important mathematical contributions do not tarnish with age. Figure 1 displays the graph of this function. Evidently there has been an enormous growth in the number of memorable mathematicians in recent times. On the other hand, it is clear that there must have been some memorable mathematicians who were born before −700, although no names have come down to us. There can be many reasons for this: the ravages of time acting on records, Struik's possible idiosyncrasies, social anonymity in early civilizations, etc. Be that as it may, Figure 1 shows a steady growth[1] in the number of memorable mathematicians from about −100 until +1400, after which there is a dramatic and still continuing increase in the rate of growth of the curve.

Notice that there were no "Dark Ages" for memorable mathematicians. The "darkness" of medieval Europe, from 600 until about 1200 (depending on the authority quoted), was compensated by the "lightness" of Arab civilization.

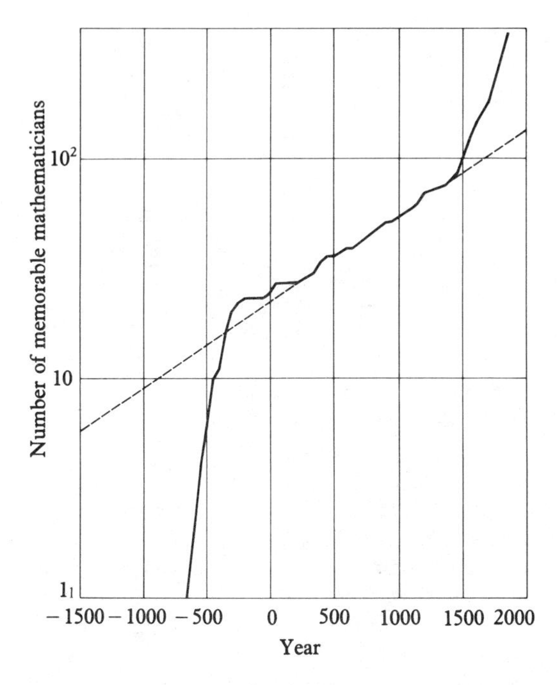

Figure 1 Memorable mathematicians (cumulative, by birthdate). Data from Struik, D. J., *A Short History of Mathematics*, 3rd ed. Dover, New York, 1968.

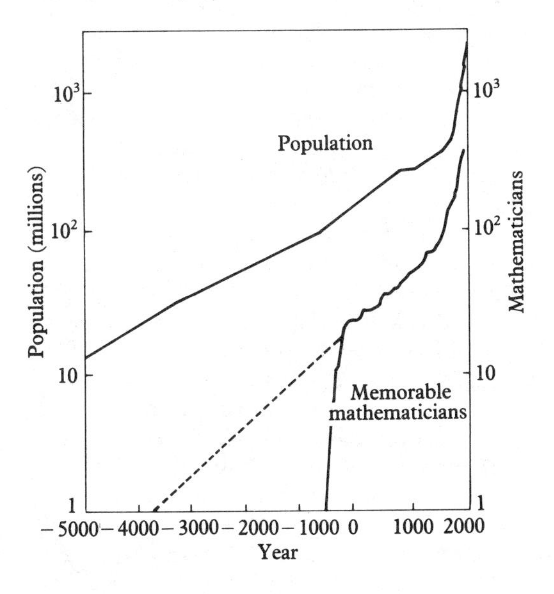

Figure 2 World population and memorable mathematicians.

Figure 2 illustrates the cumulative growth of memorable mathematicians with the portion corresponding to steady exponential growth from -100 to $+1400$ projected back in time; this projection suggests that there should have been "one" memorable mathematician about -3700. We will return to this speculation later. The same figure shows estimates of the world's population at various times. Observe that the population curve and the cumulative memorable mathematicians curve are approximately parallel. This means that memorable mathematicians have constituted about the same fraction of the population in the distant past as they have in more recent times. Inspection of the figure shows that the ratio of population to the cumulative number of memorable mathematicians has varied with time as shown in Table 1. This table cannot be brought more closely up to date because Struik considers only pre-twentieth century mathematics, and we have classified memorable mathematicians according to their date of *birth,* which is at least 20 years before their active intellectual life begins. It is remarkable that the ratios are so close in value: they all lie between 3.8 millions of population per memorable mathematician and 6.9 millions. The apparent trend seems to indicate that the number of mathematicians is growing more rapidly than population. This may be due to an increasing undernumeration of mathematicians the longer ago they lived, or to an overestimation of world population in early times, or it may simply be that the number of memorable mathematicians and world population are just not proportional. The last possibility is perhaps right, but the fact that the population growth and memorable mathematician growth curves are both so far from being simple, yet are almost parallel, suggests that the processes they represent may be closely related.

As a speculative possibility, we propose that memorable mathematician growth is actually proportional to *gross world product* (or some other index of world economic growth) rather than to world population. Gross world product is analogous to gross national product, which measures the annual value of all goods and services produced by a nation and is the best known and probably most reliable indicator of its state of

Table 1 Ratio of Population to Cumulative Number of Memorable Mathematicians

Date	*Number of Mathematicians Born Before*	*World Population*	*Ratio: Population per Number of Mathematicians (in millions)*
700	39	2.70×10^8	6.9
1049	54	2.85×10^8	5.3
1449	80	3.75×10^8	4.7
1549	97	4.20×10^8	4.3
1649	143	5.45×10^8	3.8
1749	180	7.28×10^8	4.0
1799	223	9.06×10^8	4.1
1849	296	1.17×10^9	4.0

wealth. An expanding economy increases per capita wealth with time, and the economy of the world as a whole has certainly been expanding rapidly since the industrial revolution. Because mathematicians are supported by that fraction of the world wealth that remains after food, clothing, shelter, and other "necessities" have been paid for, it seems reasonable to conclude that the number of mathematicians should increase as "excess" wealth increases. This is a possible explanation of the relatively more rapid growth of memorable mathematicians compared with population, but it is also a difficult hypothesis to verify, since there is little direct data available that would permit the calculation of gross world product for past centuries.

In any event, we assume provisionally that there will continue to be about 4 million people for every memorable mathematician, the ratio characteristic for those memorable mathematicians born before 1850. For a world population of 1.6 billion, which was the situation in 1900, there ought to have been about 400 cumulative memorable mathematicians; for a population of about 3 billion (the situation in 1960), about 750. Therefore the number of memorable mathematicians born between 1900 and 1960 should be $750 - 400 = 350$; those born since 1940 have not, for the most part, been heard from yet, although they must constitute at least $(60 - 40)/60 = 1/3$ of the total. This means that there should be (very approximately) $(2/3)350 \cong 233$ recognized memorable mathematicians alive today, which is in reasonable agreement with our earlier intuitive estimate that perhaps one in 50 working mathematicians is memorable.

Figure 2 shows that world population experienced the same kind of dramatic—one might aptly say "explosive"—growth after 1400 that was experienced by the number of memorable mathematicians. This period overlaps the humanistic Renaissance, and all three events are to a large degree responses to one critical development: the *invention of movable type* and the use of the *printing press* about 1540 by Gutenberg and Fust in Mainz, Germany. The importance of this event cannot be too greatly stressed. With a means of large-scale and rapid dissemination of information in permanent form and at low cost, it became possible to accumulate library archives for effective reference and to concentrate on the discovery of new knowledge without the need of continually reproducing what had already been done but had not been communicated to others. Again and again historical indicators point to the late fifteenth century as the most critical period in the last two millenia for the development of civilization.

The growth of the number of "memorable technologists" (Figure 3) shows one effect of the printing press in a clear way. In this case we have agreed that a technologist is memorable if he appears in the index to Derry and Williams' *Short History of Technology*, a standard work. Observe that technologists grew in number at about the same rate as mathematicians in the earliest times, from about -700 to -400, but their growth remained completely stagnant throughout the Dark Ages (is this a possible definition of the Dark Ages?) until a phoenix-like resurrection after 1400. Since 1400 the number of technologists has grown much more rapidly than the number of

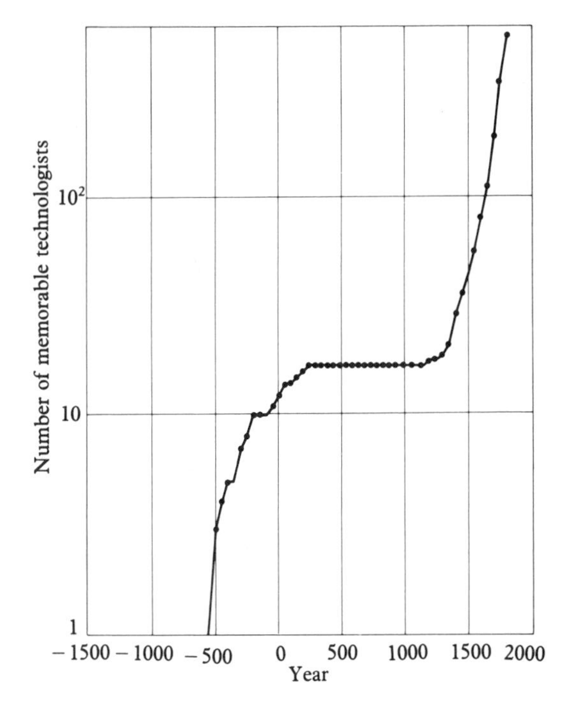

Figure 3 Memorable technologists (cumulative, by birthdate). Data from Derry, T. K., and Williams, T. J., *A Short History of Technology*, Oxford University Press, Oxford, 1960.

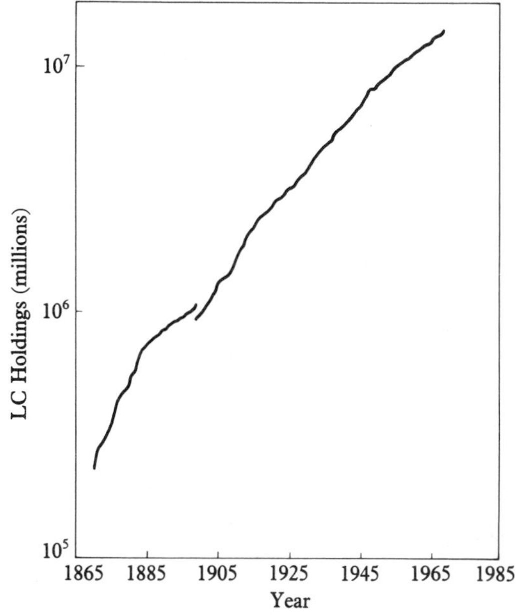

Figure 4 Library of Congress holdings, 1865–1966.

mathematicians. In the past it was much easier to find mathematicians among the scientists than it is today.

Technology depends on the timely dissemination of knowledge, even more than pure science and much more than mathematics; therefore it should come as no surprise that for hundreds of years before the invention of movable type technology was stagnant, as Figure 3 so clearly shows. How, then, can we explain the growth of the number of technologists that apparently took place from -700 on? Before we turn to this question let us look at one measure of the effect of printing that makes clear how rapidly human knowledge has accumulated since its invention.

The Library of Congress of the United States is the largest in the world. Although founded only in the early nineteenth century, it has grown with remarkable rapidity and now attempts to acquire a copy of almost every significant printed work, regardless of language; in 1966 it held nearly 14 million books. Figure 4 shows the number of books held since 1865 displayed on semi-logarithmic graph paper. The points fall nearly on a straight line; if this line were extended back in time, we would discover that the "first" book ought to have been printed about 1500, which is reasonably consistent with the facts considering the quality of the data we have used. We can therefore provisionally conclude that the sum of human knowledge, as represented in printed books, has been growing exponentially with time since the invention of printing.

With regard to the fifteenth century we have argued that it is rapid and inexpensive communication facilities, coupled with the ability to preserve information in permanent but easily retrieved form, that really set civilization going. During the period from -4000 to -700 two major advances were made which established ancient

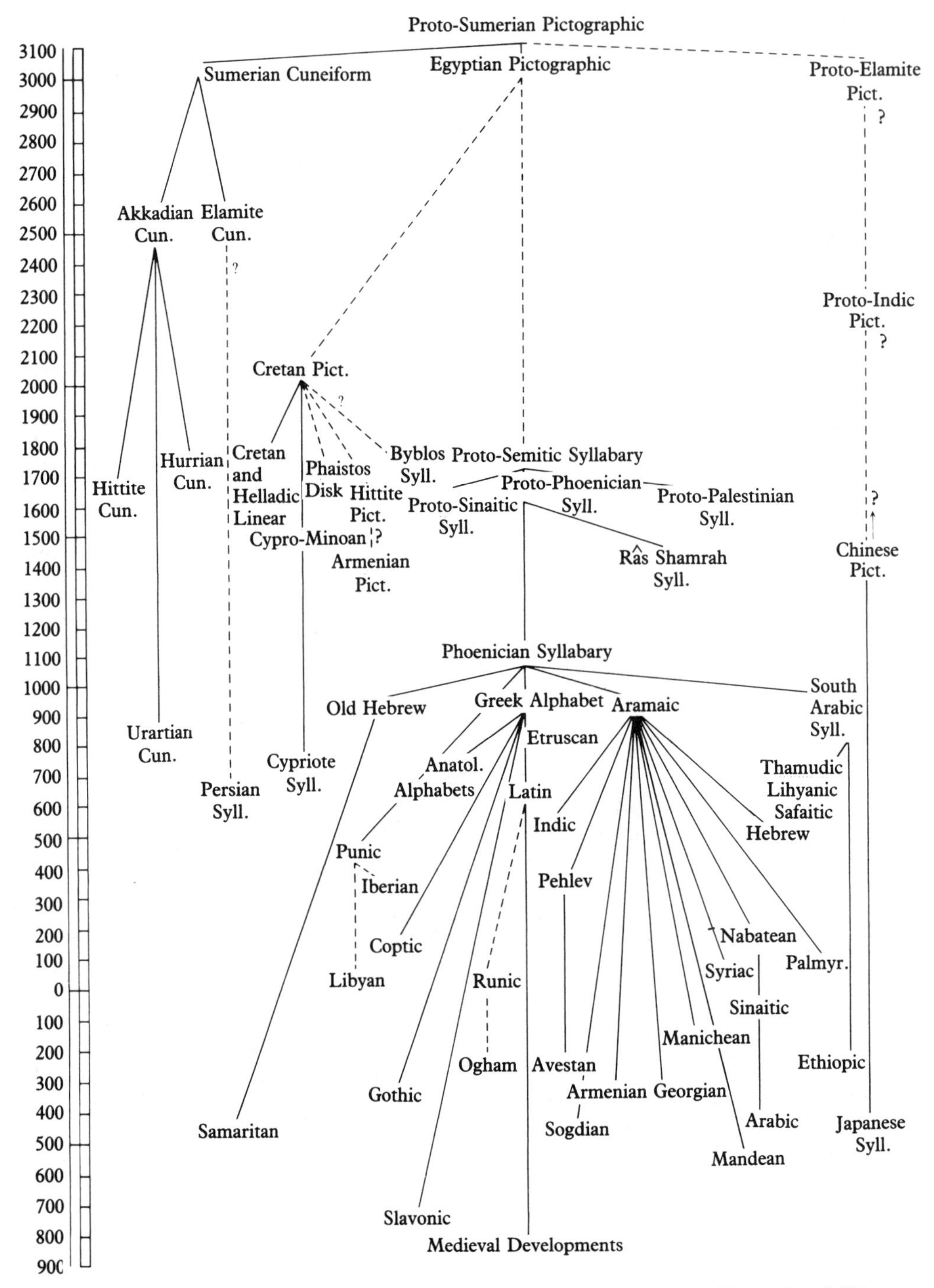

Figure 5 Origin of the alphabet. From Gelb, I. J., *A Study of Writing*, rev. ed., University of Chicago Press, Chicago, 1963.

civilizations that we easily recognize as forerunners of our own, cast in the same basic pattern. These were the *invention of writing* and the much later *invention of the alphabet.* Figure 5, taken from Gelb's *A Study of Writing,* shows that pictographic writing systems were in existence in −3100 but probably not much before that. These systems gave rise to the great near-eastern languages of antiquity, *Egyptian hieroglyphs* and *Akkadian cuneiform,* which lasted for more than 2000 years before they were displaced by the much more efficient alphabetic systems. Returning to Figure 2, we see that the straight line projection of the cumulative number of mathematicians indicates that the "first" memorable mathematician should have been born in −3743. This date is consistent with the first postneolithic developments of civilization and is some hundreds of years before the earliest known pictographic writing systems.

It is conceivable that mathematical needs for notational symbolism were later developed into full-fledged pictographic writing systems; that is, that mathematics preceded and was the catalytic agent for the formation of writing systems. Certainly primitive mathematical records are ancient. We quote from Struik:[2]

> Numerical records were kept by means of . . . strokes on a stick. . . . The oldest example of the use of a tally stick dates back to paleolithic times and was found in 1937 in Vestonice (Moravia). It is the radius of a young wolf, 7 inches long, engraved with 55 deeply incised notches, of which the first 25 are arranged in groups of 5. They are followed by a simple notch twice as long which terminates the series; then starting from the next notch, also twice as long, a new series runs up to 30.

This tally stick has been dated at about −30,000. If numbers as large as 55 were already necessary in the primitive hunting societies of the Paleolithic, it should be no surprise that settled agrarian societies would soon find it necessary to introduce much larger numbers and an efficient means for denoting them. We think it quite likely that mathematical notations are anterior to writing systems.

The invention of writing systems led to the production of massive quantities of records, some of which have survived the ravages of time and have been deciphered to reveal the nature of the civilizations that created them. Thus we know quite a bit about the Egyptian and Akkadian people, although somewhat less about the Sumerian forerunners of the Akkadians. The Summerian and Egyptian writing systems were originally pictographic; the Akkadian was a modification of the Sumerian and ultimately was simplified to a still quite complex syllabic system that utilized hundreds of different cuneiform signs to express the different syllables of the language. Pictographic systems are extremely inflexible and inefficient ways to write; large syllabary-based writing systems are only slightly more efficient. The major improvement in writing systems, which made them vastly more efficient as well as much simpler to learn, was the invention of the alphabet. Although the Phoenicians invented a small efficient syllabary about −1000, it was the Greeks who by −700 modified it to form a real alphabet. Shortly thereafter the Greeks began their phenomenal rise to political power and intellectual eminence, the latter quality persisting in its influence to this day. The chronological scheme of Figure 6 connects linguistic developments with the civilizations that produced them.

Return to Figures 2 and 3, and observe the sudden growth of both memorable mathematicians and memorable technologists that began between −800 and −700. Could this effect be due to the invention of the alphabet, an efficient tool for recording and retrieving information?

We have seen striking changes in the nature of civilization which occurred shortly after or contemporaneously with three advances in recording and communicating knowledge: the invention of *writing systems* before −3100, the invention of the *alphabet* before − 700, and the invention of *movable type* about +1450. During the last two decades we have witnessed the invention and explosive development of a similar fourth advance: the *digital computer.* Would it be unreasonable to posit a corresponding change in the fabric of our society in reponse to this novel and almost unbelievably efficient means for recording and retrieving knowledge?

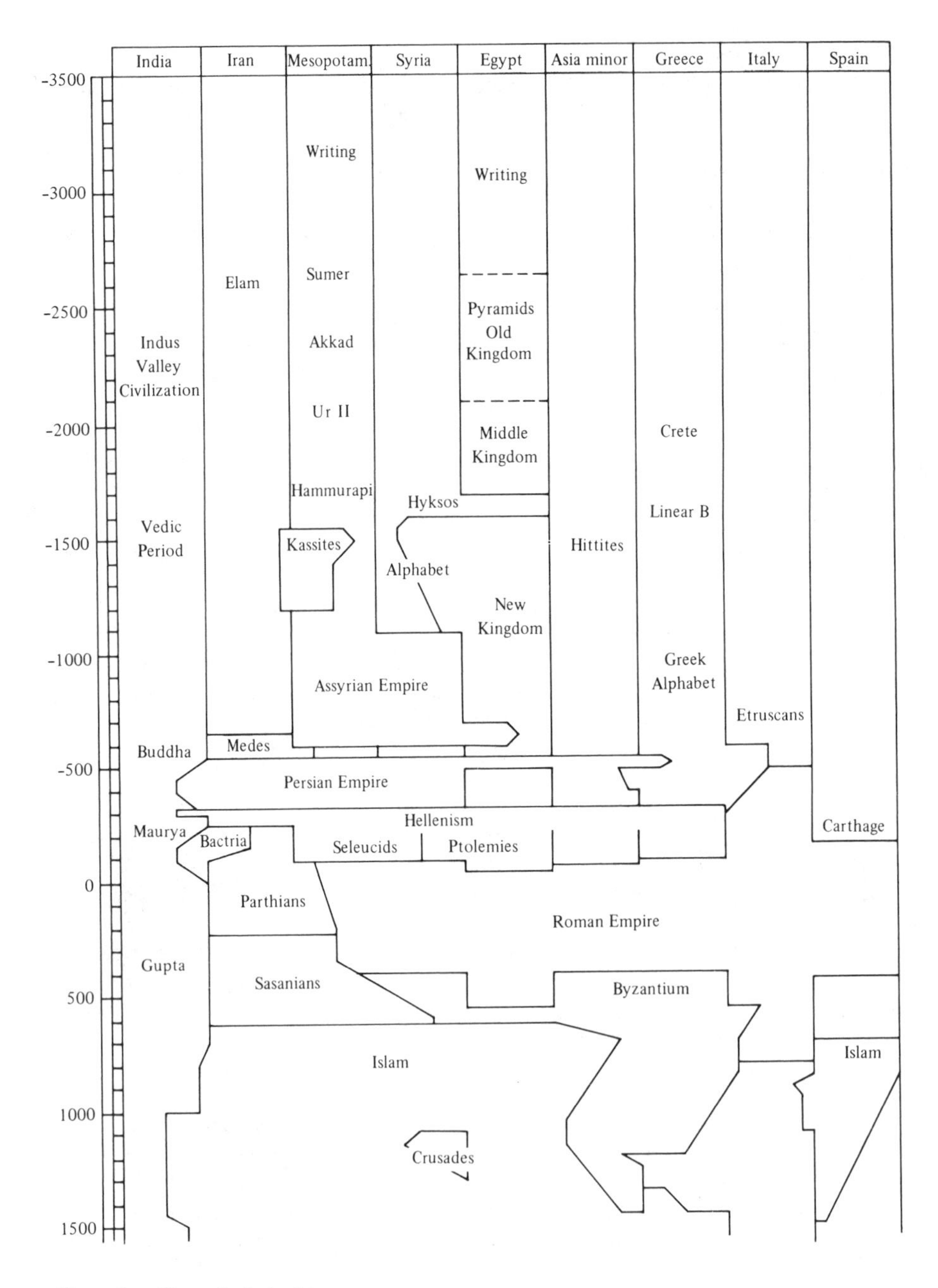

Figure 6 Chronological scheme.

Notes

1. The curve is plotted on *semilogarithmic* graph paper; that is, the ordinates are the *logarithms* of the cumulative number of mathematicians. A straight line segment corresponds to *exponential* growth.
2. D. J. Struik, *A Short History of Mathematics*, 3rd ed., Dover, New York, 1968.

Old Tripos Days at Cambridge

A. R. Forsyth

A fairly complete biographical sketch of A. R. Forsyth is given in the article "Old Cambridge Days" by Leonard Roth, which follows this essay.

Those adept at reading between the lines can get a pretty clear picture of England in the late nineteenth century from this essay. It is an utterly charming bit of nostalgia delivered by A. R. Forsyth in 1935, a reminiscence of a time a century or more remote from our own. It captures not only the experiences of an eager undergraduate sweating his way through a difficult and complicated formal education, but also the insularity, the unaffected sense of superiority, the nature of tradition, nearly everything but the smell of the England of that era. This brief work says as much about the spirit of British mathematics and physics in the nineteenth century as most formal histories.

A brief word about some terms: In the Cambridge of Forsyth's day degrees were awarded on the basis of a single series of examinations given at the end of a formal multiyear course of study. The name of the mathematics examination was the Tripos. *The individual who earned the highest score in any one year was accorded the title* Senior Wrangler. *If one person were consensus all-American, a Rhodes scholar, and Bachelor of the Year, he would not come close to commanding the lasting distinction that came to the Senior Wrangler of the Tripos. Here then is an account of a time that will never return.*

MY PURPOSE THIS evening is to submit to you an outline sketch of the general course and the circumstances of mathematical study at Cambridge in my student days between fifty and sixty years ago. It does not set out to be an estimate of the state of mathematical knowledge at that date. It is not a comparison (or would contrast be the better word?) between the empirical natural philosophy then in vogue and the rather speculative (slightly Aristotelian) physical theories that now absorb a small world of seekers after new Truth. It is only a picture of that academic life as it then appeared to an unfledged student, now drawn as faithfully as memory will allow.

I

A beginning may be made by naming some of the men of mark at Cambridge in that day.

There was Stokes, who held the chair that once was Newton's: the successor of Young, Poisson, Fresnel, in the line of mathematical physicists (to be followed, in my judgement, by Horace Lamb who died recently). We knew some of the work of Stokes, published many years earlier, on the sums of periodic series. But we could not then know how much work, silent in its anonymity, fell to him as Secretary of the Royal Society in saving the *Transactions* of that learned body from

Source: A. R. Forsyth, "Old Tripos Days at Cambridge," *Mathematical Gazette,* 19 (1935): 162–179. An address to the South-West Wales Branch of the Mathematical Association, Swansea, 2nd March, 1935.

the occasional paradoxes of less accomplished physicists. Nor could we know then that he was the scientific Master of Sir William Thomson—the title of respect with which Kelvin, to his latest days, was wont to describe Stokes in public references. We students attended one course of his lectures—on Physical Optics: it was the single professorial exception allowed, even enjoined, by tutors of all grades—and delightful the lectures were, none the worse because they were of no profit in a Tripos Examination.

There was Cayley, deemed a wonder in pure mathematics, not then held in favor at Cambridge. He had achieved marvels in analysis and high algebra, all beyond us: marvels in the geometry of curves and surfaces, in a range of which we knew nothing; also in the beginnings of the hypergeometry of ideal space: Maxwell described him as the man

> Whose soul, too large for ordinary space,
> In *n* dimensions flourished unrestricted.

Even our college manciple, a fine character in the old-world type of college servants, had his own appreciation of the great man: one day, when expounding some of the glories of Trinity to passing visitors, he halted before Cayley's picture in the dining-hall and solemnly assured them "he's that exact, he could take the earth in his hand and tell you its weight to a pound." Not for undergraduates was there to be attendance at Cayley's lectures, any one of which often contained results his research had obtained only since the preceding lecture; there, he never gave a thought to the Tripos: and rarely indeed did examiners pay the least heed to Cayley's work. But there was one rather wayward undergraduate of my year who, greatly daring, went to a course of lectures by Cayley: he had gone with the acquiescence of his college tutor, to the surprise of one or two lecturers of his college, and against the prudent warning of his coach that such expenditure of time would be unprofitable for the Tripos. That undergraduate was out of his depth after a few minutes of the first lecture: he took what notes he could: sought information somehow: delved hard (and often unwisely) in strange places, with a determination to learn something in that range of knowledge; and he then succeeded in making a beginning of the pure mathematics which, in varying forms, has absorbed a large part of a life that has not been idle.

There was Adams. He had discovered the planet Neptune thirty years earlier, by a great and long-sustained calculation first conceived by him when he was an undergraduate. We knew something of the inadequate treatment of his work in 1845 meted out by Airy, who had passed from Cambridge to Greenwich and who still was Astronomer Royal in our day. But the work of Adams was not for us: he had passed into the limited cohort of great classics, whose works are not read but to whose names men pay a willing, if uncritical, homage. Seemingly unambitious, he moved supreme in his own domain and was known to us as Father Neptune. Diligent in research, he did not publish much, but all the published work was sound: his main satisfaction appeared to be the attainment of knowledge. Thus, when G. W. Hill's work in the lunar theory first became known, Adams was found to have all the main results in his own possession, even to familiarity with those infinite determinants which long deterred even the most courageous souls.

We saw Maxwell, intellectually known to be great: we could not surmise how much greater he was yet to become even than the reverent estimate of an awakening world. His great treatise *(Electricity and Magnetism)* was still fairly recent: its range seemed to have little in common with the electricity of the blackboard and the examination paper: indeed, many of the students (I was one) could then hardly tell—except in a glib vocabulary sufficing for the examination—the difference between a conductor and a condenser. And I have heard Maxwell lecture, a declaration that now can be made by few: for he died in 1879, while still in the prime of life as men count years.

On rare occasions we saw Sir William Thomson, who professed at Glasgow: known to some of us by name, well known to students who had come to Cambridge from his classes at Glasgow. (In my day, many of the best Scots students came on to Cambridge from Glasgow, from Edinburgh, from Aberdeen, after a distinguished home-career.) I happened to know about Thomson also as the electrical engineer of the first successful Atlantic cable: for, in the mid-months of 1877 before going into residence at Trinity, I had been on the ship which was laying a cable between Marseilles and Bona; and on board that ship, at every turn, the electrical staff cited the name of Thomson. It was only years afterwards, when Stokes and J. J. Thomson and I were the official delegates of the University of Cambridge at the Kelvin Jubilee, that I learnt an earlier incident connecting three of the great men who have been mentioned: when the very young William Thomson had been a candidate for the Glasgow chair fifty years earlier, one of his testimonials came from Mr. G. G. Stokes and one from Mr. A. Cayley. Also, Thomson (with Tait) was joint author of that book on Natural Philosophy which, as "*T and T′*," was a guide to a few men in each generation of Cambridge students.

There were other names known to us, but as yet only names. Such was Sylvester, who, at the age of sixty-three and after a varied life, had just gone to the Johns Hopkins University, and who, when he was on the verge of seventy, returned to vivify Oxford disciples with his zeal for research. There was Salmon, the friend of Cayley: he had not yet completely passed from mathematics to theology where he made a second fame: he was the author of that book on Conics which, to us, was almost a mathematical Bible. We gathered that Cayley, Sylvester, Salmon, were a world-triumvirate in a dark continent of invariants: Hermite's name was mentioned occasionally: the name of Gordan must have been unknown save to the very elect. By middle life, it was my privilege to have become acquainted with all of them personally, except Hermite, and from Hermite there came letters of appreciation beyond value: all that acquaintance now is an abiding memory.

Others also there were, though not deemed of the same high rank in our youthful minds. (I am not framing an estimate of their achievements in our science: only the very youthful, and the rather omniscient, can proceed fearlessly to such a task.) There was Henry Smith of Oxford. There was Clifford, who had only recently left Cambridge for London. There was Lamb, who had only recently gone to Adelaide, and whose book on hydrodynamics (then a slight volume, being an exposition of lectures he had given at Trinity) was the first English book that revealed a use of the complex variable in mathematical physics: let me add that it was an age when the use of $\sqrt{-1}$ was suspect at Cambridge even in trigonometrical formulae. There was Greenhill, whose early papers were a development of such work revealed by Lamb. There was Glaisher of my own college, one of the best of lecturers, full of enthusiasm about differential equations and elliptic functions and the method of least squares. There was W. D. Niven, also of Trinity, an interpreter of Maxwell's work: his lectures gave every man as much labor as Cayley's course had given me. There were Burnside and Chrystal, of a then recent year: Burnside, still at Cambridge, had yet to graduate from applied mathematics into pure mathematics: Chrystal, just gone to St. Andrews and soon to succeed Kelland at Edinburgh, had achieved a reputation by his *Encyclopædia* article on "Electricity," and was yet to be a pioneer in pure mathematics.

And there were others: even of mathematicians, I do not pretend to cite all the names held in honor in the limited range of a student's knowledge: in ancient words, they were "men of learning, honored in their generations, a glory in their days: some have left a name behind them: and some there be which have no memorial." In diverse ways, and in varying degrees, they were an inspiration to the students of that epoch now more than half a century ago.

Such were the Great Ones on the heights or the upland slopes of our mathematical Olympus: what of the multitude of students on the plains below? We existed in the downland, seldom raising our eyes to the topmost ridges, hardly qualified for fit worship even if there were either wish or will. Some students of that day were, in their turn, to carry on the legendary torch of learning, though, then, all was mainly promise. We had any amount of youthful confidence, and more than any amount of exuberant prophecy, as to what was sure to happen: but sometimes the confidence waned as the hour of performance drew near; and old unfulfilled prophecies are forgotten.

Yet there were some—again let me repeat that I am recalling the young men, for the most part, within our own range of study—some who, even then, seemed assured of future greatness and would never need to abide challenge. There was J. J. Thomson—at that time, and down to this day, known as J. J., in general respect and general affection. His personality stood out: we felt that he was framed in an intellectual mold different from ours: he had our worship as completely as Alfred Lyttelton and A. G. Steel (who were our contemporaries) had secured adoration in the world of cricket. There was Charles Parsons, great son of a distinguished father: he was absorbed in his models of swiftly-moving machines almost in anticipation of his turbines, a genius not of the conventional type that obtains early academic recognition. There was Hobson: he was, of course, bound to be Senior Wrangler: could a mathematical undergraduate in that school conceive a nobler pinnacle of fame? There was Karl Pearson, looking a fair-haired Norseman, apparently ready for anything, trying many things in happy fact, not respectful to conventional thought, perhaps not always too respectful to conventional persons. There was Micaiah Hill, who might go far, in our view, though whither was beyond our prescience: he did much for our science, and more for the University of London. There was Larmor, an outstanding representative of the Irish genius, silently noted for coming greatness, an expectation amply realized in his scientific life. One other name may fitly be mentioned here: he then was merely one of the horde of mathematical students, not more noticeable than scores of others. Later, by judgment, and thought, and no little calculation, he solved the indeterminate problem of feeding a whole nation in the sternest stress of war. The name of David Thomas, Lord Rhondda, has its place of honor on the beadroll of the benefactors of his country at an epoch of danger. There were younger men too: G. B. Mathews and Whitehead in one year: William Bragg and W. H. Young in the next. But on these I do not dwell: partly because, in the year after my own graduation, there came a change in the whole system, which (though not so recognized at the time) really was the end of a long and ancient chapter in the history of English mathematics.

Between the great professors and our unfledged selves there was nothing in common, absolutely nothing, strange as such a declaration may seem. They did not teach us: we did not give them the chance. We did not read their work: it was asserted, and was believed, to be of no help in the Tripos. Probably many of the students did not know the professors by sight. Such an odd situation, for mathematical students in a University famed for mathematics, was due mainly, if not entirely, to the Tripos and its surroundings which, as undefined as is the British constitution, had settled into a position beyond the pale of accessible criticism.

II

Let me summarize, briefly, that position in the University. In early days, there was only a single test for the bachelor's degree at Cambridge. By the beginning of the eighteenth century, the test had been systematized into what was called the Senate-House Examination, largely mathematical in range, with more than doubtful Latin as the medium of expression. In 1824, a new examination in what we call classical learning (the language and literature of Greece and Rome) had been instituted; and from that date the old Senate-House Examination was called the Mathematical Tripos (the word Tripos itself being of ancient usage at Cambridge). But every candidate for admission to the new Classical Tripos must have qualified in the Mathematical Tripos: such was the ordinance. Macaulay's omniscient schoolboy (or does nobody read Macaulay's *Essays* now?) will remember how Macaulay himself failed to qualify: the Trevelyan biography contains a sympathy-begging letter to his mother in which, on the eve of some examination, he complains that Milton's descriptions of Heaven and Hell alike have been driven out of his head by two abominable trigonometrical formulae, which he quotes (one of them wrongly) as evidence of their foul character. Macaulay failed in his mathematics and had to be content with a poll degree—an instance, not a solitary instance, of the occasional ironies of academic records.

By that date the Tripos had become a venerable and venerated institution: it had acquired the characteristics of an ancient establishment: in spite of modifications, it retained some of those characteristics even in my own day. Bear in mind that the years of the eighteenth century, when the examination was settling into shape, were not far removed from the Newton-Leibnitz controversy about the differential calculus. Partisans in science can be as fierce as partisans in theology, though happily in our day they wield no spiritual weapon of excommunication. In patriotic duty bound, the Cambridge of Newton adhered to Newton's fluxions, to Newton's geometry, to the very text of Newton's *Principia:* in my own Tripos in 1881 we were expected to know any lemma in that great work by its number alone, as if it were one of the commandments or the 100th Psalm. Thus English mathematics were isolated: Cambridge became a school that was self-satisfied, self-supporting, self-content, almost marooned in its limitations; and gradually it developed a special organization the like of which has never appeared in other countries. The professors of the University did not examine: the lecturers of the colleges never examined, as teachers: the University did not, in any way, prepare its own students for its own examinations. But a new profession arose—not officially recognized: open to all, for there was neither a test of admission or means of exclusion: that of the private coach. He made it the business of his life to prepare candidates for the examination. The University, correctly impartial, chose the examiners on the successive nominations of the colleges in a statutory automatic cycle, established for the appointment of the proctors who were best known to us as the active disciplinary police of the University. Thus the examiners changed from year to year. But the private coach was continuous. He accumulated experience and skill: he sifted all examination papers, recent and old alike: he codified mathematical knowledge into small tracts or pamphlets, kept in manuscript as his own private prescription for his own set of students. Thus it came about that there were relatively few books: Euclid of course: textbooks on algebra, or trigonometry, in a limited range, with a morbid devotion to the fascinating tortuosities of fiendish problems. It is true that Griffin, who was Senior Wrangler when Sylvester was second, produced a book on geometrical optics: and that Parkinson, who was Senior Wrangler when William Thomson was second, produced an edition of that same book (it was in use in my time) with pictures of telescopes which had the simplicity of coffee-canisters. But it was upon the coaches that teaching depended: in their profession, they condensed available knowledge into potted abstracts; and so there were tabloid manuscripts on heat, on light, on sound, on lunar theory, on planetary theory, on practically every subject included in the Tripos schedule. The examination questions fell out of the unknown upon candidates: the private teaching devoted its powers to a preparation of students to face new conundrums by solving all that were old. Thus the continuous coaching gradually established a learned profession; and the Tripos became practically a game preserve for the coaches.

Yet do not imagine that the result was a mere fossilization of eighteenth-century ideas and methods. Such a tendency had shown itself at the beginning of the nineteenth century. The dot-notation and the notions of fluxions had full sway; and, in Cambridge, the differentials of the type dx, pervading continental mathematics, were avoided if not implicitly banned. One group of men—the leading spirits among whom were Whewell, Peacock, Babbage, and Herschel—fought the tendency by protest, by argument, by work; and one outcome of their action was to admit the differential calculus or, in the phrase of Babbage, to substitute pure *d*-ism for the *dot*-age of the University. How complete was the change may be gathered from even the single fact that Pollock, the Senior Wrangler in 1806, later a Fellow of the Royal Society, never acquired the differential calculus.

Gradually, however, the coaching system was firmly established: in the teaching, the coaches were all-dominant, almost the sole agents, because nearly all the college lectures were concerned with preliminary subjects. Happily for Cambridge and for mathematics, there were superb teachers engaged in the systematized round; and among all their names two survive in historic prominence, William Hopkins and Edward John Routh, both of Peterhouse. Hopkins was the leading coach through the period in which Sylvester, Green, Stokes, Cayley, Adams, William Thomson, Todhunter, Routh, Maxwell, graduated: a marvelous roll of pupils. As Hopkins retired, Routh took his place, apparently by natural merit. He had begun by taking the pupils of W. J. Steele—once described to me by Tait as the greatest mathematical teacher who ever lived—when Steele was ill: and Steele died all too soon. Routh produced a Senior Wrangler, if production be the proper word: he went on producing Senior Wranglers, some twenty-seven or twenty-eight in succession.

When I was an undergraduate, Routh was supreme: though supreme, he was not alone. There were other coaches of high repute, his contemporaries. There was Percival Frost: some of you may know his *Solid Geometry,* a few may know his *Curve Tracing,* probably none will have known his *Newton, I, II, III.* There was Isaac Todhunter, whose textbooks blazed a new development in the teaching of mathematics in the country, and whose more ambitious works (such as his *History of the Theory of Probability*) endure as authorities of reference. There was William Besant (the name of his novelist-brother, Walter Besant, was better known to the large world outside Cambridge), the last of the old school devoted to fluxions, whose re-edited book on hydrodynamics still has its vogue with students and teachers. All these had been elected Fellows of the Royal Society. There was one, much junior in standing, a superb teacher, a man (as I believe) whose powers would have taken him far as a pioneer into the domain of new knowledge had they been devoted to research rather than coaching—R. R. Webb, the earliest of the old Senior Wranglers now alive. Let me mention one other name in this hierarchy—my old friend, R. A. Herman, who died a few years ago, a man of unusual manipulative skill, a great and stimulating teacher, the last of the great coaches. Their names remain in the history of Cambridge mathematics.

III

Just as in the early years of the nineteenth century, there had been an upheaval affecting the character of the methods, so in my time as a student there was the beginning of another upheaval affecting the range of subjects to be studied. The theory of thermodynamics was coming to the front: with the beginning of its activity the name of William Thomson must be associated; and it is not yet exhausted. The theory of gases was claiming attention—the names of Maxwell, Clausius, Tait, Boltzmann, need only be mentioned: the spirits of Clausius and Tait secured a more than ample measure of resounding controversy. More silently, more surely, more steadily, and with results ranging beyond the imagination of that day, Maxwell's mathematical presentation of Faraday's work and his own mathematical developments were revolutionizing the systematic treatment of all electrical theory. The initial influence, exercised by these changes upon Tripos study and examination, was small: here a minute fragment, there a tiny snippet, just as beginnings. But all this new work was outside the range and beyond the familiar knowledge of the coaches.

Some of the older subjects remained in the vigor of full demand. Astronomy—then a range of study almost enjoined as a pious duty in the academic home of Newton, and even fortified in its claim by the achievement of Adams in the discovery of Neptune—held a double sway. On one side, there was the mathematics of the geometry of the heavens and the description of cosmical phenomena connected with time, together with all the associated instruments and their correct use. On another side, it was still the matter of active investigation in the motion of the heavenly bodies and the stability of the universe, in the figure and the pulsations of the earth, in the lunar and the planetary theories. Light found devotees (was not Stokes a living authority?): the corpuscular theory had not yet begun to recover from its old dethronement; and the wave-theory reigned unchallenged. Sound was a staple subject: in the Cambridge treatment it lent itself to varied mathematical results, picturesque in form though not contributory to progressive knowledge, and unrelated to the recent physiological work of Helmholz: though, in addition to the examination treatment, Rayleigh's great treatise had just come to our hands if we would but read. Heat (that is, conduction of heat) had pride of place: it lent itself to beautiful conundrums in the guise of mathematics; and there were attractive formulae not yet arithmetized. Sir William Thomson, in one of his frequent fervid moods, had declared Fourier's Treatise to be a poem: we mathematical undergraduates did not pretend to be appreciative critics of poetry in any disguised form, but we read unpoetic condensations in tabloid fragments; especially were we drilled in the use of Fourier series with their ever-new surprises and sudden pitfalls; and Stokes's early papers were our sole source of critical treatment, usually ignored. Rigid Dynamics (including the dynamics of a particle and, strange paradox under such a title, including also the vibrations of strings and bars and plates) claimed a general allegiance. In that range, principles were few and details were multitudinous: Cambridge examiners had an unlimited field for the practice of their ingenuity: bodies could, at will, be made to slip or roll or slide under a friction the amenable quality of which even prodigal Nature could have envied. Particles were made to describe arabesque paths under fantastic laws of force never imagined outside an examination paper. And occasionally, almost like a concession to the high-brows, there were scattered patches of theoretical dynamics which, containing little dynamics, seemed a welter of differential equations.

These were subjects on the grand scale. You may have noted that all of them belong to the domain of applied mathematics, at that epoch in Cambridge still styled Natural Philosophy.

Such pure mathematics as could be discovered in the course was usually made ancillary to this Natural Philosophy, when once beyond the rudimentary stages. Thanks to Salmon, many students knew their analytical conics. Thanks to examiners like Wolstenholme (whose volume of problems, in a more expansive and more formal continental setting, could have been deemed a varied contribution to research), many of us became even skillful in solving the wiliest conundrums. The algebra and the trigonometry (there was no "analysis" in the modern sense) belonged to the old school, the very old school: the proofs, such as were current about the Binomial Theorem, would stir the explosive contempt of the critical mathematicians of today. Nor was the calculus any better in its foundations: I can remember a college question, set years after my student time, "Define a function, and prove that every function has a differential coefficient." The marvel is that our blunders were not greater in number and more atrocious in quality. We were saved, partly by the unconscious benignity that limited the range of application of results; partly by a rough intuitive appeal to physical principles which could expose an error in application but could not construct the correction; partly, even perhaps, because our mentors shared all our blunders.

There was a special devotion to an ancient and limited geometry: to Euclid, not the old man, but to the Simson-Todhunter presentation: To Geometrical Conics, a tessellated pavement of scrappily short proofs of properties of the ellipse and the parabola, with the hyperbola rather clumsily lurking in the background: and to what was called Newton. This was not the great *Principia* which, indeed, few of us saw: certainly the first time I ever saw that work was when a copy of the Glasgow edition, duly bound and stamped, came to me as the then usual bonus additional to the Smith's Prize. What we had was an English translation, doubtless accurate, certainly overloaded with a bewildering heap of piddling minutiae, not unworthy of ancient grammarians as they dwelt on the letter of sacred texts. Even so, we never attained to the applied mathematics of that work: we were restricted to the rudiments of a geometry of infinitesimals: and we did not know that Newton had specially devised this geometry to avoid the use of the analytical calculus which had been his original weapon of research. The Anglicized form of these Newtonian Lemmas and their proofs had to be set out by us with almost the same meticulous verbal accuracy as was exacted in so-called Euclid proofs. Finally, in the earlier section of the Tripos Examination (officially described as "qualifying for honors," commonly known as "the three days"), there was a rigid rule against the explicit use of a differential coefficient and of an integration-process: we might substitute $x + h$ for x and subtract, dodging onwards to the satisfaction of the examiner: we might use a Newton curve, if we could devise it, to effect a quadrature: but never might we use d/dx or the $\int$-sign of integration which were taboo.

A few words more in amplification of a remark already made, in connection with the position of pure mathematics in the Cambridge studies in our day—that (except as occasional exercises in intellectual gymnastics) they were ancillary to applications in Natural Philosophy. Not long before my time there was a curious instance of this requirement of servitude. Airy, who had been a professor at Cambridge and had written useful Tracts, made it a public reproach against Cayley that he had used partial differential equations of the third order; and by way of driving the reproach home, declared that such equations of the second order marked the limit of the subject worthy of consideration, because they were the utmost that were required in natural philosophy. Cayley's reply—gentle, and courteous of course, but definite equally of course—was that, in such a matter and in kindred matters, he was performing the duty enjoined by the statute on the holder of his professorship, "to explain the principles of pure mathematics." Airy had been thinking of the natural philosophy of his own earlier day: his riper years, devoted to astronomy, were spent in the administration of Greenwich Observatory: and Maxwell's published papers (to cite nothing else) might have revealed to the obstinate conservative the antiquated character of his scientific dogma.

Let me return to the course of a mathematical student. In that course, pure mathematics, as a systematic range of ordered knowledge in many subjects of diverse content, received scant attention. The theory of numbers was represented by one chapter of snippets in Todhunter's large book on algebra; it was deemed to be covered by a single lecture; and we never even heard of a congruence, the very sign of which was appropriated for another purpose, let alone any work by Gauss. The theory of invariance was ignored beyond a sidling admission of some geometrical properties of conics and (for a few select adventurous minds) similar properties of quadrics. Differential equations were either catchy numerical examples of a few standard forms—mere trivialities—or they were the equations of mathematical physics, each treated by an isolated method to its own issue. The theory of functions, in any form, was absolutely unknown even in title. The imaginary *i* was suspiciously regarded as an untrustworthy intruder. The complex variable (a phrase that had not then penetrated to Cambridge) was described either as imaginary or as impossible; it compelled recognition as a root of a quadratic equation, or a cubic, or a quartic; and when I went to Cayley's lectures for one term in my third year, at the beginning the very word plunged me into complete bewilderment. A scrap of elliptic functions had been introduced to the Tripos—the scrap was too much for the coaches; we were fortunate at Trinity with Glaisher, as was St. John's with Pendlebury: but from the Tripos standpoint the subject was a cross, between a trigonometry with two cosines and Legendre exercises on elliptic integrals regarded as beyond evaluation: while double periodicity was established as a thing all by itself, and was left in the air as isolated as the coffin of Mahomet in the legend. Cayley and Salmon had made vital contributions to analytical geometry beyond conics and quadrics, and Salmon's books reposed on the shelves of college libraries. But only mathematical wanderers, the Esaus and the Ishmaels of the Tripos, knew even a fragment of such novelties; and differential geometry, save as an exercise in differential calculus, was out of bounds because Tripos questions were never set! Occasional attention, by individual students who had been pupils of Tait at Edinburgh or pupils of Barker at Manchester, was paid to quaternions: never as a noncommutative algebra: always with an introductory testimonial from Natural Philosophy. The theory of groups was unknown, except of course of Cayley; its title would have conveyed no meaning to any Cambridge student: yet nearly ten years earlier, Jordan at Paris had attracted to his lectures on that subject a German student, Felix Klein, and a Norwegian student, Sophus Lie. Perhaps the easiest indication of my estimate of the position may come from a personal confession. Something of differential equations beyond mere examples, the elements of Jacobian elliptic functions, and the mathematics of Gauss's method of least squares, I had learnt from Glaisher's lectures; and by working at the matter of the one course by Cayley which had been attended, I began to understand that pure mathematics was more than a collection of random tools mainly fashioned for use in the Cambridge treatment of natural philosophy. Otherwise very nearly the whole of such knowledge of pure mathematics as is mine began to be acquired only after my Tripos degree. In that Cambridge atmosphere we all were reared to graduation on applied mathematics. There is no regret: on the contrary, I am glad to have had the varied training in that range of knowledge, as glad as of the classical training in schooldays: my reason for seeming to dwell on the topic is that it may indicate the main trend of Cambridge studies fifty to sixty years ago.

IV

And now to turn to the actual teaching of the students by the coaches. It was excellently devised to achieve its aim, which usually was the attainment of high place in the examination; and I believe that the best of the coaches could have borne triumphant comparison with the best mathematical teachers at any epoch in any university. There must have been variations in the processes adopted by front-rank coaches in large practice: let me describe the system employed by Routh in my day.

The word system is used deliberately. Routh had fully 120 pupils in my freshman's Michaelmas term, a total which included men of four different standings because the Tripos examination was held in the Christmas vacation; and steady work on so large a scale had to be systematized. It was a marvel even of physical endurance, let alone intellectual effort. He arranged his pupils in small classes, not more than nine or ten: my own class consisted of six (being of the same year, we presumptuously called ourselves his "first" class); occasionally there was one and the same seventh member. (As an example of Routh's pre-eminent position in that little world, let me add that the said class of six were solid at the top of the list, and the occasional member was seventh in the list.) Each class was taken three times a week, on alternate days during the eight weeks of each of the three terms and the six or seven weeks of the Long Vacation; and, in all, there were ten terms and three Long Vacations in the full undergraduate course. Each attendance was for one hour exactly, never more, never less. In our class he began at 8:15 in the morning, winter or summer; and in his overcrowded Michaelmas term he would be teaching until ten at night. A very few minutes at the beginning of the hour were spent in a swift examination of exercise work; the rest of the hour was devoted entirely to continuous exposition of the current subject. The topics were treated, not in connection with general underlying principles that might characterize the subject, but in the way that a student should frame his answer in the examination. We took abundant notes, scribbling our hardest (we developed a wonderful pace which told in the Tripos); the notes rarely needed working up; and the budget of these notes would be a small useful tract on its subject. Not a moment was spent in diversion, or extraneous illustration, or side-issues, or even examples. Routh had selected the topics in the whole range of subjects as he deemed best for Tripos purposes: no oral questions were asked because we were too busy writing: passing doubts had to be resolved, later, somehow: and any lack of comprehension had to be supplied outside the hour by such enquiry among fellow-students as the student cared to make. At the end of the hour some questions, always riders or problems cognate to the subject and about six in number, were given to us: their solution had to be brought to the next lecture. There might be a manuscript to be read or to be copied—the craft of manifolding had not been developed; the manuscripts, each a little tract of proofs of propositions ready to be discharged at examiners, were left in a general pupil-room that could be rather crowded. In regard to the not very numerous books—most of them being handy textbooks of a Cambridge quality—there might be a few chosen passages suggested to the student for his reading, the information to be dovetailed into his lecture-notes to the best of his own power.

Further, and this was a special feature, in every week there was a paper of problems only, given in common to all his students whatever their standing. In one week, unlimited time was allowed for the solutions; in another, the student was expected to spend only three hours on the paper, as though he were in the mill of the Senate House. All the answers were handed in on the Friday or the Saturday of a week. On the Monday morning they had been returned to the pupil-room, together with a complete set of solutions by Routh, and with a mark-sheet which showed the mark-value of every attempt in every set of answers, as well as the respective totals. These totals were scrutinized by students, eager in their hope of ultimate high place, eager to compare themselves with one another and with men of senior standing. If a youngster scored well, in comparison with elders, he was noted; if he scored heavily in superiority to elders, his reputation blossomed—until it was nipped by his next failure. The spirit of competition had abundant stimulus, through these problem papers.

Finally, throughout the earlier years, there were occasional bookwork papers, really examination-papers of the canonical type set in the Tripos, each question consisting of bookwork and rider. In the third Long Vacation and the last term there were (I seem to remember) three of these every fortnight. The answers to these had to be written out exactly as in examination (the bookwork, be it noted, to be deliberately written out, for practice); and Routh examined these bookwork papers in class, very swiftly, spotting the deficiencies with an uncanny dexterity. Not a moment was wasted: no jesting or frivolous word over a blunder was ever spoken; and individual attention was given in public equally and impartially to all. Every subject, specified in the vast schedule for the one and undivided Tripos at that time, was covered in the first three academic years and the first two Long Vacations; the third Long Vacation and the single term of the fourth year were absorbed by revision, nothing but revision, of everything from the rudiments, upward and onward, in grim doggedness and unresting drill. Routh gave no "tips" to his students, in the usual sense of that word, meaning guesses at the minds of the examiners; and the little devices of a crammer were alien to the man and his work. He was concerned solely with the preparation of students for the examination and its demands; and anything like independence, in any direction of study or reading that led outside his scheme, was discouraged as a hazardous expenditure of precious time.

So the teaching by college lecturers counted for little over the general course compared with the coaching. Some courses of college lectures were attended, especially when related to newer subjects that had sprung up outside the old-established range. Thus in my own college W. D. Niven lectured on the Maxwell-Faraday theory of electricity and electromagnetism: though Routh and Maxwell were Senior and Second Wrangler in the same year, Maxwell's work was outside Routh's ambit. To Glaisher's lectures in pure mathematics I have already referred. Strangely, Routh's teaching in Rigid Dynamics was far from clear, falling far short of the standard he maintained elsewhere. He had written what was (and what still may be) the standard work of reference on the subject: one consequence to me was to stir a doubt (not applying to Routh alone) concerning the advisability of letting authors teach the subjects of books they have written, where presumably they have said what they have deemed important. In that subject I went to a course at St. John's by R. R. Webb, a master in its range: and his course was superb.

And finally some of us made timid ventures outside the range of Cambridge textbooks. Reference has been made to *T and T′*, a storehouse for subjects such as Potential, Attractions, Elasticity. Occasionally French books were tried, such as Verdet's *Cours d'Optique* and Pontécoulant's *Système du Monde*. Few of us had any working knowledge of German, even on a modest scale: there was Kirchhoff's *Mechanik*, and I ploughed through a large part of Durège's *Elliptische Funktionen*. But all such ventures were exceptional and rare: there was no advice, no wish, no leisure, to urge us on those paths. The coach was the autocratic director, often the sole director.

V

What of the examination itself, the goal of all the work? The examining body, five in number, had no relation to the teaching. Each year two persons were nominated by the two colleges who belonged to the pair in a rota: each nominee might happen to be a lecturer in his college, but equally he might be a non-resident. The two nominated were the moderators for that year; they served for a second year with the title of examiners. The fifth member of the body (called the additional examiner) served for only a single year: always a man of distinction, but usually unrelated to the teaching, a sort of superior external examiner whose main duty centered in the latest advanced applied subjects included in the Tripos in the early 'seventies, the last straw on that camel's back.

All the papers were set for all the candidates alike in sheer ignorance of any preferences or preparation: there were no optional subjects; no picking or choosing among sets of questions, lettered and starred and numbered: no instruction limiting the number of questions to be answered. Nothing was said about full marks: they could be obtained only by full performance (which, let me say, would have been beyond the power of any examiner to achieve in the time, even with a full knowledge of the answers desired). There were two sections of the papers: the first extended for four days: the second, following after an interval of a little more than a week, lasted for five days.

During the first three days of the earlier section there was a limitation surviving from a past age: while subjects such as statics, dynamics, astronomy, optics, hydrostatics, were included, the use of the differential and integral calculus was forbidden. There were questions in Euclid, to be written with the verbal rigor of an ancient style unknown to a modern generation. There was one question in arithmetic, a lonely poser: I have been told that, in some limited betting, long odds were always given against the Senior Wrangler obtaining any marks for that question. In preparing the list of successful candidates on this section, the examiners took account solely of the first three days' answers: the list was declared as qualifying the included candidates for honors. No standard for inclusion was fixed by the University: the unframed convention, transmitted from year to year, dared not exact a high percentage; and the competition, near the proverbial line that is never drawn beforehand, was keen indeed.

The qualifying list appeared on a Saturday; on the following Monday the second section began its five days' run. There were eight papers, of the bookwork plus rider type: there were two papers of problems, one by each of the two moderators. Some of these might be "do-able": most of them were not. They represented the inventive possibilities of the utmost range of the setter's fancy, without regard to teaching, or books, or suitability, perhaps fairly described in Pope's words

> Tricks to shew the stretch of human brain,
> Mere curious pleasure or ingenious pain.

In the "Five Days," the papers nominally covered all the subjects in the schedule: and the schedule was framed so as to include nominally all mathematics, there being no Part II or Part III or anything but simply the Tripos. Thus each subject perforce was scantily represented; all the questions ever set by an examiner who had acted before, were eagerly scrutinized as though they were clues in a crossword puzzle. The papers were very long though there was a rubric requiring them not to contain more questions than a well-prepared candidate could answer in the time—with an inevitable inference that, in my day, there were no well-prepared candidates. In this section of the examination—the division into separated Parts came into operation only in 1882—the stage was set for the grand competition in which all the interest and the excitement centered. All the training had prepared for a struggle which was an academic Grand National or Derby. Many of the questions were set at the ablest type of candidate alone, all other types being then ignored; indeed, towards the end, the very vocabulary of the questions belonged to an esoteric language strange to not a few of these successful candidates for honors. Of course, there was outside gossip as to the probable result: the dinner waiters in my college certainly ran a sweepstake in my own year: throughout the colleges there was eager speculation about the coming Senior Wrangler; and such speculation was not ill-informed, because usually the favorite won, down to 1882. I have said, down to 1882; for in those days there was only a single Tripos list, and the Senior Wrangler had his pride of place for the complete range of the study. It is true that only the men near the top of the list had answered a modest selection of the questions, but all questions were propounded unreservedly for all candidates. It often happened that two highly-placed men would be bracketed whose performances had little in common: lower down, the answers had become scattered fragments, often deserving no more than charity marks: in the humblest positions the differences became minute, and the ordered list ended in a dense ruck. But never in all the years had there been a bracket at the top of the old Mathematical Tripos.

The last paper was set on a Friday afternoon. On the succeeding Friday morning in the academic center of the University, the Senate-House, as soon as the great bell of St. Mary's in a tense silence had tolled the nine strokes for the hour, the Senior Moderator read the list to an assembled mob of alert undergraduates, who had come to the academic Senate-House in every conceivable costume that was not academic. It is the moment for yet another Senior Wrangler, an ephemeral embodiment of success and fame: it was the moment also for yet another Wooden Spoon, an equally ephemeral embodiment of the economy of labor endowed with a picturesque name. And, almost as in a day, they faded into the past, making way for another presentation of the old story, as unchanging as a nursery tale.

At this distance of time the Tripos seems interesting. As an institution it was well established; it bore high repute inside Cambridge; and it had a fascination for the sporting sense of a nonmathematical world outside. There was nothing in any university in the world that, for a student, was deemed comparable with the achievement of a Senior Wrangler. That outside world, even a large part of the small world resident in Cambridge, only saw a result in the form of an ordered list, never pausing to consider its significance and its influence.

But let me say a little about the whole course, as an insider. We students acquired good manipulative skill, possibly more expert for its purpose than is acquired today; but we became skillful mechanics rather than engineers. We were schooled indeed, even the ablest; but our education was imperfect. We were drilled in the gymnastic that led to swift answer according to rule and pattern. In the examination there was no leisure to think: even during our training there had been little leisure for thinking, because we always were being taught; and independence in reading was almost a misdemeanor in the eyes of some coaches. Further—and this is a grave charge against the style of competitive examination nominally intended for general qualification—most of the candidates had only the scantiest of tests after their long training. There would be a hundred or more candidates in any one year: many of the questions (including all that carried the highest marks) were utterly beyond the powers of all but a few of the hundred men; the vast majority had to scramble in a mark-scraping. Yet all alike were the academic offspring of the one Alma Mater: all were sent out to the world with the same badge of graduation in honors. As an instance, avowedly extreme, let me state the mere fact that, in my year, the Senior Wrangler had fully sixty-five times as many marks as the Wooden Spoon. One of the two men may have been tested adequately; the other, certainly after the very earliest stages, could not have been tested at all; and he was not the only individual, out of more than a hundred men, neglected at the close of the long training of an undergraduate career. Any school inspector, with only a scrap of experience, would have denounced the examination: and no one, not even the Recording Angel himself, could induce me to believe that, except for the purpose of piling up marks in answer to a horde of questions, any Senior Wrangler was ever worth sixty-five Wooden Spoons.

With the production of the Tripos list the association of coach and pupil ended. They passed out of one another's lives; and the coach returned to the same round of drill with the pupils who were to go through the final mill. The ablest of the men who had achieved an adequate measure of academic success had to begin life all over again, endowed with an examination facility that soon became atrophied in most instances in the absence of any practice. Here there is a temptation to embark on a discussion of the aims, the subjects, and the spirit, of a university education. There can, of course, be no single universal aim; but to my mind, the production of experts, whether mathematicians or specialists of any type, ready for research, almost driven thither by enthusiastic mentors, should not be the dominating aim of any great school of thought in a university. Such a discussion would, however, be long-drawn-out; and it would involve issues different from those which I have tried to describe.

But I must not conclude this sketch of a past era without a final remark. Do not imagine that the old system was nothing but a merciless grind. Undoubtedly, it was stern in its demands. But it earned, and it has received, tributes of gratitude for its efficiency. Among its products (in so far as they have been its products) have been men of high distinction in State and Church and University, in public life, in learned professions, in the pursuit of science, in the advancement of learning. In my own time at Cambridge there had begun (though it was hardly so recognized) a ferment of opinion against the unmitigated domination of written examinations as the sole trustworthy test of ability and powers. For many years success in one or other of the Triposes had been the only avenue to office, whether professorial inside or outside Cambridge or administrative within the several colleges. But other qualities began to be exacted. My own college had taken the initiative by requiring, from a candidate for fellowship, a thesis or a dissertation as evidence of constructive ability; and the spirit of that example spread. Nowadays there seems a tendency to rush to an opposite extreme in proceeding to the selection of individuals for responsible positions. Research (often spelled with a capital R) is made the prominent requisite; and some original production, perhaps with little enquiry as to its significance and seldom with any enquiry as to the inspiring share of a professor, becomes a dominant testimonial, occasionally to the comparative neglect of the indispensable qualities of human personality. Again I must warn myself off disputable ground beyond the boundaries of my theme; so let me end by commending to your consideration my sketch of one chapter in the mathematical history of a vanished age.

Old Cambridge Days

Leonard Roth

Leonard Roth was born and educated in England. After graduating form Cambridge in 1926 he was appointed lecturer in pure mathematics at London University where he continued until 1967. He came to the United States in 1967 to occupy the Andrew Mellon chair of mathematics at the University of Pittsburgh; he held this position until his death in 1968.

This delightful essay is another tour of Cambridge, this time from a more modern point of view. Roth offers excellent insight into very nearly the whole of the development of British mathematics and its relationship to the mathematics of continental Europe. He gives a brief but complete survey of the Newton-Leibniz calculus controversy and points out the disastrous consequences of this rather silly quarrel for British mathematics. Roth's own experiences as an undergraduate provide further clues as to what motivates and what demoralizes a budding mathematician. Finally, though it may be difficult to believe, the eccentric Professor Forsyth described in this essay by Roth is indeed the author of the preceding article "Old Tripos Days at Cambridge."

THE TOWN OF Cambridge is a rather insignificant little place situated some fifty miles to the north of London. Its only title to distinction is, and always has been, that it happens to be the site of a university. Why this should be so is a complete mystery; the historians have never been able to throw any light on it and probably never will. But how the university actually came about—that is known with some precision. The earliest universities of Christian Europe were all born in the same way: a great scholar settled in a town and attracted some students to his lectures; these disciples then stayed with him and, in their turn, began teaching the young men who continued to arrive; and so, without any forethought, the thing was done. Thus the University of Bologna, the first of the great European foundations, came into being simply because some of the local monks began giving public lectures on Roman law: this was about 1150. Today, in one of the principal squares of Bologna, there is a monument to those men who were in effect the first university lecturers of medieval Europe. (Throughout the centuries, many of Italy's greatest advocates have received their training in the Bologna law school.) The universities of Paris and of Oxford were both founded about fifty years later, in exactly the same way.

Now the universities of the second wave of foundations arose in a quite different manner. When learned men have been in one another's company for a sufficient length of time they usually begin to quarrel. From the highest motives, naturally: doctrinal questions, difficult philosophical points, and the like. In those early universities, the disputes sometimes grew so acute that the entire body of scholars split into two factions; and then the dissenting party would go away and found a rival institution elsewhere, just as a swarm of bees will leave the parent hive. Several French and Italian universities were founded in this way, by swarms from Paris and Bologna re-

Source: Leonard Roth, "Old Cambridge Days," *American Mathematical Monthly* 78 (1971): 223–226.

spectively. And Cambridge was founded, about the year 1200, by a swarm from Oxford.

Incidentally, this last migration had a curious aftermath. Until fairly recently, anyone who proceeded to a master's degree at Oxford had to sign a document affirming that he would never in any circumstances lecture to the little town of Stamford (Stamford, by the way, is on the road from London to York). The reason for this procedure was that some time later, a second migration left Oxford and set up a new center of learning at Stamford. However, the project came to nothing, and the dissidents soon returned home to Oxford. But the university authorities were so alarmed by these two flights that they took steps to prevent any more.

And so, by the end of the thirteenth century, we find two university institutions thriving in England, and situated—by Continental standards at any rate—almost within stone's throw of each other. However, although so near on the map, the climatic conditions which they enjoyed—or endured, as the case may be—were very different indeed. Oxford has a mild winter and is rather enervating on the whole. But Cambridge is much worse off; and in the early Middle Ages, when it lay on the edge of a huge undrained fen, the weather there must have been truly horrible. Even today, with its cold and damp winter, aggravated by a wind blowing across the European plain, it is nothing to joke about.

Now very soon after Oxford and Cambridge were founded they were confronted with serious problems of discipline. The presence of a large body of young men, in every respect alien to the local townsfolk, led to frequent disorders, the so-called "town and gown" riots, which often culminated in a murder. The situation was particularly galling to the magistrates because of the immunity frequently enjoyed by priests and clerks. This still persists in some institutions: thus, in Bologna university, if one student commits an offense against another which is normally punishable by law, the police are powerless to intervene. The authorities of both English universities sought a solution to the problem by putting their students into halls of residence, which later became known as colleges. The Head of the House (as he is still officially designated at Oxford) or Master of the College, to give him his usual modern title, was endowed with considerable powers over the inmates. His second in command was known as the Father of the College; he was *in loco parentis*—and he is the ancestor of the modern Dean. We shall soon see what his role was in the Cambridge scholastic world.

We may remark in passing that scarcely any other university saw fit to follow the example set by Oxford and Cambridge; this is exceedingly odd since one and all were beset by the same disciplinary problem. At both the English universities the college system gradually brought about fundamental changes in the administration. As the centuries passed, the central authority lost more and more to the individual colleges, until the latter became very nearly autonomous bodies, each of them responsible for the supervision and teaching of their own students. This shift in power was to have significant consequences for the study of mathematics in Cambridge, which is our main theme here.

There was, however, one important right which the University still retained: that of holding the degree examination. This took the form of a disputation of strict syllogistic character, in which no other kind of reasoning was permitted; it was a three-cornered affair—examiner versus candidate, with the Dean acting as buffer between the contestants. Whenever his "son" was held up for a syllogism, the Dean would slip one in; if the candidate was unable to parry the argument, the Dean would come to his assistance. One beautiful feature of these medieval examinations was that no candidate ever failed; he might do very well, in which case he received the distinction *summa cum laude,* or his performance might be execrable. But in any case, he was awarded his degree; that is what Deans were for in those days.

Echoes of this ancient ceremony are still to be heard in the Cambridge examination system of today. Because the parties to the dispute used to sit on three-legged stools, the Cambridge examination is called a Tripos; moreover, in memory of the disputation, everyone who gains a first class in the Mathematical Tripos is designated a Wrangler.

In those remote times there was no hint that mathematics was destined to play the predominant part in Cambridge university studies which it was later to be assigned. The syllabus followed the lines of the normal medieval curriculum; geometry (but not mathematics), logic, philosophy, music, with specialist courses for students of law, divinity, or medicine. We have to wait until the mid-seventeenth century for the emergence of mathematics as a major discipline, that is to say, for the period in which Isaac Barrow occupied the mathematical chair.

How important is Barrow in the history of seventeenth-century science? The answer given to this question depends largely on one's nationality. The fact is, most of the European nations have their own horse in the Calculus Stakes; thus the French have Fermat and the Italians have Torricelli; if the Russians have not yet entered a candidate, it is only a question of time before they do. Any Englishman would, I fancy, put his money on Barrow; and there are sound, not patriotic, reasons for such a choice. It seems, on reading Barrow's lectures, that he has come nearer to the general notions of derivative and integral than have any of his predecessors. Now Newton was Barrow's pupil, and he absorbed Barrow's ideas.

Newton is, of course, the greatest of all Cambridge professors; he also happens to be the greatest disaster that ever befell not merely Cambridge mathematics in particular but British mathematical science as a whole. This is a fact which the historians do not care to dwell upon; if they mention it at all, they rarely give it due emphasis.

When Newton succeeded Barrow in the Cambridge chair, he had already lost interest in mathematical studies and had turned his intellect into theological and speculative channels. Moreover, he became caught up in public affairs as well. When William and Mary came to the throne in 1688, the British coinage was in a terribly debased condition, and it was essential to commerce that this should be remedied as quickly as possible. For some years the government shrank before the magnitude and peril of the task; in the end, however, the recoinage was decided upon: Newton was called to London as Master of the Mint, and under his supervision the operation was performed with complete success.

These aspects of Newton's career—his indifference to mathematics and his absenteeism—may be termed his negative contributions to the ruin of British science. His positive contribution was far more serious: I refer to his quarrel with Leibniz concerning the origins of the infinitesimal calculus. Throughout his life, Newton shrank from every kind of controversy: by a strange irony his very reticence landed him in the bitterest dispute in mathematical history. The main facts are so well known that I need only allude to them briefly. For some time, there had been subterranean rumblings in the hitherto cordial relations between Newton and Leibniz. In 1715, however, matters came to a head; there then appeared a lengthy historical review of the work done in this field by the two rivals. The article was anonymous, but it is now generally believed to have been written entirely by Newton himself. After surveying the whole situation apparently with complete objectivity, it ends by summing up carefully but decisively against Leibniz, virtually accusing him of having stolen Newton's ideas. At this the Continental mathematicians immediately leapt to Leibniz's defense; at the same time the British mathematicians ranged themselves behind Newton. And there, for over a century, they stayed; their patriotism and solidarity were manifested in a boycott of European mathematics—and, most regrettably, during the very period when the modern science was in its full flood of development. So the Great Sulk went on; and the work of the Bernoullis, of Euler, Lagrange, Laplace, Gauss, and Cauchy remained for Britons a dead letter.

But this is not the whole story by any means; even when the period of official isolation (so to speak) was over, its disastrous consequences continued to make themselves felt. Although during the nineteenth century British applied mathematics made spectacular strides, pure mathematics was more or less neglected and, with the exception of Arthur Cayley, Great Britain produced no pure mathematician of the highest rank. It was not until the beginning of the twentieth century that research in pure mathematics, worthy of international repute, began once more to be produced in any considerable quantity. We shall have more to say of this shortly; in the meantime we continue with our story of Cambridge.

At both Oxford and Cambridge, the eighteenth century was a period of stagnation or even decay. In his autobiography Gibbon has left us a picture of one old university that will very well serve for both: idle students and still more reprehensible teachers, professors who never lectured, and some who never resided. Cambridge mathematical studies had gone the same way as the other disciplines. The University degree examinations were still held in the medieval fashion, but these disputations had by now sunk to a mere farce.

Towards the end of the century, however, the first signs of a Cambridge revival appeared. A mathematical examination with some pretensions to seriousness came into being; although at first purely optional, it gradually ousted the old degree examination, so that in the early nineteenth century it became the sole test for the B.A. It was then that the examination acquired the name of the Mathematical Tripos. The results of the examination were published in order of merit; the first man on the list was called the Senior Wrangler; after him came the other Wranglers—these were the candidates who had been deemed worthy of a first class. Next in order came the Senior Optimes—these formed the second class. Finally, there were the third class men who were called, somewhat euphemistically, Junior Optimes; one suspects that by modern standards a fair number of these would have been refused a degree altogether. On Degree Day, when the successful candidates were presented to the Vice-Chancellor in the Senate House, a curious ceremony was observed. As the last Junior Optime—the bottom man on the list—came forward to be presented, from the public gallery there was lowered an enormous wooden spoon, which he received as a consolation prize. The expression "to get the wooden spoon" has since become proverbial.

As we have said, under the new arrangements every Cambridge man who wished to graduate had to go through the mathematical examination. At this stage of events, the ease with which one could scrape a third class showed a way out of the difficulty; even so, many men of ability were compelled to waste their time, and some had to leave the University without a degree. Even after another examination, the Classical Tripos, was instituted in 1822, the degree examination remained the same. And soon after that its standard was to be stiffened to a remarkable extent. That made things much worse for the nonmathematicians.

Quite early in the nineteenth century a handful of Cambridge men began to realize that it was high time to come out of the Great Sulk. In 1821 they formed what they called the Cambridge Analytical Society, whose aim was to familiarize British mathematicians with the work of the great Europeans. One of the leaders of this movement was Charles Babbage, who has since become famous as the father of the electronic computer. The task assumed by these young men was formidable indeed: for there is only one more conservative institution in the world than Cambridge, and that is Oxford. But it may be asserted that after ten years or so of propaganda their work began to show fruit. We thus enter the modern period of Cambridge mathematical studies, and with it the utterly fantastic story of the Mathematical Tripos in its golden days.

As we have already recounted, the Tripos was from its inception a competitive examination. By some process which has never been satisfactorily described, the fresh enthusiasm for mathematics which now burst upon the University transformed this examination into a high-speed marathon whose like has never been seen before or since. It became far and away the most difficult mathematical test that the world has ever known, one to which no university of the present day can show any parallel. This is undoubtedly a sweeping statement; but the evidence for it is clear and overwhelming. The nineteenth century is, of course, the great period of Cambridge mathematical physics; it includes Ferrers, Green, Stokes, Kelvin, Clerk Maxwell, G. H. Darwin, Rayleigh, Larmor, J. J. Thomson—to mention only the top flight. Now all these men went through the Tripos mill; and it strained their abilities to the utmost.

At that time the teaching was entirely in the hands of the individual colleges, and much of it was grievously inadequate. But in any case, it could never have served the needs of the Tripos examination in its new form. Other methods of instruction were sought—and found. Any man coming to Cambridge and wishing to take a high place in the Tripos, at once put himself in the hands of a professional coach. The training was intensive and unremitting; it lasted for ten terms, with all the intervening long vacations as extra study periods. The examination was then taken early in the January of the undergraduate's fourth year of residence.

Each of the coaches divided his pupils into a number of small groups, who met him once or twice a week. At the meeting he would hand them back their solutions to the previous week's problems, and circulate in class his own set of solutions. While this was going on he would expound the new theorem or subject for study at the blackboard—which meant that a trainee had for part of the time at least to subdivide his attention between manuscript and lecture.

This relentless driving had two purposes before it: to train the candidate to the point where he could write out, with lightning speed and no hesitation, the proof of any theorem required by the syllabus; and where he could write out, at lightning speed and with almost no hesitation, the solution to any problem the examiners might set. Mere mathematical ability was not enough: rapidity of thought had to be added to it. Let me give some examples. Cayley, who afterwards occupied the Cambridge chair of pure mathematics for many years, took the Tripos in 1842. At that time a feature of the examination was a two-hour paper consisting entirely of problems; it was not expected that the candidates would do all the questions—they would naturally have a choice. On the evening after the problem paper, a friend called on Cayley to see how things were going; and, in the manner of friends, he brought bad news. "I've just seen Smith," he announced (Smith was a rival), "and do you know, he did all the questions within two hours." "Oh," remarked Cayley, "well, I cleaned up that paper in forty-five minutes."

Lord Kelvin, when he was just plain William Thomson of Peterhouse, was easily the best mathematician of his year, and was widely tipped for the Senior Wranglership. In fact, on the day that the results of the Tripos were published, he said to his college servant, "Oh, just go down to the Senate House, will you, and see who is Second Wrangler?" Soon after the man returned, and announced: "*You,* sir." Evidently there had been someone in the examination hall who could write, if not think, faster than Kelvin.

As may be imagined, coaching for the Tripos was a highly specialized profession; at any given time there were only two or three men of outstanding capacity for this odd perversion of learning. The most famous of all coaches was unquestionably E. J. Routh. Routh was a very considerable mathematician in his own right, and his books on dynamics are still standard works of reference. For a number of years he had a virtual monopoly on Senior Wranglerships, and many of the highest places were invariably taken by his pupils. Another celebrated coach was R. R. Webb of St. John's College. It may surprise most people to learn that the great mathematician W. H. Young also had a hand in the business: one hardly associates such goings-on with a man of his caliber. But it should be borne in mind that the business was very profitable.

In my student days at Cambridge, I attended some advanced lectures in differential geometry which were given by an elderly don named R. A. Herman. He was the last of the great coaches, a survival from a past epoch. Herman was a Fellow of Trinity College, of which Hardy and Littlewood also were members, and he had taught them both in his time, though their success could hardly be attributed to any of their instructors. All three of them had been Senior Wranglers. Nobody could recall when Herman had taken the Tripos—he was by then a legendary figure—but Hardy had triumphed in the year 1898 and Littlewood in 1908, the very last year of the old regime: the following year the new regulations came into force and with them the order of seniority disappeared. Herman had been a pioneer geometer in his day; it was he who introduced into England the use of the moving trihedron in differential geometry, and Hardy himself has put on record his gratitude to Herman for his inspiring lectures on this subject. But when I sat under him some thirty years later, the inspiration had all flickered out; he reminded one of an extinct volcano. Just occasionally, in a chance word or gesture, one could glimpse the man there had been.

So far we have dealt with the Tripos in its more superficial aspects. What was the examination itself like in those times? At the beginning, and indeed until the Analytical Society had brought about a change, the whole system remained tied to Newton. Out of blind loyalty to their Master, the examiners insisted as far as possible on maintaining a form and a substance of which he might have approved. Thus in problems concerning planetary motion or gravitational attraction, candidates were obliged to use the methods of classical geometry which Newton had employed in the *Principia* and which his own discoveries in the calculus had already rendered obsolete even before he composed the work. And in questions dealing with the calculus, the candidates had to adopt the bad notation of fluxions and fluants due to Newton, instead of the good notation of Leibniz which had long gained universal acceptance elsewhere.

After the reformers had had their way, such antiquated notions were discarded. All the same, the Tripos examination remained predominantly a test in applied mathematics. There was an excellent reason for this bias: the examiners themselves knew hardly any pure mathematics anyway. And the system was obviously self-perpetuating: with each generation the mantle descended from mathematical physicist to mathematical physicist. However, it should not for a moment be imagined that the examination was particularly concerned with applied mathematics in any serious sense of the term: the typical Tripos question, which has been parodied over and over again, was an unreal, often fantastically unreal, abstraction from the physical problem which had suggested it, whose sole object was to render it tractable to the candidates.

Moreover, the Tripos remained a highly conservative institution; with the passage of the years, the distinguished band of mathematicians, whom we have already named, continued to make fundamental contributions to the science; but the Tripos syllabus, generally speaking, kept a respectable distance behind them. Thus Bertrand Russell, who took the examination about 1890, has commented upon both the academic character of the courses and the time-lag in the curriculum. In his day, the Tripos examiners had not yet caught up with Maxwell's equations, which had been given to the world a generation earlier.

Clearly, then, another reformer was called for, one who would take the work on from the point where Babbage and his friends had left it. This might conceivably have been Cayley, who occupied the chair of pure mathematics from 1863 until his death in 1895. But Cayley, despite his eminence in the field of pure mathematics, was a professor of the old school, who looked upon the academic world of his time, saw that it was very bad, but continued to go his own way. Although Cayley lectured regularly, as he was strictly bound to do, upon various branches of pure mathematics, his audience was practically nonexistent. Often it consisted of no more than one pupil. But that pupil was Andrew Russell Forsyth.

This extraordinary man, whom I had the privilege to know in his capacity of Professor Emeritus at the Imperial College of Science, died in extreme old age in 1942; but his fame lives on. Everybody knows him as the author of the most successful book on differential equations that has ever appeared in any language; although it was first published as long ago as 1885, it is still being reprinted. I would venture the opinion that this work has done more than anything else to retard the true development of the subject; for over two generations it has continued to put wrong ideas into people's heads concerning the nature and scope of the theory and, thanks to the author's forceful and authoritative style, in this it has been overwhelmingly successful.

The truth is that Forsyth had the misfortune to be born a hundred years too late; in his mathematical outlook and technique, he was a man of the eighteenth century. His major work on the theory of differential equations, a colossal achievement in six volumes, is still today the only treatise in its class which is by a single hand; but a mere glance at the list of contents suffices to reveal that, on the whole, Forsyth looks backward to Lagrange rather than forward to Cauchy. However, some knowledge of the rudiments of analysis was essential to an understanding of the work; and as the Cambridge men of his generation had none, the author, who had now succeeded to Cayley's chair, set himself the task of educating them. And this is where he comes into our story.

In 1893 there appeared the first edition of Forsyth's *Theory of Functions of a Complex Variable:* another production which cannot be described as anything less than colossal—even a German professor might have quailed before such a project. The book includes fairly complete accounts of the relevant work of Cauchy, Abel, Riemann, Weierstrass, Appell, and carries on the survey right up to the then contemporary researches of Klein and Poincaré. The style of the book is magisterial, Johnsonian; the author's powers of assimilation are well-nigh incredible—and yet, strange to say, despite his intentions and his absorption of the material, he never comes within reach of comprehending what modern analysis is really about: indeed whole tracts of the book read as though they had been written by Euler.

Nevertheless, for all its shortcomings, this was the work which brought modern pure mathematics into Cambridge. The young men at once began to imbibe it; and not the young men alone. My own copy of the book once belonged to R. R. Webb, the coach whom I have mentioned above, and from his penciled notes in the margins it seems pretty clear that he was learning his function theory the hard way, much as any beginner would. The very fact that the book was written in the wrong spirit probably contributed to its great initial success. As Littlewood once put it, "Forsyth was not very good at delta and epsilon"; but neither was the public for whom he wrote: so author and readers met on common ground. In any case, it served as a stepping-stone to the real thing, which at that date was to be found only in French or German. Hardy has recorded that he himself first saw the light when he read the volumes of Jordan's *Cours d'Analyse;* and many other young men of his generation must have done likewise.

Within the space of ten years, Forsyth's treatise had achieved its aim. But it also accomplished something which its author had certainly never intended. For Cambridge now found itself equipped with a corps of modern pure mathematicians whose nominal leader was a living fossil firmly fixed in the Sadlerian chair. This grotesque situation seemed to all intents and purposes a permanent one: Forsyth's international reputation was enormous and in any case there was no possibility of removing him; it appeared as though he were there for life. But now fate took a hand in the game. In the year 1909 Forsyth, in the company of other scientists and their families, was travelling to a meeting of the British Association to be held in Canada. Among the party were the eminent physicist C. V. Boys and his wife Marion. Forsyth was then an apparently confirmed bachelor of 51; but he and Marion Boys fell in love with one another. The end of it was that she decided to leave her husband; and this meant that Forsyth was compelled to resign his professorship, for in the Cambridge of those days there was no place for even the suggestion of divorce. It is pleasant to add that everyone concerned in this affair lived happily ever after; for it was generally conceded by all his acquaintances that the bereaved husband bore his loss with remarkable fortitude. (The former Mrs. Boys was a powerful personality.)

Forsyth survived his wife by many years; in fact he contrived to outlive everything—that was his tragedy. He had to retire from his chair at the Imperial College because he had reached the extreme age limit, although he commanded enough energy to have carried on for at least another five years. He set himself to learn Arabic and Persian; he wrote several enormous volumes on what were ostensibly branches of modern mathematics, all treated from the eighteenth century point of view; the Cambridge University Press, which made a fortune out of his earlier publications, must have lost a good deal of it on these. And all the time he was filling reams of paper with formulae and calculations; I happen to possess some manuscripts of his Cambridge lectures and also of some work on which he was engaged a year or two before his death: the differences between them, from the standpoint of calligraphy, are almost negligible.

I am pleased to relate that I have been able to pay one small tribute to this remarkable son of Cambridge. Some years ago, when I was asked to rewrite the article on Cayley for the *Encyclopaedia Britannica*, I took the opportunity to slip him in; and there, for some time to come, I hope he will stay.

But to continue with our main narrative. Although it almost goes without saying that Forsyth had himself been a Senior Wrangler (he had studied with Routh and Webb) and, moreover, was temperamentally inclined towards the Tripos kind of mathematics, yet he was one of the chief promoters of Tripos reform. At this point we may conveniently put a question which must have occurred to the reader very much earlier: how did such a fantastic sort of academic contest ever take root in the university? I think that the blame for it must be laid upon the system of college autonomy which we have previously described. In the absence of any strong central authority, the examination had fallen into the hands of private individuals, owing no responsibility to anyone; and until the university was able to regain some of its lost powers, there was little chance of breaking their hold over it. This reversion to the medieval form of government took place at about the turn of the century; the time was now ripe for reform.

As I have said, the new regulations for the Tripos came into force in 1909, the very year in which Forsyth, unbeknown to himself, was preparing to quit Cambridge. When I took the examination, nearly twenty years later, the net result of all the changes could be summarized as follows. In the first place, the published order of merit had gone, and with it the rat race. Secondly, there was now a fair balance between pure and applied mathematics in both syllabus and examination questions. In the third place, a more advanced section of the old Tripos, which had been taken at a later stage by men who were seeking a fellowship, was incorporated in the undergraduate examination and entitled Schedule B, to distinguish it from the "elementary" part, known as Schedule A. (Incidentally, the University has since returned to the old practice: Schedule B, rechristened Part III, is now generally taken after a fourth year of residence.)

What was the new examination like? No doubt, if any of the high Wranglers from the past century could have returned to scrutinize our papers, they would have pronounced them pretty easy. To most of us, however, they appeared quite otherwise, particularly at first sight, in the examination room. To begin with, although the notorious Tripos trickery had now given place to a more mature outlook, there was still sufficient need for ingenuity to make the whole proceeding a highly risky business. The typical Schedule A question was a three-decker: first the candidate would be asked to prove a theorem; then would come a problem based more or less on this theorem; and thirdly, another problem even less based than the first. In fact, despite all appearances to the contrary, this last might break fresh ground: that was the sting in the tail. Everybody knew that only complete answers to questions really counted, and that the postscript usually mattered more than the rest. Hence a certain general foreboding. A candidate, even a well-prepared one, might go into the examination on the Monday morning and find himself unable to do a single complete question; if, unduly depressed by this failure, he had the same experience on Monday afternoon, then it was all over save the post-mortem.

The kind of question one had to face may be illustrated by the following example—possibly fictitious, for I do not remember seeing it anywhere. In the first part the candidate is asked to obtain the general solution to Laplace's equation in three dimensions, in terms of Legendre functions. Next, he is required to apply his results to the problem of a conducting sphere in a given field. In the third part the sphere is replaced by an ellipsoid which is nearly spherical. Now the fun begins: the answer to this part of the question is stated—not as a guide to the solver, but in order to make the question more difficult. For as the desired result is merely an approximation, anybody could produce an answer of sorts: the real difficulty is to arrive at the formula stated by the examiners. In questions of this type one might polish off the first two parts in no time at all, only to waste up to an hour on the third. And in that way madness lies.

Schedule B was a quite different affair, not without its own peculiar troubles. Whereas Schedule A was taken by everyone, and on it one's class was usually decided, Schedule B was optional, that is to say, if any undergraduate possessed the necessary tenaciousness, cunning, and indiscipline, he might be able to persuade his tutor to let him out of it. Assuming this was impracticable, he found himself confronted by a dilemma. Schedule B was based on the advanced courses in pure and applied mathematics; the questions set in it were of the longish essay type, each taking about an hour to write out. Once again, only complete answers to the questions were really of much use. Now the courses were numerous and comprehensive, while the total number of questions set was comparatively small. If a candidate chose to take a few courses only, he might find on the fatal day that he was unable to answer a single question on a particular paper. I vividly recall one session of our examination at which, ten minutes after the papers had been given out, a candidate rose in his place and walked slowly out of the hall. This act did nothing to cheer those of us who remained.

One might grasp the other horn of the dilemma by electing to take a great number of courses so as to insure against this disaster. But then there arose another serious difficulty: how could one commit all this material to memory? Here again, a fragmentary knowledge of the syllabus was only of doubtful value; for by sheer bad luck the questions set might weave in and out of the candidate's recollections and so lead inevitably to incomplete answers.

Luck certainly played a considerable part in the examination. I myself had some of each sort, though admittedly more good than bad. I took a term's course of lectures given by Littlewood on the foundations of function theory—this course, or a modified form of it, is still being reprinted as a paperback. But the printed version can give no idea of how delightful the lectures actually were: for Littlewood is one of the wittiest mathematicians that Cambridge, or indeed any other university, has ever produced. When, however, we came to the examination I found to my dismay that he had set about the most difficult question in the course. This was it: "Prove that, if a and b are any two given numbers, then one of the following possibilities must hold: either a is less than b, or a is equal to b, or a is greater than b." Perhaps this result may seem obvious to some of my readers; it certainly seemed obvious to me at the time—in the sense in which it appears obvious to them. But I distinctly recalled that, during the lectures, Littlewood had made a fearful mess of the demonstration; it had taken him the best part of an hour to write up on the blackboard. And I had the feeling that, on the present occasion, he would settle for nothing less. So, apart from the jokes, that course had to be written off as a dead loss.

One day my director of studies said to me: "I see that Mr. Pars" (they were all Misters in those days—no Cambridge man would have been seen dead with a Ph.D.) "is giving a course on general dynamics. I think you might attend." Actually there was no subject I cared less about; but this was a command, and so I attended. Now L. A. Pars was one of those lightning performers in whom Cambridge has always specialized. It is true that he wrote every single word upon the blackboard, but at such a pace that it was next to impossible to keep up with him: to understand what he was doing was quite out of the question. After two terms of this sort of treatment I had a wad of notes as bulky as Whittaker's treatise, on which the lectures were mainly based. An important difference between my account of the subject and Whittaker's was that he knew what he was writing about. As events showed, however, this was irrelevant; for in the examination I encountered two questions on the course which, although ostensibly devoted to problems of dynamics, were really the purest of pure mathematics. I managed to do them both.

This little incident brings us to the grievance which many students nourished against the Tripos system as a whole. In a reasonably balanced examination for the degree a candidate whose interests lay in pure mathematics would have been required to take all the pure mathematics courses together with a selection of the applied; and similarly for a specialist in applied mathematics. But the Cambridge plan insisted on the double dose; and for all but the highly gifted minority this laid an almost intolerable burden upon the conscientious student, to say nothing of the fact that it took no account of the use to which he would turn his knowledge in after life. Experience shows that a taste for one main branch or other of the subject is as a rule acquired fairly early, and that a change is seldom made after. Consequently most of us felt in our bones that we were wasting an awful lot of our time.

And so we arrive at the examination itself. This too could have been arranged better, one thought. Schedule A consisted of six papers: Monday, Tuesday, Wednesday, from 9 A.M. to noon, and then from 1:30 to 4:30 P.M. On the following Monday, Tuesday and Wednesday we sat for Schedule B, which likewise consisted of six papers, each of three hours duration. What went on during those thirty-six hours I cannot now recall in any detail; all I can say is that it was a kind of continuous nightmare. But there was an opening episode which still sticks in my memory. Our examination was held in the great hall of King's College, under the shadow of the famous Gothic chapel; so those candidates who were Scholars of King's had only to cross the college court. I happened to be acquainted with one of them—he was an eccentric individual hailing from Lancashire, and he was notorious for the fact that during the vacations he used to earn a living by playing the organ at the cinema in his home town. At that time vacation work was almost unheard of in England, and his conduct was generally regarded as very queer and perhaps slightly scandalous. As I have said, he had only to cross the court to reach the examination hall. He was wearing his gown, as was necessary at lectures and examinations, but he hadn't bothered to change his slippers. So the officials on duty declined to admit him because he was improperly dressed (I happen to know this as I was sitting near the door and overhead the conversation). Whether he subsequently returned, correctly shod for the occasion, or whether he remained shut out forever I really cannot say; for only a few minutes later I had many other things to worry about.

I remember clearly enough the closing scene of our Tripos. As soon as the last papers were handed in, two fellow-sufferers—myself and a friend—rushed from the hall and walked as fast as we knew how the three miles down to the river where the first of the May races were due to begin. There, in the midst of a crowd of undergraduates and their guests, we soon banished the Tripos from our thoughts. I secretly vowed never to take another examination in my life; and this vow, I am glad to say, I have kept.

Here my own reminiscences of the Tripos come to an end; but of course the story goes on. Various generations of Cambridge men have each shaped the examination according to their light, but the work is never complete and probably never will be. Other intending reformers of the Tripos are even now waiting in the wings; indeed some among them would reform it altogether. Such a notion, startling as it may appear, is by no means novel; it was held more than forty years ago, by Hardy himself, who had backed the 1909 reform as only a first stage of the program. Hardy firmly believed that the Tripos was an unmitigated evil, for which one must blame the inferior performance of British pure mathematicians *vis-à-vis* their European colleagues. So, away with the examination.

Now the weakness of this argument resides in its lack of supporting evidence. It would be very difficult to unearth any specific cases of careers which have undoubtedly been ruined or even seriously damaged by the Cambridge mathematical system: on the other hand, the supporters of the status quo can for their part point to a long line of distinguished mathematical physicists, some of whom we have already mentioned, who achieved success either because or in spite of it: looking at their record one could scarcely suppose that they would have done more or better work had they been spared the ordeal of the examination; and, in the past, what an ordeal it was!

Abolitionists are such charming people; their motives are so patently pure, and only rarely do they foresee the full consequence of their projected panaceas. All his life long Hardy moved in the highest academic circles and tutored the most talented of young men. Had he troubled to consult any lecturer from a provincial university, or even (it may be) a don from a Cambridge college less exalted than Trinity, he could easily have learned a simple but significant truth: if students know beforehand that a particular subject is not to be examined upon, they will, almost to a man—or a woman—altogether decline to study it. Even Forsyth could have told Hardy that much: for he had been a professor at London University.

Leonard Roth was a very good lecturer and had a deep and widespread knowledge of the arts, music, and literature. His many-sided gifts and his charm are only partially apparent from [this] paper found among other MSS left by him. It was probably not written for publication. Those who had the privilege of knowing him will always recall his more intimate endowments: his unusual kindness, unpretentiousness, and deeply-felt humanity. B. SEGRE.

The Work of Nicholas Bourbaki

Jean A. Dieudonné

Jean A. Dieudonné was born and educated in France and received the Doctorate of Science in Mathematics from the Ecole Normale supérieure. He taught at the Universities of Nice and Nancy, has received numerous scientific awards, and is currently retired.

In the early years of this century it was the custom at the Ecole Normale supérieure in France to subject first-year students in mathematics to a rather bizarre initiation. A senior student, in the guise of an important visitor from abroad, would give an elaborate and rather pompous lecture in which several "well-known" mathematical theorems were cited and proved. Each of the theorems bore the name of a famous or possibly not-so-famous French general, and each was wrong in some fiendishly clever way. The object of the game, naturally, was for the first-year students to spot the error in each theorem or better yet, one supposes, not to spot it and provide some comic relief. Granted that it does sound like jolly sport, one might wonder exactly what significance it has for the history of modern mathematics. The truth is that it has hardly any significance except for the fact that in the 1930s a few graduates of the Ecole Normale met and decided to put together a modest little work that would systematically describe and develop everything of importance in modern mathematics. And, quite naturally, the name they selected for this project, the name under which this effort would appear in print, was that of Nicholas Bourbaki, a not-so-famous nineteenth-century French general.

This effort, begun in the 1930s, continues to this day and the results constitute some volumes produced by the Seminaire N. Bourbaki. The participants and contributors have all been, by custom, anonymous, and there has never been official public acknowledgement of anyone's participation in this work. But despite the lack of recognition it has clearly been a labor of love and possibly the single most important unifying and directing force in the mathematics of this century. The author of the following article, Jean Dieudonné, is one of the world's distinguished mathematicians and a close personal "friend" of N. Bourbaki. He is eminently well qualified to tell the general's story.

TO UNDERSTAND THE origins of Bourbaki, we shall have to go back. . . . These were the years when we were students, the years after the 1914 war; and this war, we can very well say, was extremely tragic for the French mathematicians. I shall not try to judge or give a moral assessment of what happened at that time. In the great conflict of 1914–18, the German and French governments did not see things in the same way where science was concerned. The Germans put their scholars to scientific work, to raise the potential of the army by their discoveries and by the improvement of inventions or processes, which in turn served to augment the German fighting

Source: Excerpt from Jean A. Dieudonné, "The Work of Nicholas Bourbaki," *American Mathematical Monthly* 77 (1970): 134–145. An address before the Roumanian Institute of Mathematics, Bucharest, Oct. 1968. Translated by Linda Bennison.

power. The French, at least at the beginning of the war and for a year or two, felt that everybody should go to the front; so the young scientists, like the rest of the French, did their duty at the front line. This showed a spirit of democracy and patriotism that we can only respect, but the result was a dreadful hecatomb of young French scientists. When we open the war-time directory of the Ecole Normale, we find enormous gaps which signify that two-thirds of the ranks were mowed down by the war. This situation had unfortunate repercussions for French mathematics. We others, too young to have been in direct contact with the war, but entering the University in the years after the war ended, should have had as our guides these young mathematicians, certain of whom we are sure would have had great futures. These were the young men who were brutally decimated and whose influence was destroyed.

Obviously, people of previous generations were left, great scholars whom we all honor and respect. Masters like Picard, Montel, Borel, Hadamard, Denjoy, Lebesgue, etc., were living and still extremely active, but these mathematicians were nearly fifty years old, if not older. There was a generation between them and us. I am not saying that they did not teach us excellent mathematics: we all took first-class courses from these mathematicians . . . , but it is indubitable (and true for the matter of every period) that a 50-year-old mathematician knows the mathematics he learned at 20 or 30, but has only notions, often rather vague, of the mathematics of his epoch, i.e., the period of time when he is 50. It is a fact we have to accept such as it is, we cannot do anything about it.

So we had excellent professors to teach us the mathematics of let us say up to 1900, but we did not know very much about the mathematics of 1920. As I said before, the Germans went about things in a different way, so that the German mathematics school in the years following the war had a brilliance which was altogether exceptional. We only need to think of the mathematicians of the highest order who illustrated this point: C. L. Siegel, E. Noether, E. Artin, W. Krull, H. Hasse, etc., of whom we in France knew nothing. Not only this, but we also knew nothing of the rapidly developing Russian school, the brilliant Polish school, which had just been born, and many others. We knew neither the work of F. Riesz nor that of von Neumann, etc. We had been closed in on ourselves and, in our world, the theory of functions reigned supreme. The only exception was Elie Cartan; but being 20 years ahead of his time, he was understood by no one. (The first to understand him after Poincaré was Hermann Weyl, and for 10 years he was the only one, so how could we poor little students have known enough to understand him?) So, apart from E. Cartan, who at this time didn't count—he only started to count 20 years later, but since then his influence has grown steadily—we were entirely folded in on that theory of functions, which, while being important, represented only a part of mathematics.

Our only opening onto the outside world at this time was the seminar of Hadamard, a professor, but not a very brilliant teacher, at the Collège de France. (He was a great enough scholar for me to be able to say this without harming his reputation.) He had the idea (apparently taken from abroad, because this had never been done in France) of inaugurating a seminar of analysis of current mathematical work. At the beginning of the year he distributed, to all those who wanted to speak on the subject, what he judged to be the most important memoirs of the past year, and they had to explain them at the blackboard. It was a novelty for the time, and to us an extremely precious one, because there we met mathematicians of many different origins. Also, it soon became a center of attraction for foreigners; they came in crowds. . . . So it was for us young students a source of acquaintances and views that we did not find in the formal mathematics courses given at the University. This state of affairs lasted several years, until certain of us—starting with A. Weil, then C. Chevalley, having been out of France meeting Italians, Germans, Poles, etc.—realized that if we continued in this direction, France was sure to arrive at a dead end. We would no doubt continue to be very brilliant in the theory of functions, but for the rest, French mathematicians would be forgotten. This would break a two-hundred-year-old tradition in France, because from Fermat to Poincaré, the greatest of the French mathematicians had always had the reputation of being universal mathematicians, as capable in arithmetic as in algebra, or in analysis, or in geometry. So we had this warning of the bubbling of ideas that was beginning to be seen outside, and several of us had the chance to go and see and learn at first hand the development that was going on outside our walls. After Hadamard retired in 1934, the seminar was carried on, in a slightly different form, by G. Julia. This consisted of studying in a more systematic manner the great new ideas which were coming in from all directions. This is when the idea of drawing up an overall work which, no longer in the shape of a seminar, but in book form, would encompass the principal ideas of modern mathematics. From this was born the Bourbaki treatise. I must say that the collaborators of Bourbaki were very young at the time and doubtless they would never have started this job had they been older and better informed. In the first meetings for the project, the idea was that it would be finished in three years, and in this time we should draft the basic essentials of mathematics. Events and history decided differently. Little by little, as we became rather more competent and more aware, we realized the enormity of the job that had been taken on, and that there was no hope of finishing it as quickly as that.

It is true that there were already excellent monographs at the time and, in fact, the Bourbaki treatise was modeled in the beginning on the excellent algebra treatise of Van der Waerden. I have no wish to detract from his merit, but as you know, he himself says in his preface that really his treatise had several authors, including E. Noether and E. Artin, so that it was a bit of an early Bourbaki. This treatise made a great impression. I remember it—I was working on my thesis at that time; it was 1930 and I was in Berlin. I still remember the day that Van der Waerden came out on sale. My ignorance in algebra was such that nowadays I would be refused admittance to a university. I rushed to those volumes and was stupefied to see the new world which opened before me. At that time my knowledge of algebra went no further than *mathématiques spéciales,* determinants, and a little on the solvability of equations and unicursal curves. I had graduated from the Ecole Normale and I did not know what an ideal was, and only just knew what a group was! This gives you an idea of what a young French mathematician knew in 1930. So we tried to follow Van der Waerden, but in effect he only covered algebra, and even then just a small part of algebra. (Since then, algebra has developed considerably, partly because of Van der Waerden's treatise, which is still an excellent introduction. I am often asked for advice on how to start out studying algebra, and to most people I say: First read Van der Waerden, in spite of what has been done since.)

So we intended to do something of this kind. Now Van der Waerden uses very precise language and has an extremely tight organization of the development of ideas and of the different parts of the work as a whole. As this seemed to us to be the best way of setting out the book, we had to draft many things which had never before been dealt with in detail. General topology could only be found in a few memoirs and in Fréchet's book, which was, in effect, a compilation of an enormous quantity of results, without any kind of order. I can say the same of Banach's book, which is admirable for research but completely disorganized; in other subjects such as integration (as presented by Bourbaki) and certain algebra questions, there was nothing. Before the chapter of Bourbaki on multilinear algebra, I don't think there was a didactic work in the world that explained what exterior algebra was. We had to refer to the work of Grassmann, which is not particularly clear. Thus we quickly realized that we had rushed into an enterprise which was considerably more vast than we had imagined, and you know that this enterprise is still far from finished. In my briefcase I have the proofs on the 34th volume, which is devoted to the three chapters of the theory of Lie groups. There are others, many others, being prepared; there are already three or four editions of preceding volumes, and the end of the work is not in sight.

We had to have a starting point—we had to know what we wanted to do. Of course, there was the idea of the Encyclopedia, which, in fact, already existed. As you know, it had been started by the Germans in 1900, and despite their proverbial tenacity and ardor for work, in 1930, after several editions and alterations, etc., it was hopelessly behind in comparison to the mathematical science of that time. Nowadays, nobody would think of starting on such an impossible enterprise, knowing the vast amount of mathematical publications released every year. I believe that we shall have to wait for the day when computers have minds and are able to assimilate all that in a few minutes. For the time being we have not progressed that far, nor had we gone that far in 1930. Moreover, it would have been useless to redo something which despite its merits had failed. The Encyclopedia, even at that period, was above all useful as a bibliographical reference, to find out where such and such a result could be found. But naturally, it contained no proofs, because if the Encyclopedia, already gigantic with its 25–30 volumes, had included proofs it would have been ten times larger. No, we did not want to produce a work of bibliographic reference, but one which would be a demonstrative mathematical text from beginning to end. And this forced us into making an extremely strict selection. What selection? Well, that is the crucial part in Bourbaki's evolution. The idea which soon became dominant is that the work had to be primarily a *tool*. It had to be something usable not only in a small part of mathematics, but also in the greatest possible number of mathematical places. So if you like, it had to concentrate on basic mathematical ideas and essential research. It had to reject completely anything secondary that had no immediately known application and that did not lead directly to conceptions of known and proved importance. There was much sifting, which started innumerable discussions among the collaborators, and which also earned Bourbaki a great deal of hostility. Because as the works of Bourbaki became known, all those who found that their favorite subject was not included were not inclined to do much propaganda in his favor. So I think that we can attribute much of the hostility that has been shown toward Bourbaki at certain periods, and which is still widespread in certain countries, to this extremely strict selection.

So how do we choose these fundamental theorems? Well, this is where a new idea came in: that of *mathematical structure*. I do not say it was an original idea of Bourbaki—there is no question of Bourbaki's containing anything original. Bourbaki does not attempt to innovate mathematics, and if a theorem is in Bourbaki, it was proved 2, 20, or 200 years ago. What Bourbaki has done is to define and generalize an idea which already was widespread for a long time. Since Hilbert and Dedekind, we have known very well that large parts of mathematics can develop logically and fruitfully from a small number of well-chosen axioms. That is to say, given the bases of a theory in an axiomatic form, we can develop the whole theory in a more comprehensible way than we could otherwise. This is what gave the general idea of the notion of mathematical structure. Let us say immediately that this notion has since been superseded by that of category and functor, which includes it under a more general and convenient form. It is certain that it will be the duty of Bourbaki, who, as I shall explain later, never fears change, to incorporate the valid ideas of this theory in his works.

Once this idea had been clarified, we had to decide which were the most important mathematical structures. Naturally, this was the root of many discussions before we found ourselves in agreement. I might say that Bourbaki does not pretend to be infallible; he has been mistaken several times about the future of structures, and apologized when it was necessary, withdrawing his original ideas. Successive editions trace some changes clearly. Bourbaki does not pretend to want to fix or nail down mathematics; that would be exactly contrary to his original purpose. But if one does not recoil from new ideas, even when they go beyond Bourbaki, one has no respect for tradition. Consequently this open systematic attitude of Bourbaki has also been a cause of hostility, this time on the part of people of previous generations, who criticized the liberties Bourbaki took with the mathematics of their time. In particular, the choice of definitions and the order in which the subjects were arranged were decided according to a logical and rational scheme. If this did not agree with what was done previously, well, it means that what was done previously had to be thrown overboard, without sparing even long-established traditions. To give you an example: Bourbaki refuses to say *non-decreasing* when referring to an increasing function because this would be a total absurdity. We know that this term means what we want to say only when talking about linear (total) order relations. (If one says non-decreasing in the setting of a non-linear order relation, this hardly means *increasing but not strictly increasing*.) So Bourbaki purely and simply abolished this terminology, as he did many others. He also invented terminology, using Greek when it was necessary, but also using many words from ordinary speech, which made traditionalists wince. They did not admit easily that what we now call *boule* or *pavé* used to be called *hypersphéroide* or *parallélotope,* and their reaction was: "This work is not to be taken seriously." A little book came out recently, which we liked very much. It is called *"Le Jargon des Sciences"* by Etiemble, vigilant guardian of the French language. He insists on preserving it in its original purity and is up in arms against the gibberish of most scientists. Happily, he makes an exception of French mathematicians, saying that they had the good sense to take simple, authentic French words from ordinary speech, sometimes changing their meaning. He cites attractive examples, recent titles such as *Platitude et privilège* and *Sur les variétés riemanniennes non suffisamment pincées.* This is the style in which Bourbaki is written—in a recognizable language and not in a jargon sprinkled with abbreviations, as in Anglo-Saxon texts where you are told about the C.F.T.C. which is related to an A.L.V. unless it is a B.S.F. or a Z.D., etc. After ten pages of this you have no idea what they are talking about. We think that ink is cheap enough to write things in full, with a well-chosen vocabulary.

I told you then that we made a selection. I shall explain this choice in more detail, using a metaphor. We realized very quickly that despite introducing the idea of structure, which was meant to clarify and separate things, mathematics refused to separate into small pieces. On the other hand, it was clear that the old divisions, Algebra, Arithmetic, Geometry, Analysis were out of date. We had no respect for them and abandoned them from the start, to the fury of many. For example, it is well known that euclidean geometry is a special case of the theory of hermitian operators in Hilbert spaces. The same goes for the theories of algebraic curves and numbers, which come essentially from the same structures. I compare the old mathematical divisions with the divisions of the ancient zoologists, who, seeing that a dolphin and a shark or a tuna-fish were similar animals, said: These are fish because they all live in the sea and have similar shapes. It was quite a while before they realized that the structures of these animals were not at all similar, and they had to be classified very differently. Algebra, Arithmetic, Geometry and all that nonsense compare easily to this. One has to look at the structure of each theory and classify it in this way. In spite of everything though, it does not take long to make one realize that despite this effort towards the isolation of structures, they have a way of mixing very quickly and extremely fruitfully. One could say that the great ideas in mathematics have come when several very different structures met. So here is my picture of mathematics now. It is a ball of wool, a tangled hank where all mathematics react one upon another in an almost unpredictable way. Unpredictable, because a year almost never passes without our finding new reactions of this kind. And then, in this ball of wool, there are a certain number of threads, coming out in all directions and not connecting up with anything else. Well, the Bourbaki method is very simple—we cut the threads. What does this mean? Let us look at what remains; then we make a list of what remains and a list of what is eliminated. What remains: The archiclassic structures (I don't speak of sets, of course), linear and multilinear algebra, a little general topology (the least possible), a little topological vector spaces (as little as possible), homological algebra, commutative algebra, noncommutative algebra, Lie groups, integration, differentiable manifolds, riemannian geometry, differential topology, harmonic analysis and its prolongations, ordinary and partial differential equations, group representation in general, and in its widest sense, analytical geometry. (Here of course I mean in the sense of Serre, the only tolerable sense. It is absolutely intolerable to use *analytical geometry* for linear algebra with coordinates, still called analytical geometry in the elementary books. Analytical geometry in this sense has never existed. There are only people who do linear algebra badly, by taking coordinates and this they call analytical geometry. Out with them! Everyone knows that analytical geometry is the theory of analytical spaces, one of the deepest and most difficult theories of all mathematics.) Algebraic geometry, its twin sister, is also included, and finally the theory of algebraic numbers.

This makes an imposing list. Let us now see what is excluded. The theory of ordinals and cardinals, universal algebra (you know very well what that is), lattices, non-associative algebra, most general topology, most of topological vector spaces, most of group theory (finite groups), most of number theory (analytical number theory, among others). The processes of summation and everything that can be called hard analysis—trigonometrical series, interpolation, series of polynomials, etc.; there are many things here; and finally, of course, all applied mathematics.

There I wish to explain myself a little. I absolutely do not mean that in making this distinction Bourbaki makes the slightest evaluation on the ingeniousness and strength of theories catalogued in this way. I am convinced that the theory of finite groups, for example, is at the present time one of the deepest and richest in extraordinary results, while theories like non-commutative algebra are of medium difficulty. And if I had to make an evaluation I should probably say that the most ingenious mathematics is excluded from Bourbaki, the results most admired because they display the ingenuity and penetration of its discoverer.

We are not talking about classification then, the good on my right, the bad on my left—we are not playing God. I just mean that if we want to be able to give an account of modern mathematics which satisfies this idea of establishing a center from which all the rest unfolds, it is necessary to eliminate many things. In group theory, despite the extraordinary penetrating theorems which have been proved, one cannot say that we have a general method of attack. We have several of them, and one always has the impression that one is working like a craftsman, by accumulating a series of stratagems. This is not something which can be set forth by Bourbaki. Bourbaki can only and only wants to set forth theories which are rationally organized, where the methods follow naturally from the premises, and where there is hardly any room for ingenious stratagems.

So, I repeat, those which Bourbaki proposes to set forth are generally mathematical theories almost completely worn out already, at least in their foundations. This is only a question of foundations, not details. These theories have arrived at the point where they can be outlined in an entirely rational way. It is certain that group theory (and still more analytical number theory) is just a succession of contrivances, each one more extraordinary than the last, and thus extremely anti-Bourbaki. I repeat, this absolutely does not mean that it is to be looked down upon. On the contrary, a mathematician's work is shown in what he is capable of inventing, even new stratagems. You know the old story—the first time it is a stratagem, the third time a method. Well, I believe that greater merit comes to the man who invents the stratagem for the first time than to the man who realizes after three or four times that he can make a method from it. The second step is Bourbaki's aim: to gather from the diverse processes used by mathematicians whatever can be shaped into a coherent theory, logically arranged, easily set forth and easily used.

The work method used in Bourbaki is a terribly long and painful one, but is almost imposed by the project itself. In our meetings, held two or three times a year, once we have more or less agreed on the necessity of doing a book or chapter on such and such a subject (generally, we foresee a certain number of chapters for a book), the job of drafting it is put into the hands of the collaborator who wants to do it. So he writes one version of the proposed chapter or chapters from a rather vague plan. Here, generally, he is free to insert or neglect what he will, completely at his own risk and peril, as you will see. After one or two years, when the work is done, it is brought before the Bourbaki Congress, where it is read aloud, not missing a single page. Each proof is examined, point by point, and is criticized pitilessly. One has to see a Bourbaki Congress to realize the virulence of this criticism and how it surpasses by far any outside attack. The language cannot be repeated here. The question of age does not come into it. The ages of the Bourbaki members vary considerably—later I shall tell you the maximum age limit—but even when two men have a 20-year age difference, this does not stop the younger from hauling the elder, who he feels has understood nothing of the question, over the coals. One has to know how to take it, as one should, with a smile. In any case, the reply is never late in coming, no one can boast of being infallible before Bourbaki members, and in the end, everything works out fine, despite the very long and extremely animated arguments.

Certain foreigners, invited as spectators to Bourbaki meetings, always come out with the impression that it is a gathering of madmen. They could not imagine how these people, shouting—some times three or four at the same time—about mathematics, could ever come up with something intelligent. It is perhaps a mystery but everything calms down in the end. Once the first version has been torn to pieces—reduced to nothing—we pick a second collaborator to start it all over again. This poor man knows what will happen because although he sets off following the new instructions, meanwhile the ideas of the Congress will change and next year *his* version will be torn to bits. A third man will start, and so it will go on. One would think it was an unending process, a continuous recurrence, but in fact, we stop for purely human reasons. When we have seen the same chapter come back six, seven, eight, or ten times, everybody is so sick of it that there is a unanimous vote to send it to press. This does not mean that it is perfect, and very often we realize that we were wrong, in spite of all the preliminary precautions, to start out on such and such a course. So we come up with different ideas in successive editions. But certainly the greatest difficulty is in the delivery of the first edition.

An average of 8–12 years is necessary from the first moment we set to work on a chapter to the moment it appears in the bookshop. The ones that are coming out now are the ones that were discussed for the first time about 1955.

I said earlier that there is a maximum age limit. This was recognized quite quickly for the reason I was speaking about at the start of this talk—a man of over 50 can still be a very good and extremely productive mathematician but it is rare for him to adapt to the new ideas, to the ideas of people 25 and 30 years younger than he. Now, an enterprise like Bourbaki seeks to be permanent. There is no question of saying that we nail down mathematics to such or such a period. If the mathematics set forth by Bourbaki no longer corresponds to the trends of the period, the work is useless and has to be redone. This has already happened, for that matter, with several volumes of Bourbaki. If there were elderly members of Bourbaki, they would tend to put a brake on this healthy tendency, believing that everything being fine at the time of their youth, there is no reason for change. This would be disastrous. So, to avoid tensions such as this, which sooner or later would cause Bourbaki's break-up, it was decided at the time the question arose, that all the Bourbaki collaborators retire at 50.

And it is so; the present Bourbaki collaborators are all under 50. The founder-members, of course, retired almost ten years ago, and even those who not long ago were considered young are already past—or about to reach—retiring age. So it is a question of replacing the members who leave. How do we do that? Well, there are no rules, because in Bourbaki the only formal rule is the one I have just told you, retirement at 50. Apart from this, we can say that the only rule is that there are no rules. There are no rules in the sense that there is never a vote, we have to have unanimity on every point. Each member has the right to veto any chapter he feels is bad. The veto simply signifies that we do not allow the printing of the chapter and we have to go back and restudy it. This explains the lengthiness of the process—the fact that we have such a hard time agreeing on a final version.

We are concerned then with replacing members affected by the age limit. We do not replace them formally (this would be a rule and there are no rules). There is no vacant seat, as with an academy. As most of the members of Bourbaki are professors—many in Paris—they have a chance to see at close range the young mathematicians, the youths who are just starting mathematical research. A youth of value who shows promise of a great future is quickly noticed. When this happens, he is invited to attend one of the Congresses as a guinea pig. This is the traditional method. You all know what a guinea pig is—the small animal that we use to test all viruses. Well, it is much the same thing; the wretched young man is subjected to the ball of fire which constitutes a Bourbaki discussion. Not only must he understand, but he must also participate. If he is silent, he is simply not invited again.

He must also show a certain quality. The absence of this tendency has stopped many great and valuable mathematicians from joining Bourbaki. During a Congress, the chapters come up in the order of the day, in no particular order, and we never know in advance if we shall be doing only differential topology at this Congress, or if at the next one we shall be doing commutative algebra. No, everything is mixed—I cite the same example, the symbol that could be thought of as the Bourbaki symbol, the ball of wool. Consequently a Bourbaki member is supposed to take an interest in everything he hears. If he is a fanatical algebraist and says "I am interested in algebra and nothing else," fair enough, but he will never be a member of Bourbaki. One has to take an interest in everything at once. Not to be capable of creating in all fields, that is all right. There is no question of asking everyone to be a universal mathematician; this is reserved for a small number of geniuses. But still, one should take an interest in everything, and be able, when the time comes, to write a chapter of the treatise, even if it is not in one's speciality. This is something which has happened to practically every member, and I think most of them have found it extremely beneficial.

In any case, in my personal experience, I believe that if I had not been submitted to this obligation to draft questions I did not know a thing about, and manage to pull through, I should never have done a quarter or even a tenth of the mathematics I have done. When one starts to write on questions one does not know and if one is a mathematician, one is forced to put questions to oneself. This is characteristic of the mathematician. Consequently one tries to solve them, and this leads to personal work, independent of Bourbaki, and more or maybe less valuable, but which was born of Bourbaki. So one cannot say that this is a bad system. But there are excellent minds which cannot adapt to this sort of obligation, profound minds which are first-class in their field, but to whom one must not mention other fields. There are unbending algebraists who will never be made to swallow analysis, and analysts for whom the field of quaternions is a monstrosity. These mathematicians may be first-class mathematicians, superior to most Bourbaki members—we admit it freely and I could give you illustrious examples—but they could never be members of Bourbaki.

To return to the guinea pig. When he is invited, we start by looking for this quality of adaptation. Often it is not there, so we wish him luck and he goes on his way. Fortunately one finds from time to time among the youths, this tendency, this appetite for universal knowledge of mathematics and adaptation to diverse theories. After a very short time, if we find that he gives a good return, he becomes a member without any voting, election, or ceremony. Bourbaki, I repeat, has one rule, which is not to have rules, except for retirement at 50.

To end, I should like to reply to a recent attack on Bourbaki by certain young men of a certain country. Bourbaki is accused of sterilizing mathematical research. I must say that I completely fail to comprehend this, since Bourbaki has no pretension of being a work stimulating to research. I was saying earlier that Bourbaki can only allow himself to write on dead theories, things which have definitely settled and which only need to be gleaned (except for the unexpected, of course). Actually one must never speak of anything dead in mathematics, because the day after one says it, someone takes this theory, introduces a new idea into it, and it lives again. Rather let us say theories dead at the time of writing, that is to say, nobody has made any significant discoveries in these theories Bourbaki develops for 10, 20, or 50 years, whereas they are in the part judged important and central, serving as tools for research elsewhere. But they are not necessarily stimulants for research. Bourbaki is concerned with giving references and support to anyone who wants to know the essentials in a theory. He is concerned with knowing that when one wants to work, for example, on topological vector spaces there are three or four theorems one has to know: Hahn-Banach, Banach-Steinhaus, the closed graph; it is a question of finding them somewhere. But nobody has the idea of ameliorating the theorems; they are what they are, they are extremely useful (this is the fundamental point) so they are in Bourbaki. This is the important thing. As for stimulating research, if open problems exist in an old theory, obviously they are pointed out, but this is not the aim of Bourbaki.

The aim is, I repeat, to provide worktools, not to give stimulating speeches on the open problems of the new mathematics, because these open problems are in general much farther than Bourbaki can go. This is living mathematics and Bourbaki does not touch living mathematics. He cannot when, by definition, it changes each year. If one wrote a book on that, following Bourbaki's method, i.e., taking eight or ten years to work it out, you can imagine the book after twelve years. It would represent absolutely nothing. It would have to be modified continually and would be like the old Encyclopedia, never finished.

Those are the few explanations I wanted to give you. Now I shall be very happy to answer questions, to add to what I have said.

Answers to Questions

. . . Bourbaki sets off, if you like, from a basic belief, an unprovable metaphysical belief we willingly admit. It is that mathematics is fundamentally simple and that for each mathematical question there is, among all the possible ways of dealing with it, a best way, an optimal way. We can give examples where this is true and examples where we cannot say, because up to now we have not found the optimal method.

I cited, for example, group theory and analytical number theory, which are characteristic. In both one has a quantity of methods, each one more clever than the last. This is splendid and ingenious and of a complexity never before known, but we are sure that this is not the final way to deal with the question. On the other hand, take algebraic number theory. Since Hilbert, it is so systematized that we know there is a right way to handle its questions. We change them sometimes, but in the end, little by little, we manage to find one way which is better than the others. This is only a belief, I repeat, a metaphysical belief.

. . . On foundations we believe in the reality of mathematics, but of course when philosophers attack us with their paradoxes we rush to hide behind formalism and say: "Mathematics is just a combination of meaningless symbols," and then we bring out Chapters 1 and 2 on set theory. Finally we are left in peace to go back to our mathematics and do it as we have always done, with the feeling each mathematician has that he is working with something real. This sensation is probably an illusion, but is very convenient. That is Bourbaki's attitude toward foundations.

part two

Some Mathematical Lives

IT WOULD BE tempting to suggest that the development of mathematics is revealed through the lives of great mathematicians. But that just isn't so. One can learn all the significant ideas of mathematics and still know nothing of the lives and affairs of the creators of those ideas. One could, in fact, waste a great deal of time in surveying the lives of great mathematicians, trying to discover some connecting thread. The truth is that some famous mathematicians were rich and others poor, some members of the nobility and others children of anonymous peasants, some died in pale consumption in their twenties and others enjoyed good health into their nineties, some were religious zealots and others dedicated skeptics, some were celibate and others sybarite. There is nothing to distinguish mathematicians as a group except their genius.

Since so little in the lives of these elect aids in the understanding of mathematics, the reader may rightly wonder why this text includes a section on biographies. The answer is, "We care because we are human." It would be less than human not to wonder about the life of someone such as Newton. His marvelous insights into the fundamental nature of mechanical physical phenomena and his creation of a mathematics suitable for explaining those events are a monument to the human spirit. We are curious about Newton the man not because we hope to profit from the minutiae of his life. We could not duplicate his achievements simply by dressing as he dressed or eating as he ate. No, we care because in his life and achievement we are all affirmed. He is of our kind. His life is not so removed from our own that we cannot find a commonality that, if only by reflection, lends some degree of majesty to us all.

What follows in this part, then, are some sketches of the lives and times of a few mathematicians. The articles are by design a varied collection by almost any measure of similarity. In such brief essays one can hardly expect to come to know the persons. But one may catch a brief view of something about the nature of these mathematicians and perhaps of one's own nature. For us, that will have to suffice.

On the Seashore: Newton

E. T. Bell

E. T. Bell was born in Scotland and educated in the United States. He received a Ph.D. in mathematics from Columbia University and taught at the University of Washington and the California Institute of Technology. His books on mathematics history are among the most popular works on that subject in the English language. He died in 1960.

It would indeed be a curious volume of mathematics lore that failed to take note of Sir Isaac Newton, the greatest scientist-mathematician of the English-speaking world. This work would not presume such eccentricity and thus this essay makes its appearance. Written by E. T. Bell for Men of Mathematics, *his widely acclaimed volume of biographies, this article first appeared in print in 1937. While the intervening years have shed a bit more light on some of the dimmer corners of seventeenth-century science, the major events of Newton's life remain exactly as Bell related them almost half a century ago.*

Bell was not one to couch his opinions of historical persona or events in the polite equivocation of scholarship. If he felt that someone was a villain or that some deed was despicable, then he said so in unmistakable terms. Unquestionably, some of his historical asides are debatable and others merely false, but his writing is never dull. Bell's capacity to accurately and clearly inject mathematical ideas into a narrative that is witty, biting, and insightful has made his volume one of the most widely known works about mathematicians in the English language.

The method of Fluxions [the calculus] is the general key by help whereof the modern mathematicians unlock the secrets of Geometry, and consequently of Nature.—BISHOP BERKELEY

I do not frame hypotheses.—ISAAC NEWTON

"I DO NOT know what I may appear to the world; but to myself I seem to have been only like a boy playing on the seashore, and diverting myself in now and then finding a smoother pebble or a prettier shell than ordinary, whilst the great ocean of truth lay all undiscovered before me."

Such was Isaac Newton's estimate of himself toward the close of his long life. Yet his successors capable of appreciating his work almost without exception have pointed to Newton as the supreme intellect that the human race has produced—"he who in genius surpassed the human kind."

Isaac Newton, born on Christmas Day ("old style" of dating), 1642, the year of Galileo's death, came of a family of small but independent farmers, living in the manor house of the hamlet of Woolsthorpe, about eight miles south of Grantham in the county of Lincoln, England. His father, also named Isaac, died at the age of thirty-seven before the birth of his son. Newton was a premature child. At birth he was so frail and puny that two women who had gone to a neighbor's to get "a tonic" for the infant expected to find him dead on their return. His mother said he was so undersized at birth that a quart mug could easily have contained all there was of him.

Not enough of Newton's ancestry is known to interest students of heredity. His father was described by neighbors as "a wild, extravagant, weak man"; his mother, Hannah Ayscough, was thrifty, industrious, and a capable manageress. After her husband's death Mrs. Newton was recommended as a prospective wife to an old bachelor as "an extraordinary good woman." The cautious bachelor, the Reverend Barnabas Smith, of the neighboring parish of North Witham, married the widow on this testimonial. Mrs. Smith left her three-year-old son to the care of his grandmother. By her second marriage she had three children, none of whom exhibited any remarkable ability. From the property of his mother's second marriage and his father's estate Newton ultimately acquired an income of about £80 a year, which of course meant much more in the seventeenth century than it would now. Newton was not one of the great mathematicians who had to contend with poverty.

As a child Newton was not robust and was forced to shun the rough games of boys his own age. Instead of amusing himself in the usual way, Newton invented his own diversions, in which his genius first showed up. It is sometimes said that Newton was not precocious. This may be true so far as mathematics is concerned, but if it is so in other respects a new definition of precocity is required. The unsurpassed experimental genius which Newton was to exhibit as an explorer in the mysteries of light is certainly evident in the ingenuity of his boyish amusements. Kites with lanterns to scare the credulous villagers at night, perfectly constructed mechanical toys which he made entirely by himself and which worked—waterwheels, a mill that ground wheat into snowy flour, with a greedy mouse (who devoured most of the profits) as both miller and motive power, workboxes and toys for his many little girl friends, drawings, sundials, and a wooden clock (that went) for himself—such were some of the things with which this "un-precocious" boy sought to divert the interests of his playmates into "more philosophical" channels. In addition to these more noticeable evidences of talent far above the ordinary, Newton read extensively and jotted down all manner of mysterious recipes and out-of-the-way observations in his notebook. To rate such a boy as merely the normal, wholesome lad he appeared to his village friends is to miss the obvious.

The earliest part of Newton's education was received in the common village schools of his vicinity. A maternal uncle, the Reverend William Ayscough, seems to have been the first to recognize that Newton was something unusual. A Cambridge graduate himself, Ayscough finally persuaded Newton's mother to send her son to Cambridge instead of keeping him at home, as she had planned, to help her manage the farm on her return to Woolsthorpe after her husband's death when Newton was fifteen.

Before this, however, Newton had crossed his Rubicon on his own initiative. On his uncle's advice he had been sent to Grantham Grammar School. While there, in the lowest form but one, he was tormented by the school bully who one day kicked Newton in the stomach, causing him much physical pain and mental anguish. Encouraged by one of the schoolmasters, Newton challenged the bully to a fair fight, thrashed him, and, as a final mark of humiliation, rubbed his enemy's cowardly nose on the wall of the church. Up till this young Newton had shown no great interest in his lessons. He now set out to prove his head as good as his fists and quickly rose to the distinction of top boy in the school. The Headmaster and Uncle Ayscough agreed that Newton was good enough for Cambridge, but the decisive die was thrown when Ayscough caught his nephew reading under a hedge when he was supposed to be helping a farmhand to do the marketing.

While at Grantham Grammar School, and subsequently while preparing for Cambridge, Newton lodged with a Mr. Clarke, the village apothecary. In the apothecary's attic Newton found a parcel of old books, which he devoured, and in the house generally, Clarke's stepdaughter, Miss Storey, with whom he fell in love and to whom he became engaged before leaving Woolsthorpe for Cambridge in June, 1661, at the age of nineteen. But although Newton cherished a warm affection for his first and only sweetheart all her life, absence and growing absorption in his work thrust romance into the background, and Newton never married. Miss Storey became Mrs. Vincent.

Before going on to Newton's career at Trinity College we may take a short look at the England of his times and some of the scientific knowledge to which the young man fell heir. The bullheaded and bigoted Scottish Stuarts had undertaken to rule England according to the divine rights they claimed were vested in them, with the not uncommon result that mere human beings resented the assumption of celestial authority and rebelled against the sublime conceit, the stupidity, and the incompetence of their rulers. Newton grew up in an atmosphere of civil war—political and religious—in which Puritans and Royalists alike impartially looted whatever was needed to keep their ragged armies fighting. Charles I (born in 1600, beheaded in 1649) had done everything in his power to suppress Parliament; but in spite of his ruthless extortions and the villainously able backing of his own Star Chamber through its brilliant perversions of the law and common justice, he was no match for the dour Puritans under Oliver Cromwell, who in his turn was to back his butcheries and his roughshod march over Parliament by an appeal to the divine justice of his holy cause.

All this brutality and holy hypocrisy had a most salutary effect on young Newton's character: he grew up with a fierce hatred of tyranny, subterfuge, and oppression, and when King James later sought to meddle repressively in University affairs, the mathematician and natural philosopher did not need to learn that a resolute show of backbone and a united front on the part of those whose liberties are endangered is the most effective defense against a coalition of unscrupulous politicians; he knew it by observation and by instinct.

To Newton is attributed the saying "If I have seen a little farther than others it is because I have stood on the shoulders of giants." He had. Among the tallest of these giants were Descartes, Kepler, and Galileo. From Descartes, Newton inherited analytic geometry, which he found difficult at first; from Kepler, three fundamental laws of planetary motion, discovered empirically after twenty-two years of inhuman calculation; while from Galileo he acquired the first two of the three laws of motion which were to be the cornerstone of his own dynamics. But bricks do not make a building; Newton was the architect of dynamics and celestial mechanics.

As Kepler's laws were to play the role of hero in Newton's development of his law of universal gravitation they may be stated here.

1. *The planets move around the Sun in ellipses; the Sun is at one focus of these ellipses.*

[If S, S′ are the foci, P any position of a planet in its orbit, SP + S′P is always equal to AA′, the major axis of the ellipse.]

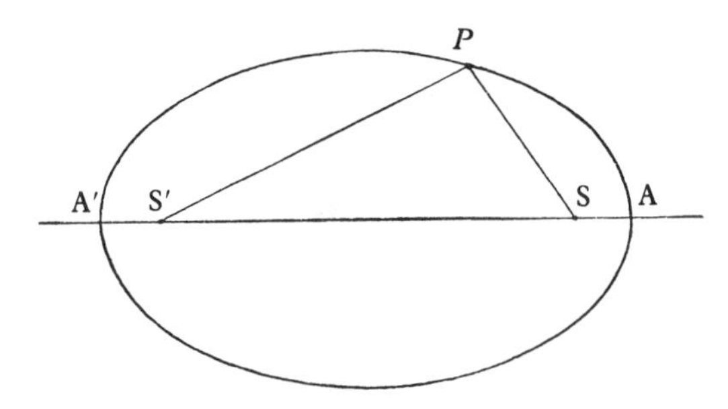

2. *The line joining the Sun and a planet sweeps out equal areas in equal times.*
3. *The square of the time for one complete revolution of each planet is proportional to the cube of its mean* [or average] *distance from the Sun.*

These laws can be proved in a page or two by means of the calculus applied to Newton's law of universal gravitation:

Any two particles of matter in the universe attract one another with a force which is directly proportional to the product of their masses and inversely proportional to the square of the distance between them. Thus if m, M are the masses of the two particles and d the distance between them (all measured in appropriate units), the force of attraction between them is $k \times m \times M/d^2$, where k is some constant number (by suitably choosing the units of mass and distance k may be taken equal to 1, so that the attraction is simply $m \times M/d^2$).

For completeness we state Newton's three laws of motion.

1. *Every body will continue in its state of rest or of uniform* [unaccelerated] *motion in a straight line*

except in so far as it is compelled to change that state by impressed force.

2. *Rate of change of momentum* ["mass times velocity," mass and velocity being measured in appropriate units] *is proportional to the impressed force and takes place in the line in which the force acts.*
3. *Action and reaction* [as in the collision on a frictionless table of perfectly elastic billiard balls] *are equal and opposite* [the momentum one ball loses is gained by the other].

The most important thing for mathematics in all of this is the phrase opening the statement of the second law of motion, *rate of change*. What is a rate, and how shall it be measured? Momentum, as noted, is "mass times velocity." The masses which Newton discussed were assumed to remain constant during their motion—not like the electrons and other particles of current physics whose masses increase appreciably as their velocity approaches a measurable fraction of that of light. Thus, to investigate "rate of change of momentum," it sufficed Newton to clarify *velocity*, which is rate of change of position. His solution of this problem—giving a workable mathematical method for investigating the velocity of any particle moving in any continuous manner, no matter how erratic—gave him the master key to the whole mystery of rates and their measurement, namely the *differential* calculus.

A similar problem growing out of rates put the *integral* calculus into his hands. How shall the total distance passed over in a given time by a moving particle whose velocity is varying continuously from instant to instant be calculated? Answering this or similar problems, some phrased geometrically, Newton came upon the integral calculus. Finally, pondering the two types of problems together, Newton made a capital discovery: he saw that the differential calculus and the integral calculus are intimately and reciprocally related by what is today called "the fundamental theorem of the calculus." . . .

In addition to what Newton inherited from his predecessors in science and mathematics he received from the spirit of his age two further gifts, a passion for theology and an unquenchable thirst for the mysteries of alchemy. To censure him for devoting his unsurpassed intellect to these things, which would now be considered unworthy of his serious effort, is to censure oneself. For in Newton's day alchemy *was* chemistry and it had *not* been shown that there was nothing much in it—except what was to come out of it, namely modern chemistry; and Newton, as a man of inborn scientific spirit, undertook to find out *by experiment* exactly what the claims of the alchemists amounted to.

As for theology, Newton was an unquestioning believer in an all-wise Creator of the universe and in his own inability—like that of the boy on the seashore—to fathom the entire ocean of truth in all its depths. He therefore believed that there were not only many things in heaven beyond his philosophy but plenty on earth as well, and he made it his business to understand for himself what the majority of intelligent men of his time accepted without dispute (to them it was as natural as common sense)—the traditional account of creation.

He therefore put what he considered his really serious efforts on attempts to prove that the prophecies of Daniel and the poetry of the Apocalypse make sense, and on chronological researches whose object was to harmonize the dates of the Old Testament with those of history. In Newton's day theology was still queen of the sciences and she sometimes ruled her obstreperous subjects with a rod of brass and a head of cast iron. Newton however did permit his rational science to influence his beliefs to the extent of making him what would now be called a Unitarian.

In June, 1661 Newton entered Trinity College, Cambridge, as a subsizar—a student who (in those days) earned his expenses by menial service. Civil war, the restoration of the monarchy in 1661, and uninspired toadying to the Crown on the part of the University had all brought Cambridge to one of the low-water marks in its history as an educational institution when Newton took up his residence. Nevertheless young Newton, lonely at first, quicky found himself and became absorbed in his work.

In mathematics Newton's teacher was Dr. Isaac Barrow (1630–1677), a theologian and mathematician of whom it has been said that brilliant and original as he undoubtedly was in mathematics, he had the misfortune to be the morning star heralding Newton's sun. Barrow gladly recognized that a greater than himself had arrived, and when (1669) the strategic moment came he resigned the Lucasian Professorship of Mathematics (of which he was the first holder) in favor of his incomparable pupil. Barrow's geometrical lectures dealt among other things with his own methods for finding areas and drawing tangents to curves—essentially the key problems of the integral and the differential calculus respectively, and there can be no doubt that these lectures inspired Newton to his own attack.

The record of Newton's undergraduate life is disappointingly meager. He seems to have made no very great impression on his fellow students, nor do his brief, perfunctory letters home tell anything of interest. The first two years were spent mastering elementary mathematics. If there is any reliable account of Newton's sudden maturity as a discoverer, none of his modern biographers seems to have located it. Beyond the fact that in the three years 1664–66 (age twenty-one to twenty-three) he laid the foundation of all his subsequent work in science and mathematics, and that incessant work and late hours brought on an illness, we know nothing definite. Newton's tendency to secretiveness about his discoveries has also played its part in deepening the mystery.

On the purely human side Newton was normal enough as an undergraduate to relax occasionally, and there is a record in his account book of several sessions at the tavern and two losses at cards. He took his B.A. degree in January, 1664.

The Great Plague (bubonic plague) of 1664–65, with its milder recurrence the following year, gave Newton his great if forced opportunity. The University was closed, and for the better part of two years Newton retired to meditate at Woolsthorpe. Up till then he had done nothing remarkable—except make himself ill by too assiduous observation of a comet and lunar halos—or, if he had, it was a secret. In these two years he invented the method of fluxions (the calculus), discovered the law of universal gravitation, and proved experimentally that white light is composed of light of all the colors. All this before he was twenty-five.

A manuscript dated May 20, 1665, shows that Newton at the age of twenty-three had sufficiently developed the principles of the calculus to be able to find the tangent and curvature at any point of any continuous curve. He called his method "fluxions"—from the idea of "flowing" or variable quantities and their rates of "flow" or "growth." His discovery of the binomial theorem, an essential step toward a fully developed calculus, preceded this. . . .

The second of Newton's great inspirations which came to him as a youth of twenty-two or three in 1666 at Woolsthorpe was his law of universal gravitation (already stated). In this connection we shall not repeat the story of the falling apple. . . .

Most authorities agree that Newton did make some rough calculations in 1666 (he was then twenty-three) to see whether his law of universal gravitation would account for Kepler's laws. Many years later (in 1684) when Halley asked him what law of attraction would account for the elliptical orbits of the planets Newton replied at once the inverse square.

"How do you know?" Halley asked—he had been prompted by Sir Christopher Wren and others to put the question, as a great argument over the problem had been going on for some time in London.

"Why, I have calculated it," Newton replied. On attempting to restore his calculation (which he had mislaid) Newton made a slip, and believed he was in error. But presently he found his mistake and verified his original conclusion.

Much has been made of Newton's twenty years' delay in the publication of the law of universal gravitation as an undeserved setback due to inaccurate data. Of three explanations a less romantic but more mathematical one than either of the others is to be preferred here.

Newton's delay was rooted in his inability to solve a certain problem in the integral calculus which was crucial for the whole theory of universal gravitation as expressed in the Newtonian law. Before he could account for the motion of both the apple and the Moon Newton had to find the total attraction of a solid homogeneous sphere on any mass particle outside the sphere. For *every* particle of the sphere attracts the mass particle outside the sphere with a force varying directly as the product of the masses of the two particles and inversely as the square of the distance between them: how are all these separate attractions, infinite in number, to be compounded or added into one resultant attraction?

This evidently is a problem in the integral calculus. Today it is given in the textbooks as an example which young students dispose of in twenty minutes or less. Yet it held Newton up for twenty years. He finally solved it, of course: the attraction is the same as if the entire mass of the sphere were concentrated in *a single point* at its center. The problem is thus reduced to finding the attraction between two mass particles at a given distance apart, and the immediate solution of this is as stated in Newton's law. If this is the correct explanation for the twenty years' delay, it may give us some idea of the enormous amount of labor which generations of mathematicians since Newton's day have expended on developing and simplifying the calculus to the point where very ordinary boys of sixteen can use it effectively.

Although our principal interest in Newton centers about his greatness as a mathematician we cannot leave him with his undeveloped masterpiece of 1666. To do so would be to give no idea of his magnitude, so we shall go on to a brief outline of his other activities without entering into detail (for lack of space) on any of them.

On his return to Cambridge Newton was elected a Fellow of Trinity in 1667 and in 1669, at the age of twenty-six, succeeded Barrow as Lucasian Professor of Mathematics. His first lectures were on optics. In these he expounded his own discoveries and sketched his corpuscular theory of light, according to which light consists in an emission of corpuscles and is not a wave phenomenon as Huygens and Hooke asserted. Although the two theories appear to be contradictory both are useful today in correlating the phenomena of light and are, in a purely mathematical sense, reconciled in the modern quantum theory. Thus it is not now correct to say, as it may have been a few years ago, that Newton was entirely wrong in his corpuscular theory.

The following year, 1668, Newton constructed a reflecting telescope with his own hands and used it to observe the satellites of Jupiter. His object doubtless was to see whether universal gravitation really was universal by observations on Jupiter's satellites. This year is also memorable in the history of calculus. Mercator's calculation by means of infinite series of an area connected with a hyperbola was brought to Newton's attention. The method was practically identical with Newton's own, which he had not published, but which he now wrote out, gave to Dr. Barrow, and permitted to circulate among a few of the better mathematicians.

On his election to the Royal Society in 1672 Newton communicated his work on telescopes and his corpuscular theory of light. A commission of three, including the cantankerous Hooke, was appointed to report on the work on optics. Exceeding his authority as a referee Hooke seized the opportunity to propagandize for the undulatory theory and himself at Newton's expense. At first Newton was cool and scientific under criticism, but when the mathematician Lucas and the physician Linus, both of Liège, joined Hooke in adding suggestions and objections which quickly changed from the legitimate to the carping and the merely stupid, Newton gradually began to lose patience.

A reading of his correspondence in this first of his irritating controversies should convince anyone that Newton was not by nature secretive and jealous of his discoveries. The tone of his letters gradually changes from one of eager willingness to clear up the difficulties which others found, to one of bewilderment that scientific men should regard science as a battleground for personal quarrels. From bewilderment he quickly passes to cold anger and a hurt, somewhat childish resolution to play by himself in future. He simply could not suffer malicious fools gladly.

At last, in a letter of November 18, 1676, he says, "I see I have made myself a slave to philosophy, but if I get free of Mr. Lucas's business, I will resolutely bid adieu to it eternally, excepting what I do for my private satisfaction, or leave to come out after me; for I see a man must either resolve to put out nothing new, or become a slave to defend it." Almost identical sentiments were expressed by Gauss in connection with non-Euclidean geometry.

Newton's petulance under criticism and his exasperation at futile controversies broke out again after the publication of the *Principia*. Writing to Halley on June 20, 1688, he says, "Philosophy [science] is such an impertinently litigious Lady, that a man had as good be engaged to lawsuits, as to have to do with her. I found it so formerly, and now I am no sooner come near her again, but she gives me warning." Mathematics, dynamics, and celestial mechanics were in fact—we may as well admit it—secondary interests with Newton. His heart was in his alchemy, his researches in chronology, and his theological studies.

It was only because an inner compulsion drove him that he turned as a recreation to mathematics. As early as 1679, when he was thirty-seven (but when also he had his major discoveries and inventions securely locked up in his head or in his desk), he writes to the pestiferous Hooke: "I had for some years last been endeavoring to bend myself from philosophy to other studies in so much that I have long grudged the time spent in that study unless it be perhaps at idle hours sometimes for diversion." These "diversions" occasionally cost him more incessant thought than his professed labors, as when he made himself seriously ill by thinking day and night about the motion of the Moon, the only problem, he says, that ever made his head ache.

Another side of Newton's touchiness showed up in the spring of 1673 when he wrote to Oldenburg resigning his membership in the Royal Society. This petulant action has been variously interpreted. Newton gave financial difficulties and his distance from London as his reasons. Oldenburg took the huffy mathematician at his word and told him that under the rules he could retain his membership without paying. This brought Newton to his senses and he withdrew his resignation, having recovered his temper in the meantime. Nevertheless Newton thought he was about to be hard pressed. However, his finances presently straightened out and he felt better. It may be noted here that Newton was no absent-minded dreamer when it came to a question of money. He was extremely shrewd and he died a rich man for his times. But if shrewd and thrifty he was also very liberal with his money and was always ready to help a friend in need as unobtrusively as possible. To young men he was particularly generous.

The years 1684–86 mark one of the great epochs in the history of all human thought. Skilfully coaxed by Halley, Newton at last consented to write up his astronomical and dynamical discoveries for publication. Probably no mortal has ever thought as hard and as continuously as Newton did in composing his *Philosophiae Naturalis Principia Mathematica* (Mathematical Principles of Natural Philosophy). Never careful of his bodily health, Newton seems to have forgotten that he had a body which required food and sleep when he gave himself up to the composition of his masterpiece. Meals were ignored or forgotten, and on arising from a snatch of sleep he would sit on the edge of the bed half-clothed for hours, threading the mazes of his mathematics. In 1686 the *Principia* was presented to the Royal Society, and in 1687 was printed at Halley's expense.

A description of the contents of the *Principia* is out of the question here, but a small handful of the inexhaustible treasures it contains may be briefly exhibited. The spirit animating the whole work is Newton's dynamics, his law of universal gravitation, and the application of both to the solar system—"the system of the world." Although the calculus has vanished from the synthetic geometrical demonstrations, Newton states (in a letter) that he used it to *discover* his results and, having done so, proceeded to rework the proofs furnished by the calculus into geometrical shape so that his contemporaries might the more readily grasp the main theme—the dynamical harmony of the heavens.

First, Newton deduced Kepler's empirical laws from his own law of gravitation, and he showed how the mass of the Sun can be calculated, also how the mass of any planet having a satellite can be determined. Second, he initiated the extremely important theory of *perturbations*: the Moon, for example, is attracted not only by the Earth but by the Sun also; hence the orbit of the Moon will be perturbed by the pull of the Sun. In this manner Newton accounted for two ancient observations due to Hipparchus and Ptolemy. Our own generation has seen the now highly developed theory of perturbations applied to electronic orbits, particularly for the helium atom. In addition to these ancient observations, seven other irregularities of the Moon's motion observed by Tycho Brahe

(1546–1601), Flamsteed (1646–1719), and others, were deduced from the law of gravitation.

So much for lunar perturbations. The like applies also to the planets. Newton began the theory of planetary perturbations, which in the nineteenth century was to lead to the discovery of the planet Neptune and, in the twentieth, to that of Pluto.

The "lawless" comets—still warnings from an angered heaven to superstitious eyes—were brought under the universal law as harmless members of the Sun's family, with such precision that we now calculate and welcome their showy return (unless Jupiter or some other outsider perturbs them unduly), as we did in 1910 when Halley's beautiful comet returned promptly on schedule after an absence of seventy four years.

He began the vast and still incomplete study of planetary evolution by calculating (from his dynamics and the universal law) the flattening of the earth at its poles due to diurnal rotation, and he proved that the shape of a planet determines the length of its day, so that if we knew accurately how flat Venus is at the poles, we could say how long it takes her to turn completely once round the axis joining her poles. He calculated the variation of weight with latitude. He proved that a hollow shell, bounded by concentric spherical surfaces, and homogeneous, exerts no force on a small body anywhere inside it. The last has important consequences in electrostatics—also in the realm of fiction, where it has been used as the motif for amusing fantasies.

The precession of the equinoxes was beautifully accounted for by the pull of the Moon and the Sun on the equatorial bulge of the Earth causing our planet to wobble like a top. The mysterious tides also fell naturally into the grand scheme—both the lunar and the solar tides were calculated, and from the observed heights of the spring and neap tides the mass of the Moon was deduced. The First Book laid down the principles of dynamics; the Second, the motion of bodies in resisting media, and fluid motion; the Third was the famous "System of the World."

Probably no other law of nature has so simply unified any such mass of natural phenomena as has Newton's law of universal gravitation in his *Principia*. It is to the credit of Newton's contemporaries that they recognized at least dimly the magnitude of what had been done, although but few of them could follow the reasoning by which the stupendous miracle of unification had been achieved, and made of the author of the *Principia* a demigod. Before many years had passed the Newtonian system was being taught at Cambridge (1699) and Oxford (1704). France slumbered on for half a century, still dizzy from the whirl of Descartes' angelic vortices. But presently mysticism gave way to reason and Newton found his greatest successor not in England but in France, where Laplace set himself the task of continuing and rounding out the *Principia*.

After the *Principia* the rest is anticlimax. Although the lunar theory continued to plague and "divert" him, Newton was temporarily sick of "philosophy" and welcomed the opportunity to turn to less celestial affairs. James II, obstinate Scot and bigoted Catholic that he was, had determined to force the University to grant a master's degree to a Benedictine over the protests of the academic authorities. Newton was one of the delegates who in 1687 went to London to present the University's case before the Court of High Commission presided over by that great and blackguardly lawyer the Lord High Chancellor George Jeffreys—"infamous Jeffreys" as he is known in history. Having insulted the leader of the delegates in masterly fashion, Jeffreys dismissed the rest with the injunction to go and sin no more. Newton apparently held his peace. Nothing was to be gained by answering a man like Jeffreys in his own kennel. But when the others would have signed a disgraceful compromise it was Newton who put backbone into them and kept them from signing. He won the day; nothing of any value was lost—not even honor. "An honest courage in these matters," he wrote later, "will secure all, having law on our sides."

Cambridge evidently appreciated Newton's courage, for in January, 1689, he was elected to represent the University at the Convention Parliament after James II had fled the country to make room for William of Orange and his Mary, and the faithful Jeffreys was burrowing into dunghills to escape the ready justice of the mob. Newton sat in Parliament till its dissolution in February,

1690. To his credit he never made a speech in the place. But he was faithful to his office and not averse to politics; his diplomacy had much to do with keeping the turbulent University loyal to the decent King and Queen.

Newton's taste of "real life" in London proved his scientific undoing. Influential and officious friends, including the philosopher John Locke (1632–1704) of *Human Understanding* fame, convinced Newton that he was not getting his share of the honors. The crowning imbecility of the Anglo-Saxon breed is its dumb belief in public office or an administrative position as the supreme honor for a man of intellect. The English finally (1699) made Newton Master of the Mint to reform and supervise the coinage of the Realm. For utter bathos this "elevation" of the author of the *Principia* is surpassed only by the jubilation of Sir David Brewster in his life of Newton (1860) over the "well-merited recognition" thus accorded Newton's genius by the English people. Of course if Newton really wanted anything of the sort there is nothing to be said; he had earned the right millions of times over to do anything he desired. But his busybody friends need not have egged him on.

It did not happen all at once. Charles Montagu, later Earl of Halifax, Fellow of Trinity College and a close friend of Newton, aided and abetted by the everlastingly busy and gossipy Samuel Pepys (1633–1703) of diary notoriety, stirred up by Locke and by Newton himself, began pulling wires to get Newton some recognition "worthy" of him.

The negotiations evidently did not always run smoothly and Newton's somewhat suspicious temperament caused him to believe that some of his friends were playing fast and loose with him—as they probably were. The loss of sleep and the indifference to food which had enabled him to compose the *Principia* in eighteen months took their revenge. In the autumn of 1692 (when he was nearly fifty and should have been at his best) Newton fell seriously ill. Aversion to all food and an almost total inability to sleep, aggravated by a temporary persecution mania, brought on something dangerously close to a total mental collapse. A pathetic letter of September 16, 1693 to Locke, written after his recovery, shows how ill he had been.

Sir,

Being of the opinion that you endeavored to embroil me with women and by other means,* I was so much affected with it that when one told me you were sickly and would not live, I answered, 'twere better if you were dead. I desire you to forgive me for this uncharitableness. For I am now satisfied that what you have done is just, and I beg your pardon for having hard thoughts of you for it, and for representing that you struck at the root of morality, in a principle you laid down in your book of ideas, and designed to pursue in another book, and that I took you for a Hobbist. I beg your pardon also for saying or thinking that there was a design to sell me an office, or to embroil me.

I am your most humble
And unfortunate servant,
IS. NEWTON

The news of Newton's illness spread to the Continent where, naturally, it was greatly exaggerated. His friends, including one who was to become his bitterest enemy, rejoiced at his recovery. Leibniz wrote to an acquaintance expressing his satisfaction that Newton was himself again. But in the very year of his recovery (1693) Newton heard for the first time that the calculus was becoming well known on the Continent and that it was commonly attributed to Leibniz.

The decade after the publication of the *Principia* was about equally divided between alchemy, theology, and worry, with more or less involuntary and headachy excursions into lunar theory. Newton and Leibniz were still on cordial terms. Their respective "friends," ignorant as Kaffirs of all mathematics and of the calculus in particular, had not yet decided to pit one against the other with charges of plagiarism in the invention of the calculus, and even grosser dishonesty, in the most shameful squabble over priority in the history of mathematics. Newton recognized Leibniz' merits, Leibniz recognized Newton's, and at this peaceful stage of their acquaintance neither for a moment

*There had been gossip that Newton's favorite niece had used her charms to further Newton's advancement.

suspected that the other had stolen so much as a single idea of the calculus from the other.

Later, in 1712, when even the man in the street—the zealous patriot who knew nothing of the facts—realized vaguely that Newton had done something tremendous in mathematics (more, probably, as Leibniz said, than had been done in all history before him), the question as to who had invented calculus became a matter of acute national jealousy, and all educated England rallied behind its somewhat bewildered champion, howling that his rival was a thief and a liar.

Newton at first was not to blame. Nor was Leibniz. But as the British sporting instinct presently began to assert itself, Newton acquiesced in the disgraceful attack and himself suggested or consented to shady schemes of downright dishonesty designed to win the international championship at any cost—even that of national honor. Leibniz and his backers did likewise. The upshot of it all was that the obstinate British practically rotted mathematically for all of a century after Newton's death, while the more progressive Swiss and French, following the lead of Leibniz, and developing his incomparably better way of merely *writing* the calculus, perfected the subject and made it the simple, easily applied implement of research that Newton's immediate successors should have had the honor of making it.

In 1696, at the age of fifty-four, Newton became Warden of the Mint. His job was to reform the coinage. Having done so, he was promoted in 1699 to the dignity of Master. The only satisfaction mathematicians can take in this degradation of the supreme intellect of ages is the refutation which it afforded of the silly superstition that mathematicians have no practical sense. Newton was one of the best Masters the Mint ever had. He took his job seriously.

In 1701–2 Newton again represented Cambridge University in Parliament, and in 1703 was elected President of the Royal Society, an honorable office to which he was re-elected time after time till his death in 1727. In 1705 he was knighted by good Queen Anne. Probably this honor was in recognition of his services as a money-changer rather than in acknowledgement of his preeminence in the temple of wisdom. This is all as it should be: if "a riband to stick in his coat" is the reward of a turncoat politician, why should a man of intellect and integrity feel flattered if his name appears in the birthday list of honors awarded by the King? Caesar may be rendered the things that are his, ungrudgingly; but when a man of science, *as* a man of science, snaps up the droppings from the table of royalty he joins the mangy and starved dogs licking the sores of beggars at the feast of Dives. It is to be hoped that Newton was knighted for his services to the money-changers and not for his science.

Was Newton's mathematical genius dead? Most emphatically no. He was still the equal of Archimedes. But the wiser old Greek, born aristocrat that he was—fortunately, cared nothing for the honors of a position which had always been his; to the very last minute of his long life he mathematicized as powerfully as he had in his youth. But for the accidents of preventable disease and poverty, mathematicians are a long-lived race intellectually; their creativeness outlives that of poets, artists, and even of scientists, by decades. Newton was still as virile of intellect as he had ever been. Had his officious friends but let him alone Newton might easily have created the calculus of variations, an instrument of physical and mathematical discovery second only to the calculus, instead of leaving it for the Bernoullis, Euler, and Lagrange to initiate. He had already given a hint of it in the *Principia* when he determined the shape of the surface of revolution which would cleave through a fluid with the least resistance. He had it in him to lay down the broad lines of the whole method. Like Pascal when he forsook this world for the mistier if more satisfying kingdom of heaven, Newton was still a mathematician when he turned his back on his Cambridge study and walked into a more impressive sanctum at the Mint.

In 1696 Johann Bernoulli and Leibniz between them concocted two devilish challenges to the mathematicians of Europe. The first is still of importance; the second is not in the same class. Suppose two points to be fixed at random in a vertical plane. What is the shape of the curve down which a particle must slide (without friction) under the influence of gravity so as to pass from the upper point to the lower in the *least time*? This is the problem of the *brachistochrone* (= "shortest time"). After the problem had baffled the mathematicians of Europe for six months, it was proposed again, and Newton heard of it for the first time on January 29, 1696, when a friend communicated it to him. He had just come home, tired out, from a long day at the Mint. After dinner he solved the problem (and the second as well), and the following day communicated his solutions to the Royal Society anonymously. But for all his caution he could not conceal his identity—while at the Mint Newton resented the efforts of mathematicians and scientists to entice him into discussions of scientific interest. On seeing the solution Bernoulli at once exclaimed, "Ah! I recognize the lion by his paw." (This is not an exact translation of B's Latin.) They all knew Newton when they saw him, even if he did have a moneybag over his head and did not announce his name.

A second proof of Newton's vitality was to come in 1716 when he was seventy-four. Leibniz had rashly proposed what appeared to him a difficult problem as a challenge to the mathematicians of Europe and aimed at Newton in particular. Newton received this at five o'clock one afternoon on returning exhausted from the blessed Mint. He solved it that evening. This time Leibniz somewhat optimistically thought he had trapped the Lion. In all the history of mathematics Newton has had no superior (and perhaps no equal) in the ability to concentrate all the forces of his intellect on a difficulty at an instant's notice.

The story of the honors that fall to a man's lot in his lifetime makes but trivial reading to his successors. Newton got all that were worth having to a living man. On the whole Newton had as fortunate a life as any great man has ever had. His bodily health was excellent up to his last years; he never wore glasses and he lost only one tooth in all his life. His hair whitened at thirty but remained thick and soft till his death.

The record of his last days is more human and more touching. Even Newton could not escape suffering. His courage and endurance under almost constant pain during the last two or three years of his life add but another laurel to his crown as a human being. He bore the tortures of "the stone" without flinching, though the sweat rolled from him, and always with a word of sympathy for those who waited on him. At last, and mercifully, he was seriously weakened by "a persistent cough," and finally, after having been eased of pain for some days, died peacefully in his sleep between one and two o'clock on the morning of March 20, 1727, in his eighty-fifth year. He is buried in Westminster Abbey.

Gauss and the Present Situation of the Exact Sciences

Richard Courant

Richard Courant was born in 1888 in Germany. He received his Ph.D. from Göttingen in 1910. He was associated with Göttingen from 1912 until 1933. With the rise of the Nazis he emigrated to the United States. From 1934 until his death in 1972, he worked to try to make New York University's Institute of Mathematical Sciences a center for mathematics, as Göttingen had been before World War II.

There has circulated an apocryphal story concerning Garrett Birkhoff, well-known son of the first famous American mathematician, G. H. Birkhoff. Birkhoff is a tenured faculty member of the Harvard mathematics department yet does not possess a Ph.D. When asked why not, supposedly he responded, "Where could I find someone to examine me?" Similar variants are known for other distinguished professional mathematicians who simply bypassed the doctorate to immediately begin contributing research articles.

Gauss was one of the three most famous mathematicians in history, in company with Archimedes and Newton. In reading the following article by Richard Courant one quickly realizes that there really was no one who could examine Gauss either. His singular accomplishments in number theory, physics, astronomy, geology, actuarial mathematics, mechanics, numerical analysis, geometry, and magnetism boggle the imagination. Courant has undertaken a difficult task merely in surveying the highlights of this extraordinary career.

One should keep in mind three points when reading Courant's essay. First, Courant had been invited to give this talk at the University of Göttingen from which he had had to flee twenty years earlier at the start of the Nazi era, the same university at which Gauss lived out his life. Courant pointedly observes that Gauss's creative work had actually been done before Göttingen, at Braunschweig under the generous sponsorship of the Duke of Braunschweig. Courant remarks that the Göttingen period subjected Gauss to "a firm professional obligation with administrative responsibility and with a plethora of activities imposed from the outside," together with their attendant toll on creative energy. When the generous sponsorship of the University of Göttingen ended with the death of the Weimar republic, Courant escaped the Nazi tyranny by accepting a position at New York University. In his biography by Constance Reed, Courant states that the obligations, responsibilities, and outside activities that were necessary to recreate the Göttingen idea at New York University took their toll on his creative energy. The unconscious parallelism is striking.

The second point to remember is that Courant's address was made two years before the launching of Sputnik, the event that caused massive amounts of money to be pumped into the American university community. Unfortunately, the plea that Courant makes in the essay to recognize that Gauss's mighty theoretical strides were made from applied problems was generally ignored. With the sudden influx of money researchers could and did go after the most

Source: Richard Courant, "Gauss and the Present Situation of the Exact Sciences," in Thomas L. Saaty and F. Joachim Weyl (eds.), *The Spirit and the Uses of the Mathematical Sciences* (McGraw-Hill, 1969), pp. 141–155. Address given at the Gauss Centennial of the Academy and University of Göttingen, Germany, February 19, 1955.

arcane, abstruse, ethereal topics without giving a second thought to the significance of the research. Finally, Courant's essay on the life of Gauss is timely for every mathematician. It particularly belies the idea that a mathematician is washed up after the age of twenty-five.

ONE HUNDRED YEARS after his death, Gauss' scientific greatness remains as much an enigma as it must have been to his contemporaries. The visionary originality, the depth and many-sidedness of Gauss' achievements, combined with demonstrations of extraordinary power and persistence—these are seen not only in his purely theoretical studies but also in his work in applied fields such as physics, astronomy, geodesy, insurance mathematics, mechanics, and earth magnetism. We know of no other scientist who so steadfastly led a long life full of concentrated work. The enormous contribution of Gauss to the insight and artistry of science has been absorbed only very gradually into the mainstream of developments.

When Gauss began his career, science was still the exalted occupation of a relatively small, isolated intellectual elite, comparatively protected from the political excursions of the time. Since then science has become inextricably interwoven with our complex civilization and has drawn into its service a great many people of all kinds. Thus more and more tension and problems have arisen which must be resolved if science within the framework of human society, and human society with the help of science, are to continue their development without catastrophe.

A retrospective look at the gigantic apparition of Gauss, towering over the level plain, can help us in our search for orientation among the conflicts that threaten the course of scientific development, present and future. It would be an impossible task to attempt even a superficial appreciation of Gauss' scientific merits within the span of a short essay; I shall, however, try to illuminate the phenomenon of Gauss' life in the light of some searching questions suggested by the present situation of science.

Above all, I think, is the question of what society can and ought to do in order to discover and encourage talent. How should we go about rearing scientific leaders, not just specialists and experts, but educated, responsible, and humble members of human society? The influence of narrow-minded specialists is almost as dangerous as the influence of those who haven't learned anything deeply, but believe that their formal training enables them to make decisions on everything.

Another significant question of the present arises from the great scientific projects which call for cooperation among different specialists, while new and original ideas only come to the mind of single individuals and are sometimes scared off by organized and disciplined labor. Teamwork and office hours do not mix well with top productivity. How do we find a balance here?

Furthermore, many questions spring from the confrontation between the defenders of purely theoretical science that has no external motivation, and the representatives of the applied sciences who will give precedence every time to mission-oriented research. Still another set of problems involves the relationship of research to teaching and scientific education.

Let us briefly look at the work of Gauss from the perspective of a present which is searching the past for answers to such questions. Gauss, who came from a lower middle-class community, early developed obvious mental powers for which his home environment could certainly not have provided the stimulus. Yet only after he entered the advanced grade in the elementary school of Braunschweig was his extraordinary mathematical talent discovered. It was a lucky coincidence that assistant teacher Bartels, then all of sixteen years old, himself had a strong interest in mathematics and quickly became a counselor of the ten-year-old prodigy. Going through advanced arithmetic books as well as mathematical papers with Gauss, he opened the way not only for the child but also for himself to the works of Euler, Newton, and other scientists of the then still recent past.

It was also fortunate that Bartels's enthusiasm succeeded in getting men at higher echelons in this small-scale, feudal, yet closely knit society interested in the young Gauss. It was against the will of his father, who wanted his son to become a craftsman, but with the support of his somewhat more farseeing mother, that Gauss was sent by aristocratic benefactors to the gymnasium, where he acquired the ancient languages with incredible speed and reached the graduating class at the age of fourteen.

In the end, the shy prodigy was presented to his provincial ruler, Carl Wilhelm Ferdinand, Duke of Braunschweig. The boy made a favorable impression on the Duke, who from then until his death in 1806 assumed all financial burdens for Gauss' further education and his later existence as an independent scholar. For a start, he financed his studies at the Collegium Carolinum of Braunschweig and subsequently (1795–1797) at the University of Goettingen, to which Gauss was drawn by its excellent library. His stipend amounted to 158 talers per year and was sufficient for a young man who had been brought up in frugality.

At the university Gauss was disappointed and even repelled by the pretentious and uncomprehending manner in which Professor and Court Counselor Kaestner conducted his mathematics courses. On the other hand, he was attracted by philology, which was brilliantly taught by Heyne. Nevertheless, Gauss finally decided in favor of mathematics, after he had made an exciting mathematical discovery in his nineteenth year which filled him with pride to the end of his life. He found that the regular 17-sided polygon can be constructed by ruler and compass. Ever since classical antiquity, such construction had been known only for the regular triangle, quadrangle, and pentagon. His discovery was based on a deep algebraic insight concerning an equation of sixteenth degree. He managed to solve this so-called "cyclotomic equation" of degree 16 in terms of a sequence of quadratic equations; the reduction to a chain of extracting square roots is the exact algebraic equivalent of the geometric constructability. Gauss' result for the regular 17-gon (as well as for any n-gon where n is a prime number of the form $2^{2^n} + 1$) later became the starting point for some of his far-reaching theoretical investigations in algebraic number theory.

Soon he was occupied with the preparation of his first great masterpiece, *Disquisitiones Arithmeticae* which, after years of difficulties, went to press in 1801 with the support of the Duke. Not only does it arrange in orderly fashion the earlier unsystematic beginnings of the theory of numbers, but it presents a vast collection of new discoveries and may be said to have established number theory as an organic branch of mathematics. The exposition—couched in excellent Latin—is of a classically systematic rigor, which made study of the book difficult and greatly delayed its impact. Throughout his life Gauss used such a fully matured, classically crystallized form of exposition as a matter of principle. Readers found it difficult and often impossible to penetrate the underlying structure of the ideas which finally emerged, but he countered criticism by saying that any builder removes the scaffolding after the building has been completed.

During his adolescent years in Goettingen and later in Braunschweig, where he returned in 1799 with the Duke's stipend, Gauss was nearly overwhelmed by the continuous flow of his ideas; the comparison to Mozart comes to mind. Although incredibly industrious, he was incapable of writing down all his ideas in a communicable manner.

Gauss did not receive his doctorate from Goettingen but in absentia from Helmstedt, where he found an understanding professor of scientific stature in Johann Friedrich Pfaff. Gauss' thesis was another landmark in the history of mathematics. After many unsuccessful and misleading starts by the great eighteenth-century mathematicians, the fundamental theorem of algebra was proved here for the first time. This is the theorem which states the existence of either real or complex roots for arbitrary algebraic equations of any degree. The paper initiated the era of existence proofs, which played an important part in the mathematics of the nineteenth century. Gauss' accomplishment, by the way, was based on his clear comprehension of the then hardly understood nature of imaginary and complex numbers.

Gauss' studies of astronomy had led him early to the method of least squares. This provides a sound mathematical approach to the problem of bringing the different values of repeated observations of the same physical unit into harmony with each other, so that a well-defined, trustworthy result can be drawn from apparently contradictory measurements. The least squares technique, along with other insights which Gauss had gained during early youth, proved to be of great advantage in one of his major accomplishments in the field of applied mathematics, which turned him into a world-famous man overnight at the age of 25.

On January 1, 1801, a new planet, the tiny Ceres, was detected, where only seven planets had been known since time immemorial. After a few days of observation the planet, which had very low luminosity, disappeared near the sun and seemed lost once more in the overwhelming abundance of the star-studded sky. The observations were published, and astronomers the world over made efforts to determine the ellipse of the planet's orbit from the scarce data, so that they could later relocate Ceres under more favorable circumstances. With the methods known at the time, such calculations based on Kepler's laws presented an insoluble problem, since they could be practically applied only to orbits that were almost circular.

It was summer before this news and the observational data percolated through to Gauss in Braunschweig. Immediately he set himself the task of determining the orbit. His method of least squares, his other unpublished findings, his incomparable skill in numerical computation, and his tremendous energy all combined to lead him to a precise determination of the orbit. This differed substantially from the inadequate results which many other scientists were publishing, and soon Ceres was located at precisely the place predicted by the computations of Gauss. The impression on the scientific world and the public at large was sensational. Men of learning from all countries sought contact with him. He received many attractive offers—for instance, from the imperial observatory of St. Petersburg—to the displeasure of the Duke, who increased Gauss' pension and kept his grateful and loyal protégé in Braunschweig. It was only after the Duke's death from wounds received while serving as Prussian commanding officer in the battle of Jena and Auerstaedt that Gauss accepted a long-standing offer to become professor of astronomy and director of the not yet existing observatory at Goettingen. He had married shortly before, and the transfer of his family to Goettingen in 1807 ended a period in his scientific development of overwhelming and varied productivity whose impact has made itself felt only gradually.

In summary, Gauss' career from his final years in school until he accepted the professorship at Goettingen was completely conditioned by the generosity of the Duke, a man of good sense and judgment, yet a typical representative of the dying feudal era. The Duke faced the People's Army of the French Revolution as commander of the monarchistic coalition army and led the retreat through the bombardment of Valmy (which was impressively described by Goethe). His death after the defeat at Jena is almost symbolic. Although his support of Gauss may be counted as peripheral in the sphere of a prince's interests, it is this action which gained him an honorable place in history.

It is remarkable that the discovery and sponsorship of what was probably the greatest scientific talent of the time occurred within the framework of dying feudalism. For it was at this time that the decisive turn toward furthering the progress of talented young men from all social levels had already been taken by the French Revolution. At the Ecole Normale and the Ecole Polytechnique of Paris, founded in 1794, the principles were proclaimed for the first time in the history of science that instruction and research belong closely together, that the best, productive scholars must take part as professors in the instruction of gifted students and must inspire them, and that all talented individuals should be provided with the best possible conditions for their education, regardless of status or fortune. These doctrines were set forth as being in the highest interest of human society.

The direct and, even more, the indirect success of such ideas of education has been extraordinary. They contributed much to strengthening revolutionary and postrevolutionary France and later had an essential share in the tremendous development of science and technology in the rest of nineteenth-century Europe. In Germany it was Alexander von Humboldt who, after extended studies in Paris, used these ideas in the university reform which he effected with the support of his brother Wilhelm, Prussia's great minister of education. Men like the youthful chemist Liebig or the equally young mathematician Jacobi were put into university positions in spite of resistance from the faculties. There they formed intellectual centers which, through a combination of instruction and research, had the effect of a chain reaction. Until today this tradition which was so successfully carried out by men like Poincaré and Hilbert has remained a live force. In the wake of the ideas which had their origin in France and were further developed in Prussia, the social basis for the discovery and education of gifted men was greatly widened, and new careers were created which gave a professional future to trained scientists.

The world in which Gauss grew up was untouched by this development. One might wonder what would have become of that gifted child if good fortune had not brought him to the attention of a powerful patron. We can only conjecture that many great talents have been lost to the world because of less favorable circumstances. Perhaps it is true that even under adverse conditions the very greatest and rarest talents may find their way to the top along irregular channels. Surely there were cases like this in Gauss' time, such as Faraday, Bessel, and Fraunhofer. Yet for modern society it is vital that the great number of outstanding, even if not unique, talents be discovered and encouraged. To achieve this goal is of decisive significance for civilization; its importance has steadily increased since the times of Monge and Humboldt.

Gauss never participated in the task of stimulating the younger generation, either through direct influence as a teacher or in personal contact. Ever since childhood he had paid for his superiority with a frightening intellectual isolation. He found hardly anybody capable of understanding the flight of his thoughts and interchanging ideas with him. Repelled by the mathematical instruction at Goettingen during the period of Kaestner, he abhorred the idea of being a professor of mathematics. Public recognition did not interest him. Therefore he published only a small part of his discoveries, and even those only after he had achieved such a finished presentation, so thoroughly polished, that it all but ceased to be comprehensible. Many of Gauss' mathematical discoveries remained unknown or unnoticed until they were later rediscovered by others. Despite the admiration of his contemporaries, Gauss' scientific influence was in many respects an "effect at a distance"—in time.

A characteristic example of the lack of psychological contact between Gauss and younger mathematicians is the episode with Wolfgang von Bolyai, son of Gauss' only close friend from student days. The father presented the aging Gauss with Wolfgang's revolutionary work, which contained the discovery of non-Euclidean geometry. Instead of replying with enthusiastic encouragement, Gauss sent a brief note saying that he was sorry not to be able to praise the work because he would then be praising himself, having found this non-Euclidean geometry years ago himself in a similar manner. The depth of disappointment which siezed the young Bolyai can be imagined.

Alexander von Humboldt, an early admirer and later friend of Gauss, made continuous attempts to move him to Berlin, where plans were being developed to found something like an Ecole Polytechnique in connection with the Academy. Gauss showed little inclination to take on a teaching load or to burden himself more than necessary with a position of influence and responsibility. Berlin offered attractive conditions, however, and it appeared at times as though he would accept in the end. But after negotiations which stretched over many years, Gauss finally remained at Goettingen.

In his relation to the scientific world around him, Gauss never completed the transition from the essentially aristocratic attitude of the eighteenth century to the broader life of the nineteenth. He exchanged scientific ideas, mostly about small sections of his work, by correspondence with a few men of stature. There were no international meetings, and he maintained frugality in everyday life as an undisputed virtue into old age. Politically he remained a conservative legitimist, loyal and grateful to his noble patron. After the protest of "Goettingen's Seven" against the arbitrary rule of the Hanoverian king, Gauss' son-in-law was expelled, while Wilhelm Weber, his closest and possibly his only really kindred and indispensible co-worker, lost his appointment at Goettingen. Yet by inner conviction, Gauss was probably on the king's side and distinctly kept his distance from the rebels.

The younger generation of mathematicians in the early nineteenth century looked at the cool and inaccessible giant with admiration, but at times with frightened distrust. They always had to anticipate that their own discoveries would turn out to have been made long ago by the uncommunicative Gauss. It is significant that the young and gifted Niels Henrik Abel avoided visiting Gauss while he was making his important journey from Norway to the "mathematical continent" of Europe. Jacobi made open and critical comments, as in the perhaps apocryphal but significant story of the occasion when Gauss informed him that he had long possessed Jacobi's revolutionary discoveries on elliptic functions. Jacobi replied: "Sir, my discoveries are so important that you would not have needed to be ashamed of publishing them, if you had them that much earlier." It is sad to think how great an impetus Gauss could have given to younger scientists if he had delivered or published numerous lectures—for instance, like those of Dirichlet or of Henri Poincaré, the greatest mathematician after the era of Gauss and Riemann. But it would be unjust to criticize this overpowering scientific personality for a limitation which was deeply founded in his nature.

The Goettingen professorship, which Gauss began under difficult political and economic conditions in 1809, subjected him for the first time to a firm professional obligation with administrative responsibility and with a plethora of activities imposed from the outside. During the decades in Goettingen until his death, Gauss met these obligations with great conscientiousness. He supervised the construction of the observatory, recorded many astronomic observations, invented ingenious precision instruments for optical and magnetic measurements, and spent much time on practical work in the field as director of the Hanoverian geodetical survey.

Except for the determination of the orbit of Ceres, his early scientific development had centered on theoretical and pure mathematics; now his work during the rest of his life seemed to cast applications in the principal role. Tasks presented from outside resulted in a stream of new practical successes, but at the same time they led Gauss to deep theoretical insights which evolved, apparently as a matter of course, as by-products of his applied scientific activity.

First Gauss completed his fundamental work on the determination of planetary orbits, a summary and continuation of methods which had led to the recovery of Ceres. This book, for which he had difficulty finding a publisher, remained the bible of computing astronomers for over a century. Its important original ideas for numerical analysis have not been completely exhausted to this day.

Gauss then proceeded with another project of astronomical mathematical computation; the determination of perturbations for the small planet Pallas, discovered after Ceres, whose long, extended Kepler orbit is unusually sensitive to gravitational influences of heavy celestial bodies. Ever since Gauss' time, the theory of perturbations has remained a pivotally important element of theoretical mechanics and physics. The case of Pallas presented exceptional difficulties. Even Gauss' ingenious methods of approximation failed to approach the complicated reality, although he returned time and again with almost superhuman effort to the perturbations of Pallas. Somehow, perhaps for psychological reasons (like many scientists, Gauss seems to have undergone periods of depression), this task was never concluded. Yet it produced valuable corollary results of a pure scientific theoretical nature.

In France the outstanding mathematician Legendre, twenty-five years older than Gauss and in many respects his forerunner, had dealt with theoretical and practical geodesy. Now, around 1820, the government of Hanover requested Gauss to take over the survey of the kingdom, after Gauss' former astronomical assistant, Schumacher, had already carried out surveys around Kiel and Altona. Almost a decade of Gauss' life was dedicated to this complex mission. Perhaps the outdoor activity with the surveying crews brought him some respite from the strains of long years of mental concentration.

Gauss completed the task with his usual excellence, in the process establishing geodesy as a legitimate science. Furthermore, he derived a rich yield of important new mathematical insights; Gauss simply could not help but penetrate to the deepest foundations, even in applied pursuits. His paper on the theory of surfaces is one purely mathematical work which sprang from his survey of the surface of the earth. This treatise on curved surfaces is a classical masterpiece, both in content and form. It culminates in the clearly formulated concept of curvature and the insight that the "Gauss surface curvature" does not change when the surface is subjected to arbitrary bending. Gauss called this result the *theorema egregium*.

These ideas which were soon to be generalized by Bernhard Riemann, have profoundly influenced scientific development. The Gauss-Riemann concept of curvature, for example, plays an important part in the modern theory of relativity. Closely connected with it, and evidently inspired by his practical involvement in geodetic measurements, are Gauss' insights into the nature of space, the foundation of geometry, and non-Euclidean, or as he put it, anti-Euclidean geometry. It is said that as an old man, he mumbled a few words of understanding admiration when he heard Riemann's habilitation address on the foundations of geometry. Yet he himself never published any of his ideas; only in his personal correspondence did he refer to them. Strangely enough, he was intent on keeping his discoveries in non-Euclidean geometry from the public; he was afraid of the "clamor of the Böotians," he said.

Gauss' work in the field of geodesy was followed by a fertile period during which he concentrated on physics, especially terrestrial magnetism, and electricity. For the first and only time, he found a partner of sufficient stature with whom he could work as a peer—the esteemed, though much younger, physicist Wilhelm Weber. Gauss met him at Humboldt's house and had him called from a lectureship at Leipzig to Goettingen. Their investigations in terrestrial magnetism were carried on over years while extensive series of observations accumulated, ingenious instruments were invented, and theoretical calculations were executed to characterize the earth's magnetic field as well as its daily and long-period variations. This work, too, yielded advances in pure mathematics, especially in the development of modern potential theory.

Gauss' collaboration with Weber also led to the construction of the electromagnetic telegraph, with which the two friends exchanged signals over a distance of several hundred yards. The tremendous possibilities for the future of telegraphy were obvious to both of them, and with practical foresight but without much success, they attempted to interest a newly founded railroad company in installing a telegraph connection between its two terminals. It is significant that the telegraph experiment made Gauss and Weber more famous than many of their profoundest discoveries, although this technical achievement was in the air at the time and was being independently attained by others.

Another of Gauss' remarkable contributions in the field of applied mathematics was accomplished during the long period of his tenure at Goettingen. Asked by the university to investigate the administration of the widows and orphans pension fund and work out reform proposals, he fulfilled this assignment so thoroughly that his memorandum served as the beginning of modern actuarial mathematics. It is hoped that these examples of his many-sided achievements will give the reader some idea of the extraordinary range of Gauss' work. In reviewing several aspects of his intellectual development, we may find that they suggest answers to the questions which are urgent in mathematics today.

Gauss came to mathematics in a natural way through the theory of numbers. As a child he played with numbers almost before he could read and write. He gained a surprising skill in numerical arithmetic and noticed various regularities, especially in connection with prime numbers, which he began to turn over in his mind. Thus his keen interest developed of itself; the unsystematic, elementary knowledge of mathematics, which he acquired quickly, was enough to start him on research of his own. As a very young man, with no knowledge of the literature, he made many of the number theoretical discoveries which Euler, Legendre, and others had published previously, and even then his originality frequently led him beyond his predecessors.

Only when he was a student in Goettingen did

he become familiar with the existing literature. In no time he had outstripped it with his *Disquisitiones Arithmeticae,* which created modern number theory as a scientific discipline. He never later recanted his youthful enthusiasm for the theory of numbers, which he titled the "queen of mathematics" while he called mathematics "queen of the sciences." In fact, however, as he matured Gauss turned more and more to applied mathematical problems which come to the scientist from the outside rather than springing directly from introspective meditation on self-generated interests. Although self-motivated, or so-called "pure," mathematics occupied him most intensively throughout his life, it was nevertheless the applications which formed the chief professional substance of his production.

Gauss, however, was never aware of any contrast, not even of a slight line of demarcation, between pure theory and applications. His mind wandered from practical applications, undaunted by required compromise, to purest theoretical abstraction and back, inspiring and inspired at both ends. In the light of Gauss' example, the chasm which was to open in a later period between pure and applied mathematics appears as a symbol of limited human capability. For us today as we suffocate in specialization, the phenomenon of Gauss serves as an exhortation. The representatives of both camps should not be proud of their limitation, but should do everything to understand each other and to bridge the chasm. In my judgment it is critical for the future of our science that mathematicians adopt this course, both in research and in education. In mathematics (and this is a personal opinion, considered not objectionable, yet often objected to) we must tear down the walls of professional separation between pure and applied work, and differentiate only between good and bad mathematics and perhaps between introspective and outward-oriented mathematicians. Young people should have an opportunity to mature and widen their knowledge rather than being forced to become highly specialized.

The mathematical sciences are in a critical period of transition. As mentioned before, Gauss' dissertation inaugurated the era of existence proofs. In many cases, brilliant ingenuity and great effort were concentrated not on finding the solution of a problem, but only on proving that such a solution exists. The success of such a simplification of the problem is often astounding; yet the voices that declare an existence proof without construction to be unacceptable have grown louder and louder.

Gauss, as the pioneer of existence proofs, was at the same time the greatest artist of mathematical construction, of the completed concrete details. No one since has shown the same virtuosity in producing the solutions by numerical computation. Gauss committed the entire arsenal of theoretical insight and practical skill to the numerical execution of problems. Today more than ever, mathematical problems in physics, technology, and statistics demand methods for obtaining numerical answers. If mathematicians were to insulate themselves from these demands, the fate of their discipline as a responsible science in human society would be in serious jeopardy. It is therefore only natural that in recent years interest in numerical analysis, as it is called, has greatly increased.

A tremendous stimulus, even a completely new turn, came from a significant technical development: the modern electronic digital computer. Only in recent years has a series of these machines been produced with the necessary reliability for tackling complicated mathematical problems. Even at maximum accuracy, these machines can calculate a thousand times, even a million times, faster than a skilled human computer. This has opened dazzling possibilities for the computational treatment of problems in number theory as well as in physics, mechanics, aerodynamics, chemical engineering, oceanography, and meteorology.

Gauss would doubtless have been extremely pleased if such resources had been available to him. But for us today it is important to understand that the proliferation of such machines also may have a disruptive impact on the fabric of our society. If the development of objective insight and foresight, no less than the education of the young generation, cannot keep up with these far-reaching shifts of technology, if technical achievements are permitted to become not just means but an end in themselves, then computers will not be tools but tyrants which make men their slaves.

During Gauss' youth there were few professional opportunities for mathematicians or scientists in related fields. This situation improved after the reforms of the early nineteenth century, especially as additional institutions of higher learning were established. Up into modern times, however, the teaching profession remained the only normal occupational goal in the mathematical sciences. This traditional restriction to the academic labor market has disappeared now. There is a rapidly developing demand of unpredictable dimension for mathematicians and theoretical physicists at all levels, up to that of the most accomplished scientists, and thousands of well-trained scientists are needed to operate these technical monsters for the benefit of society and science. Universities must continue unrelentingly in their efforts to remain in step with these developments.

After this most technical advance we are probably more than ever in need of the guidance which we can gain from a retrospective view of Gauss.

William Rowan Hamilton—An Appreciation

C. Lanczos

C. Lanczos is the author of a number of works on mathematics and the history of mathematics. Associated with the Dublin Institute for Advanced Studies, he has been a visiting professor at Yale and at several other universities in the United States.

When the Dublin Institute for Advanced Studies was given an endowment in Hamilton's name, the institute's only obligation (besides spending the money!) was to honor Hamilton with a public lecture at routine intervals. In 1965 the lecture, given by C. Lanczos, was to commemorate the hundredth anniversary of Hamilton's death.

In 1827 Hamilton was appointed Professor of Astronomy and Astronomer Royal of Ireland, yet he hardly did anything in astronomy during his thirty-eight-year term of office. So there would have been a certain ironic justice if Lanczos's lecture had been a routine, perfunctory tribute. However, this brief review of the man and his work is anything but that. Lanczos divides the article into three topics: Hamilton's contributions to mathematical physics, Hamilton's contributions to pure mathematics, and Hamilton's "failure" to do anything significant in the last twenty years of his life.

WILLIAM ROWAN HAMILTON, the greatest genius in the field of modern Irish science, was from his early youth an extraordinary phenomenon. Born in 1805 in Dublin, he was in his seventh year already conversant with nine languages, with a bias toward oriental tongues, since his father (a practicing attorney, residing at 29 Dominick Street, Dublin), had the wish that his son should some day make his career with the East India Company. Fate decided differently. The professor of Hebrew at Trinity College examined the boy of seven years and found, to his amazement, that the young Hamilton knew more Hebrew than many candidates for fellowship. When a new ambassador from Persia arrived, the boy, now fourteen, formulated a welcoming letter in faultless Persian.

These amazing linguistic gifts were paralleled, however, by an early proficiency in mathematics, in which field he was essentially self-taught. He read the original texts of Euclid (in Greek), then Newton (in Latin), and finally Laplace (in French). His astonishing precocity is best demonstrated by the fact that, in his seventeenth year, he found an error in the celebrated Mécanique Céleste of Laplace, which he communicated to Dr. John Brinkley, President of the Irish Academy. The Academy thus became aware of the extraordinary gifts of this young scientist, who in the same year submitted a long essay, entitled "Theory of Systems of Rays," which contained the germs of his later optical discoveries. The refereeing committee was impressed but found comprehension difficult and suggested a further elaboration of the theme. The essay was rewritten and submitted again in 1827, when it was printed.

He entered Trinity College in 1823; taking his B.A. in 1827, M.A. in 1837, LL.B. and LL.D. in 1839. He was still a student when, in 1826, Brinkley, having been appointed to a Bishopric,

Source: C. Lanczos, "William Rowan Hamilton—An Appreciation," *American Scientist,* 55 (1967): 129–143. By permission of *American Scientist.*

nominated the young Hamilton to become his successor as Professor of Astronomy at Trinity College and Astronomer Royal of Ireland. He was elected to this post in 1827 and moved to Dunsink Observatory, where he remained to the end of his life. In 1832 he was elected a member of the Academy and served between 1837 and 1845 as its President.

These are the external landmarks in the life of an extraordinary individual. Hiding behind them are the intellectual struggles of a man who was destined to establish one of the most decisive ideas in the history of mathematical physics. Optical phenomena interested Hamilton from his early youth, and he was soon intrigued by the fact that two fundamental approaches were possible: one could start with a bundle of rays which emanate from a light source and study their path, but one could also start with the wave fronts which were perpendicular to these rays. This dualism is much more than a matter of mathematical method. It was the central issue of Hamilton's investigations, the true significance of which did not come to light until much later, when the wave mechanics of our days demonstrated that particles and waves are two facets of the same thing, the particle picture corresponding to the rays, the wave picture to the associated surfaces. Hamilton recognized the remarkable analogy which exists between two apparently widely separate fields: mechanics and optics. Mechanics deals with the motion of particles or their conglomeration (the name "dynamics" is also often used, since forces are involved which cause the motion): optics deals with the phenomena of light. Newton thought of light as caused by the emission of minute material particles which moved with a very large velocity from point to point, whereas Huygens, and later Fresnel, thought of light as a propagation of wave surfaces. Hamilton found the common theoretical basis of both viewpoints.

"Characteristic Function" and "Principal Function"

The first ingenious idea occurred to him in optics, when he found that there exists a surprising analogy between optics and geometry. In geometry the remarkable fact holds that the distance between two points (x_1, y_1) and (x_2, y_2) is given by the law of Pythagoras as $[(x_1 - x_2)^2 + (y_1 - y_2)^2]^{1/2}$, and this expression is enough to derive from it all the theorems of Euclidean geometry. Something similar happens in optics, except that the ordinary geometrical distance has to be replaced by the optical distance. Hamilton called this fundamental quantity, associated with two points of space, the "characteristic function." Of course, this "optical distance" of two points, expressed in terms of their coordinates, is generally a much more complicated expression than their simple Pythagorean distance. But Hamilton found a way of obtaining it by solving a certain partial differential equation; in fact *two* such equations had to be solved. In practice, it will happen only under specially simple circumstances that we can actually obtain Hamilton's "characteristic function" in explicit mathematical form; but the existence of this function and the exploration of its properties is of greatest significance.

A few years later, he succeeded with a similarly major discovery in the realm of mechanics. The connecting link was the "principle of least action," which already played such a fundamental role in the Méchanique Analytique of Lagrange (which came out fifty years earlier). In the realm of optics a similar principle had existed since the days of Fermat, called the "principle of shortest time." Hamilton realized that his "characteristic function" was not a specific property of optics, but something that can always be constructed if a minimum principle is involved. (Let us not forget that the same happens in geometry, where the "distance between two points" is obtained by finding the shortest communication between those two points.) Hence, it is understandable that Hamilton could carry over his partial differential equation from optics to the field of mechanics and there again establish a fundamental function, which he now called "principal function." In this reinterpretation, he discovered a new form of the mechanical equations of motion which were first formulated in full generality by Lagrange; but the Hamiltonian formulation was actually of universal significance, destined to play a vital role in the physics of the twentieth century. These investigations were published by the Irish Academy in the years 1834–1835, under the title "A General Method of Dynamics." On the basis of these investigations, it was now possible to work out problems of geometrical optics and problems of mechanics interchangeably. An optical problem could be reformulated as a problem of mechanics, and any mechanical problem had an optical counterpart.

Hamilton's great papers on rays in 1827, on dynamics in 1834, and later on quaternions in 1843 contain his three most fundamental discoveries, around which his other mathematical ideas were clustered. Peculiarly enough, his early researches, for which he is so justly famous today (all done before his thirtieth year), were not properly recognized in Hamilton's own time. The German mathematician, Jacobi, was in fact the only scientist of high rank who wholeheartedly adopted and further developed the Hamiltonian dynamical method. Jacobi was much more than a mere mouthpiece of Hamilton. He gave a much broader interpretation of Hamilton's partial differential equation by showing that any complete solution of this equation (without solving a second equation) contains the solution of the dynamical problem, although we have not obtained that special function that Hamilton called the "principal function." It was in fact this generalization of Hamilton's procedure which became so eminently fruitful for the later development of theoretical physics. Hence, it seems entirely justified that Hamilton's equation is now usually quoted as the "Hamilton-Jacobi partial differential equation."

It seems a peculiar paradox that the Director of Dunsink Observatory and the Royal Astronomer of Ireland has hardly done anything specific in the field of astronomy. Yet Hamilton was far from pursuing mathematics for its own sake. All his efforts were directed toward a deeper understanding of the physical universe. Today we would call him a "mathematical physicist." But the physicists of his own time considered his theories as an elegant mathematical scheme, which did not lead to any new physical results. The Lagrangian formalism was good enough for the solution of dynamical problems encountered in physics, and, for the majority of physicists, mathematics was no more than a mere tool, which should not be pursued beyond the minimum essentials. Under these circumstances, it is well understandable that the highly advanced and refined ideas of Hamilton fell by the wayside.

Even in the first quarter of the present century it was customary to treat Hamilton's work only as an appendix to the Lagrangian theory of mechanics, in the form of a brief and entirely inadequate survey (usually not understood by the teacher himself), and if the student of physics failed to see the point, this was not considered as very important, since the whole Hamiltonian theory seemed to be more mathematics than physics.

Hamilton and Quantum Theory

This changed suddenly in 1916, when Paul Epstein, a young student of A. Sommerfeld, succeeded with a brilliant theoretical explanation of a phenomenon called the "Stark effect," i.e., the splitting of the spectral lines due to the presence of a strong electric field. Hamilton's method, modified by Jacobi, demanded the solution of a difficult partial differential equation, which seemed a practically insuperable task. It so happened that in the case of the Stark effect an exceptionally lucky circumstance prevailed. It was possible to separate the variables, with the result that Hamilton's equation was reducible to a few easily solved ordinary differential equations. That was not all. The second much more decisive circumstance was that exactly in the separable case, one could properly formulate those peculiar and rather mysterious "quantum conditions," which seemed to operate inside the atom, according to the model designed by the great Danish physicist, Niels Bohr. Thus it emerged that one could calculate with full accuracy the very intricate motions that a tiny electron performed around the hydrogen nucleus when submitted to an external electric field. The results of the calculations were in excellent agreement with the spectroscopic observations, thus demonstrating the soundness of Bohr's atomic model. At the same time, Hamilton's fundamental researches came suddenly into the focus of interest and were lifted from that relative obscurity which surrounded them for almost a century.

This was only the first instance in which the quickly developing quantum theory benefited from Hamilton's work. The French physicist, de Broglie, was struck by the remarkable analogy that Hamilton discovered between optics and mechanics. He succeeded, with a remarkably simple interpretation of the mysterious quantum conditions, by assuming an optical field accompanying the atom, the so-called "matter waves." Still more fundamental was the discovery of Schrödinger in 1925, who took cognizance of the optical nature of the matter waves and described them with the help of the celebrated Schrödinger equation. It is no accident that Schrödinger repeatedly and emphatically referred to his outstanding teacher at the University of Vienna, Fritz Hasenöhrl. Hasenöhrl was one of the few theoretical physicists who had recognized the importance of Hamilton's work and gave full account of it in his lectures. Thus, it was completely in line with Schrödinger's theoretical background to take de Broglie's geometrical optics and change it into a physical form of optics, with the result that he arrived of necessity at his famous equation.

Algebraic Concepts

Hamilton, himself, did not develop further his early dynamical investigations. From his thirtieth year, he drifted more and more into purely algebraic subjects and when, in 1843, he discovered the calculus of quaternions, he considered this achievement of such fundamental importance that he devoted the rest of his life, i.e., the time between 1843 and 1865, to the exploitation of this discovery.

Let us briefly survey the stand of algebra as it existed in Hamilton's time. It seems almost self-evident to assume that the development of science follows a natural order in the sense that the simple comes before the complicated and the more fundamental before the detail. And indeed, if we watch the history of geometry, we observe that the natural order of things is preserved. Euclidean geometry comes before non-Euclidean geometry and that again precedes Riemannian geometry. If our aim is to introduce the concepts of geometry on an elementary level, we can hardly do better than follow closely the historical development. The Greeks succeeded in establishing the pattern of an exact science in their geometrical investigations and that pattern has never been abandoned. We start our discussions with definitions, axioms, and postulates. Then, by logical arguments, we draw conclusions from the basic postulates. In every new step we make use of what has been found before and we thus erect an imposing building by putting brick upon brick. Algebra is also an exact science and, in fact, we know since Descartes' analytic geometry that algebra and geometry are but two different languages for the same essence. Hence, we might think that algebra developed similarly to geometry, starting with definitions, axioms, and postulates, and continuing from there in a purely logical manner. Strangely enough, the history of algebra shows a completely different picture. Even today, if we look around in our schools and examine the way in which algebra is taught, we find to our amazement that, while geometry is usually pursued in a more or less logical fashion, the procedure in algebra is totally different. The student learns about fractions, negative numbers, rational and irrational numbers, he memorizes rules about the handling of parentheses and the multiplication of positive and negative numbers, yet he is constantly groping in the dark and his knowledge is not more than the mere remembering of rules. It is difficult to understand why this pattern of teaching continues, when in actual fact the processes of algebra could be made just as comprehensible and logically satisfactory as the theorems of geometry.

The greatest surprise is that the clear realization of the true significance of algebraic operations came so late. Although algebra was invented by the Hindus almost five hundred years ago, algebra, as we understand it today, is not more than one hundred years old. It is quite astonishing to see that Hamilton was struggling with such things as the nature of negative numbers—something one can easily explain to a student of a secondary school. Hamilton liked to ask fundamental questions and deal with these questions in his own manner, unhampered by the dictum of other scientists, but often hampered by philosophical and even metaphysical speculations, which makes the reading of his publications sometimes excessively tiresome.

It seems, in particular, that the German philosopher, Kant, had a strong influence on his thinking. Kant's philosophy assigned to space and time a particularly important role as the two fundamental categories of our cognition. Now space is reflected in geometry, which by its very nature is the science of space. What about the other category, time? Is it possible that the duality of space and time is reflected in the duality of geometry and algebra? In that case one should conceive algebra as the "Science of Pure Time." This peculiar and perhaps misguided thought had a fascinating hold on Hamilton, started him on his algebraic investigations, and undoubtedly led to very constructive results. An immediate consequence was that he took exception to the so-called "imaginary numbers," such as the square root of minus one, usually denoted by the symbol i: $\sqrt{-1} = i$. Such numbers and the so-called "complex numbers," which are constructed with the help of these imaginary numbers, could not be subordinated to the flow of time. Hamilton found, however, a solution by considering them as a *number pair*, composed of two ordinary (real) numbers. Exactly the same idea occurred several years before Hamilton to the great German mathematician Gauss. The remarkable feature of these number pairs was that all the ordinary laws of algebra, which hold for real numbers, could be transferred to these more elaborate numbers, by using a single symbol, let us say c, for their notation.

These laws, called the basic "postulates of algebra," look harmless enough and we might think that they are self-evident, if we restrict ourselves to the case of whole numbers. But algebra operates with a much wider class of numbers than mere integers. To all these numbers we extend the laws which operate in the case of integers (whole numbers). We accept these basic laws as postulates. They are just as basic for the operations of algebra as Euclid's postulates are for the constructions of geometry. We do not try to *prove* the postulates. We accept their validity and base our conclusions on the assumption that the postulates hold.

Can we be sure that these postulates are free of inner contradictions? The answer is No; such a proof has never been given. We know, however, that if the postulates of Euclid's geometry are free of contradictions, then the same is true of the postulates of algebra, which, in disguised form, express some simple facts of geometry.

Although the basic ideas of algebra were invented by the Hindus as early as the sixth century A.D., and the operations of algebra have been in use for hundreds of years, the realization that the operations of algebra are based on these fundamental postulates came very late. This is best documented by the fact that the word "associative law of algebra" was coined by Hamilton, although the words "commutative" and "distributive" existed before. Gauss certainly gave extensive attention to the axiomatization of algebra, but did not publish his results.

Hamilton was much impressed by the idea of a "number pair." A pair of numbers could be handled algebraically as one single number, called a "complex number." Such a number could be represented by a point in the plane. In Descartes' analytic geometry a point of the plane has two coordinates a and b. The fact that these two numbers could be united in the form $c = a + bi$, had the great advantage that *two* quantities could be replaced by *one* and where Descartes needed *two* equations, now *one* would be enough. What great simplification is thus obtained in our calculations! And that is not all. The coordinates of Descartes are ingenious constructions; but we have to set up axes, which do not belong to the problem, and we have to deal with these two numbers a and b, although they have not more than accidental significance. If we can dispense with these coordinates, then we come nearer to the real spirit of the problem by not bringing in anything from the outside and operating only with elements which are in the nature of the problem.

Now physical space has three and not two dimensions. It was a natural idea—and Hamilton was not the first one to attempt it—to try something similar for space. A point in space needs *three* coordinates for its characterization, usually called length, width, and height. We have to set up *three* mutually perpendicular axes. For example, if an airplane flies over Dublin, it is not enough to give the geographic longitude and latitude of Dublin, we also have to say, how *high* the airplane flies above the ground. Thus we need *three* numbers—or, as Hamilton called it, a "number triplet"—to characterize the position of the airplane in space, at a certain instant. If before we have talked of a number *pair,* we now have to talk of a number *triplet.* If, before, our "complex number" c could be written in the form $a + bi$, we might think that perhaps we can invent a "hyper-complex number" q by putting $q = a + bi + cj$, where j is a new imaginary unit, to be added to the previous imaginary unit i. Hamilton was completely fascinated by this idea and tried his luck for years, without arriving at the desired goal. He was so imbued with these speculations that he could not refrain from talking even to his children about them and frequently his sons teased him at the breakfast table, receiving him with the greeting: "Daddy, can you multiply triplets?" to which Hamilton had to answer ruefully: "I can add and subtract triplets, but I cannot multiply them."

Today, we know that Hamilton's program as originally conceived is unsolvable. The eminent algebraist Frobenius showed, in 1878, that no numbers exist beyond the ordinary complex numbers, which could satisfy all the postulates of ordinary algebra. The hyper-complex numbers as originally envisaged by Hamilton, simply did not exist.

Quaternions

After years of fruitless groping in the dark, the solution came to Hamilton suddenly, in a flash of inspiration. He describes vividly, in a letter to his son how the discovery was made. (This letter was written many years after the event.) It happened in October 1843 that he was walking from the Observatory to town, in order to preside at a Council meeting of the Academy. His wife was walking with him and tried to make conversation, but in all probability he gave the wrong answers, since his mind was glued on the multiplication problem of triplets. Suddenly he had an illumination which showed him at what point he went astray in his previous speculations. *Two* imaginary units were not enough, *three* were in fact needed: i, j, and k. These three units had to play a completely symmetric role, to take cognizance of the fact that there are no preferential directions in space, one axis being just as good as any other axis. Hence, the number triplet had to appear in the form $ai + bj + ck$; but to this an ordinary number d has to be added, so that a full-fledged quaternion (this name was already settled in his mind, when he arrived at the Academy) must appear in the form

$$q = ai + bj + ck + d.$$

It was not enough to operate with a three-dimensional world, one had to go into a world of *four* dimensions, in which the position of a point is characterized by *four* numbers: a, b, c, d.

It is astonishing to see how the quaternions of Hamilton foreshadowed our four-dimensional world, in which space and time are united into a single entity, the "space-time-world" of Einstein's Relativity. Today, we are inclined to call d the "time-part," and $ai + bj + ck$ the "space-part" of the quaternion q.

But this is not all. On that memorable occasion, Hamilton clearly recognized the fundamental multiplication laws that the three units i, j, k must satisfy. In analogy to the ordinary imaginary unit i, whose square is -1, the three Hamiltonian units satisfy the relations $i^2 = j^2 = k^2 = -1$. To this, however, has to be added one more relation, in the form

$$ijk = -1.$$

Hamilton, in his letter to his son, recounts that he had just arrived at Broome Bridge, when in a flash of insight he suddenly saw clearly in front of his eyes the basic multiplication laws of the quaternions and he adds: "Nor could I resist the impulse—unphilosophical as it may have been—to cut with a knife on a stone of Brougham Bridge, as we passed it, the fundamental formula with the symbols i, j, k, namely

$$i^2 = j^2 = k^2 = ijk = -1$$

which contains the solution of the problem, but of course, as an inscription, has long since moldered away."

As a consequence of the above formula, we can

immediately deduce the peculiar result that $ij = -ji$. It was in that moment that Hamilton realized that the problem of hyper-complex numbers is only solvable if we sacrifice one of the fundamental laws of algebra, viz. the *commutative law of multiplication*. For Hamilton's quaternions, it is not true that *ab* and *ba* are the same. Hamilton's quaternion algebra gave the first example of a consistent algebra, in which one of the fundamental postulates of algebra was violated. It happened quite similarly in the realm of geometry that in the first half of the nineteenth century one of Euclid's postulates was abandoned, without any logical upheaval. Hence, about the same time, Kant's assumption of the absoluteness of both geometry and algebra was shattered by showing that neither the postulates of geometry, nor the postulates of algebra are of a unique and absolute character.

That Hamilton kept the associative law and sacrificed the commutative law, demonstrates his excellent mathematical instinct, since the associative law is actually far more important than the commutative law. In 1878, Frobenius had shown that Hamilton's quaternion algebra is entirely unique inasmuch as there exists no other algebra which preserves all the postulates of algebra, with the exception of the commutative law.

Hamilton was so enthused by the discovery of the quaternions that he devoted the remaining twenty-two years of his life to the exploitation of this discovery. He thought that the quaternions would give the universal clue for the understanding of the physical universe. He demonstrated the usefulness of quaternion calculus by applying it to a very large number of geometrical and mechanical problems. He showed, for example, in what a simple and elegant manner one could obtain the formulae of spherical trigonometry with the help of the quaternions. Likewise, he showed how a simple quaternion equation leads to a strikingly beautiful theorem concerning the rotations of a sphere.

Robert P. Graves, the biographer of Hamilton, characterizes in an interesting way the contrasting scientific tastes of Hamilton and his friend and occasional rival J. MacCullagh, eminent professor of mathematics at Trinity College. If MacCullagh conceived a new idea, he first of all formulated in a few lines a brief summary of the things he thought could be done with the new concepts. Then he sat in front of the paper for hours, without adding another word. To stoop down and try out the idea on a simple case, that any undergraduate student of mathematics could have done, was below the dignity of a first-rank scientist. One had to move in a rarified air of sophistication and refinement, impressing the reader by ingenious tricks and surprising turns. Hamilton was of a totally different ilk. If he hit on a new inspiration, he was exuberant and immediately started to see how the general principles worked out in a special case, even if it was a case which was completely elementary and well investigated by other methods. This exuberance had its occasional drawbacks, by leading to excessive verbosity. It is difficult to see how the students of Trinity College could learn anything from his "Lectures on Quaternions," whereas a condensation of the same work to perhaps one tenth of the original volume could have resulted in a splendid textbook of the newly discovered quaternion calculus.

Hamilton was possibly mistaken in his belief that the quaternions can do everything under the sun. He imbued them with cosmic significance and thought that they would revolutionize the edifice of mathematical physics. Today, we believe that quaternions are exactly the right tools in all problems which involve some kind of rotation. Since quaternions belong to a world of four dimensions, they can be used with great success in problems involving rotations in four dimensions, to which Hamilton did not even apply them, since he could not foresee the unification of space and time which came through the discoveries of Einstein and Minkowski in the first decade of our century. The so-called "Lorentz transformations," which play such a fundamental role in modern physics and which are in fact rotations of the space-time-world, can be represented in the most elegant and adequate manner with the help of quaternions. Moreover, Maxwell's electromagnetic equations can likewise be written in an exceptionally simple fashion in terms of quaternions. Maxwell himself was well aware of the usefulness of quaternion calculus. But he broke up the quaternion operator into a space and time part, not realizing that the unbroken quaternion operator is in fact much more in line with his equations. This, however, could only be realized after special relativity came into existence. Even Dirac's famous equation of the electron can be formulated in quaternionic terms.

In view of these posthumous victories, we may ask whether the quaternions of Hamilton may not, after all, be the fundamental building blocks of the physical universe? This possibility stood open up to the time of 1916. But the great revolution which came through Einstein's general relativity, opened a new chapter of mathematical physics, which required tools with which Hamilton's quaternions could not cope. The theory of relativity demonstrated that the fundamental building blocks of the universe are not vectors but "tensors." Tensors cannot be handled in terms of vectors. Tensor calculus demands a form of algebra which goes far beyond the world of vectors. It is a curious historical coincidence that exactly at the time of the discovery of quaternions, around the years between 1840 and 1844, a little known but unusually ingenious German mathematician with the name of Hermann Grassmann developed a form of algebra which he called "Ausdehnungslehre," i.e., the Theory of Extension. Grassmann's book, published in 1844, was overgrown with unnecessary philosophical contemplations, which made it practically unreadable. A second edition in 1862 was not much more successful. In fact, he developed in his work a space structure which went much deeper than Hamilton's own algebraic investigations. Grassmann's ideas were of too great generality, but it was he who first introduced the "inner product"—now called "scalar product"—of two vectors, and the "outer product"—now called "vector product"—of two vectors. Furthermore, the entire scheme of tensor calculus fits easily into Grassmann's calculus. We may wonder, how that is possible in view of our previous claim that Hamilton's quaternions are the *only* example of a noncommutative algebra. The answer is that Grassmann's algebra is less complete than Hamilton's, because he sacrificed not only the commutative law of multiplication, but another commonly accepted postulate of algebra, according to which zero has no factors. If the product of two numbers gives zero, this is only possible if at least one of the two factors is zero. This is true also in Hamilton's algebra. It does not hold, however, in Grassmann's algebra, where the product of two numbers may come out as zero, although neither of the two numbers vanishes. Under such conditions the operation of division has to be sacrificed. In Grassmann's algebra we can add, subtract, and multiply, but we cannot divide. In Hamilton's algebra all the four operations of algebra exist: addition and subtraction, multiplication and division.

For the practical purposes of physics and engineering, the operations with quaternions were not sufficiently flexible. The eminent physicist J. W. Gibbs in the U.S. (and O. Heaviside in England) simplified the operations with quaternions by making the whole calculus more pliable, but at the cost of the theoretical beauty of the original scheme. The quaternion product of two ordinary vectors resulted in a quaternion which had a time part and a space part. Now the time part was taken separately and called "scalar product" or "dot product," while the space part, taken separately, was called "vector product" or "cross product." Hence, *two* operations took the place of the single Hamiltonian product. The new form of vector algebra and vector analysis had the added advantage that it is much more closely related to the demands of tensor calculus than Hamilton's original scheme. Moreover, Hamilton himself was often compelled, in the applications of quaternion calculus, to separate the time part and the space part of the quaternion. This separation is more natural in the Gibbs' type of operations.

Hamilton's Other Activities

Hamilton has often been criticized for the exaggerated views he held about the importance of quaternions. Why did he not apply himself to something more useful in the last twenty years of his life? Similar charges have been brought against other great men of science. Many physicists believe that Einstein wasted the last thirty years of his life by his efforts directed toward a unified field theory. The great Newton spent many years on theological disputations, which to us appear completely senseless. But who are we to tell the genius what he should or should not do? The genius is not free to act because he is relentlessly driven by his "daimonion," as Socrates expressed it. Is what Hamilton accomplished in the span of a lifetime not enough? His optical and dynamical investigations were prophetic and foreshadowed the quantum theory of our days. His quaternions foreshadowed the space-time world of relativity. The quaternion algebra was the first example of a non-commutative algebra, which released an avalanche of literature in all parts of the world. Indeed, his professional life was fruitful beyond measure.

We should not forget, however, in the evaluation of Hamilton that he was much more than a professional scientist; and certain features of his scientific career cannot be properly understood without coordinating his scientific achievements to the entire field of activities that his life encompassed. We have seen before that he was a splendid linguist who mastered Latin, Greek, and Hebrew as a child. He retained these abilities throughout his life, because of his fabulously retentive memory. His love of literature was profound. Occasionally he said that, although he made his living as a mathematician, in his heart he was a poet. His friendship with the poet Wordsworth was based on mutual admiration, and Wordsworth confessed that he felt inferior only in the company of Coleridge and Hamilton. He wrote a large number of sonnets which were of high order according to the taste of his time, although measured by present standards they appear wooden, pompous, and void of genuine poetic inspiration. He was likewise interested in religion and theology and was well-versed in philosophical and metaphysical speculations. Thus, we see in him a man of broadest education and many interests, who differs markedly from the highly specialized scientist of our own days. In the world in which he lived, it was still possible to pursue science without the heavy obligations which plague the scientist of today. At the incredibly young age of twenty-two he was made Director of the Dunsink Observatory and, in recognition of his exceptional gifts, he was exempted from the usual duties of a professor. He had practically all the free time he wanted for his research and could choose the subjects of his investigations without qualms of conscience. Nor did the pressure exist to publish all the time, since the craze of measuring a scientist's value by the number of his publications did not yet exist. Consequently, Hamilton could work unhampered by outside pressure and for his recreation he could read profusely. It is indeed probable that he read more in the fields of literature and philosophy than in the field of mathematics, since he had the urge to work out his mathematical discoveries from first principles, rather than to follow other people's ideas. These were work conditions which fitted his personality to perfection.

We are all tainted, of course, by the time in which we live. Hamilton lived in the Victorian era, which in so many points contrasts with our present historical epoch. One lived in the best of all possible worlds, which one tried to improve by virtue and valor. The Empire set heavy obligations on the shoulders of a gentleman. One was saddled by the ghosts of honor and dignity. One was faced with the serious task of contributing, to the best of one's abilities, to the glory of the nation. One aspired to the stars and loved the grand gestures; (Disraeli: "I lay India at your feet, Majesty"). Thus, one had to walk around on cothurns like the tragic actors on the Greek stage. If one visited the local "pub" on the way to an Academy meeting and arrived there in somewhat animated mood, the danger existed that one might have ruined one's reputation forever. Under these circumstances little place was left for laughter and humor. Hamilton had a pet dog, which sometimes indulged in mischief. Like all young pups, he loved to chew up things and when he found a book around, he had to try his teeth on it. Unfortunately, it so happened that the book was the Bible, of whose holiness the dog was apparently unaware, but this was more than a Christian gentleman could tolerate. Hamilton lost his temper and gave the poor dog a sound thrashing.

But are we so much better? Will not people in a hundred years time laugh about *our* foibles and immaturities? The American mathematician E. T. Bell wrote a collection of highly romanticized and partly fictionalized biographies of great mathematicians (published under the title "Men of Mathematics"), which had great popular success.

The chapter on Hamilton is entitled "An Irish Tragedy." If we investigate what was so tragic in Hamilton's life, we find first of all the usual charge that he overestimated the value of quaternions by spending twenty years of his life on this limited subject. But Hamilton had so many fundamental discoveries to his credit that this somewhat one-sided predeliction would hardly deserve the name "tragedy." Hamilton was no grabber, he did not want to do everything himself. He was satisfied with doing a few fundamental things extremely well. Then we hear about the usual domestic difficulties, so frequent in the lives of men of imagination—and lack of imagination; but to pry into the emotional conflicts of a great man of science cannot be our business. From the intellectual viewpoint, we observe that Hamilton overcompensated for his lack of marital happiness and threw himself into work with double intensity. What remains then, are his occasional drinking bouts, and that is something that we would hardly call a "tragedy" here in Ireland.

We have no reason to assume that Hamilton's life was not on the whole a happy life. There is, however, a point which is worth discussing. This is his remarkable isolation in intellectual matters. He had little contact with his immediate colleagues. Living in the country at the Dunsink Observatory, and seldom coming to town, except to the occasional meetings of the Academy, he did not see much of his colleagues at Trinity College. The twelve lectures he had to deliver annually did not contribute much to an exchange of scientific ideas. To this has to be added his personal misfortune—entirely unaided by his own efforts—that he was knighted by the British Crown in his thirtieth year—certainly a distinction which hardly increased his popularity with his Irish colleagues.

It is true, of course, that he had an elaborate correspondence with people of all walks of life—poets, writers, theologians, philosophers, humanists, and scientists—but there was not a single first-class mathematician among his correspondents. His exchange of letters reminds one of the dialogues of Plato, which are in fact essentially monologues, since the other participants are to a large extent yes-men, who constantly praise the wonderful wisdom of Socrates, without contributing much of their own. The result was that Hamilton worked in almost complete isolation. He did not try to coordinate his results to the current stream of science. And yet, so many outstanding names were engaged in analysis about the same time, for example, Gauss and Riemann and Weierstrass in Germany, Cauchy, Legendre, Liouville in France, and many others. Hamilton took little interest in their efforts. If somebody pointed out to him certain analogies between his own discoveries and those of others, he was satisfied with defending his own priority in a few footnotes, without examining the question of how his approach compared with that of others. This, of course, is well understandable. Fame and glory were the primary prerequisites of the Victorian times. (This craving for fame was entirely *l'art pour l'art,* without pragmatic overtones. The Nobel prize was not yet established and famous physicists were not offered yearly salaries of $100,000.) The question of priority was thus a matter of utmost importance.

And yet, there lurked in the dark a dangerous rival, whom Hamilton did not even suspect. Gauss, the prodigious genius, anticipated all the mathematical discoveries of his time and threw his shadow ahead up to a hundred years. Gauss was aware of the beautiful theorem concerning the rotations of a sphere that we have quoted above. He was familiar with the idea of number pairs and number triplets, and he came across the fundamental rules of quaternion calculus as early as 1819, about 25 years before Hamilton. Fortunately for the rest of humanity, Gauss did not care to publish. While other people rushed to publication when they found something new, Gauss tucked away his remarkable discoveries in a black notebook for further exploration, without any desire to get recognition on the part of his contemporaries. But as far as Hamilton is concerned, nothing is subtracted from his merits by the fact that somebody found the basic ideas earlier, since Hamilton made his discoveries quite independently and applied them to a wide range of problems in the most versatile fashion.

In conclusion, we have little reason to line up with those critics who objected to Hamilton's too heavy involvement with the formal intricacies of the quaternions, nor with the panegyrics of the

Scotch mathematician P. G. Tait, who wanted to solve the riddles of the world by quaternions. If we survey the remarkable achievements of this great man of science, we see a man of extraordinary talents and extraordinary personality. A man of so many gifts and interests could easily have been lost to the world, had he not received the proper recognition and appreciation for his unique qualities. It is no accident that a small country like Ireland provided that recognition. We may well ponder, whether any other country but Ireland, with its unhurried and individualistic intellectual atmosphere, could have provided the proper background to a man so self-willed and autistic as Hamilton, a man who had to fight it out with his own demons, a man who loved to walk on the lonesome paths of country roads and yet craved for the company and recognition of his fellowmen. Ireland's part in the successful completion of Hamilton's scientific endeavors is ever-present.

John Edensor Littlewood

J. C. Burkill

John Charles Burkill, FRS, was born in 1900 and received his Science Doctorate from Cambridge University. He was the Master of Peterhouse, Cambridge, and a Fellow of the Royal Society of London.

J. E. Littlewood, along with G. H. Hardy, was instrumental in establishing a school of pure mathematics in twentieth-century Great Britain. Compared with the mathematics of continental Europe, nineteenth-century British mathematics was rather barren of significant figures and was most emphatically subordinate to the natural sciences in practice. That it experienced a modest renaissance beginning at the turn of the century is in large measure due to the efforts of a few individuals such as Littlewood. Since Hardy and Littlewood were collaborators the reader will enjoy comparing this biography with the contrasting flavor of Mordell's article describing Hardy (following this article).

A mathematician is certainly more than the theorems he proves. Yet so often a mathematician's biography is reduced to a list of significant theorems. Burkill's biography allows us to see Littlewood as a human who happened to be doing mathematics as he lived a full and varied life.

IN 1900 PURE mathematics in this country was at a low ebb. Since the days of Newton mathematics had come to be regarded as ancillary to natural philosophy. In the nineteenth century this attitude had been confirmed by the prestige of Stokes, Clerk Maxwell, Kelvin and others. On the continent of Europe the nineteenth century was as fruitful in pure mathematics as England was barren. The central property of functions of a complex variable was found by Cauchy, and further light was shed on the theory by Riemann and Weierstrass. France, Germany and Italy had many pure mathematicians of the first rank. The leading British scholars, notably Cayley, had been solitary figures and had not led young men into research.

After 1900, the principal architect of an English school of mathematical analysis was G. H. Hardy (1877–1947). In strengthening the foundations and building on them he found a partner in J. E. Littlewood (1885–1977). The inspiration of their personalities, their research and their teaching established by 1930 a school of analysis second to none in the world.

1885–1900: Family

John Edensor Littlewood was born at Rochester on 9 June, 1885, the eldest son of Edward Thornton Littlewood and Sylvia Maud, daughter of Dr. William Henry and Sophia Ackland (*née* Lott). In recent reminiscences J. E. L. wrote that the name Littlewood is not uncommon in the North of England; his ancestors were squires of Baildon Hall

Source: J. C. Burkill, "John Edensor Littlewood," *Bulletin of the London Mathematical Society* 11 (1979): 59–70. Reprinted by permission of the author and The London Mathematical Society. (Some references omitted.)

near Bradford, and a platoon of Littlewood archers fought at Agincourt. More recent Littlewoods have been farmers, land-owners, ministers, schoolmasters, printers, publishers, editors and doctors.

J. E. L.'s grandfather, the Reverend William Edensor Littlewood (1831–86) (named after his grandmother Sarah Edensor of Edensor in Derbyshire), went to Pembroke College, Cambridge. Having been thirty-fifth wrangler in the Mathematical Tripos and a third class Classic, he won the Chancellor's medal for English verse in 1851. He became a theologian and a schoolmaster. He is entered in the *Dictionary of National Biography* as a miscellaneous writer; a representative title (1876) is "Lovely in their lives, a book for earnest boys."

Edward Thornton Littlewood (1859–1941), the eldest son of W. E. L., went to Peterhouse and was ninth wrangler in the Mathematical Tripos of 1882. He later accepted the headmastership of a newly founded school at Wynberg near Cape Town, taking his family there in 1892.

J. E. L.'s reminiscences run:

> He was also offered a Fellowship at Magdalene College, but quixotically felt bound to the Cape post. Our lives would have been very different in Cambridge. There is much to be said for being born in the purple, and children of dons acquire an easy self-confidence, much as Etonians do. Incidentally there is now a good deal of intermarriage in dons' families, presumably of high eugenic value. As it was I had a happy childhood among mountains, the ocean, and a beautiful climate. Education was inevitably meager—adequate staff in a Colony was then very hard to get. I remember failing in arithmetic (my only failure in examinations), and that algebra as taught was quite unintelligible. I had a period at the University of Cape Town, but my father soon realized that I was doing no good there, and I left the University at 14 to go to St. Paul's School in England.

E. T. Littlewood stayed at Wynberg until he retired in 1920, when he and his wife came to live in Trinity Street, Cambridge. Their second son, Martin Wentworth, qualified in medicine at St. Thomas's Hospital and, in 1913, joined his uncle Charles Kingsley Ackland in practice at Bideford. He returned there after war service 1914–19 succeeding a line of four Drs. Ackland. Martin Littlewood's elder son, Mark John, continued the medical tradition as a doctor on the staff of the British Petroleum Company. The third son of E. T. Littlewood died as a boy of eight by falling off a bridge into a lake.

1900-1911: A Mathematical Education

This is the title of the only segment of autobiography that Littlewood put into print. It is tantalizingly short. . . . It starts: "It is *my* education. It illustrates conditions before 1907, but has some oddities of its own."

The force and economy of the style preclude either paraphrase or précis.

Littlewood was at St. Paul's from 1900 to 1903 during the reign of the famous High Master, F. W. Walker. The school's reputation for excellence in classics and mathematics was outstanding. From 1885 to 1911 the master of the highest mathematical form was F. S. Macaulay, a creative mathematician who was elected F.R.S. in 1928. The boys in that form were encouraged to work independently and by discussion with one another. Anyone in a difficulty could go up to Macaulay, but on the whole it was not done. This was education with a university atmosphere; at no school could Littlewood have acquired a better foundation of self-reliance, knowledge and judgement. He understood uniform convergence, and he could discriminate between basic ideas and tricks of manipulation. He got an Entrance Scholarship at Trinity College, Cambridge, in December 1902 and went into residence in October 1903.

Let him continue the story: "I coached for Part I of the Tripos with R. A. Herman, contemporary and friend of my father, and the last of the great coaches." (The reform of 1910, abolishing the order of merit, drove the coaches out of business.) "To be in the running for Senior Wrangler one had to spend two-thirds of the time practising how to solve difficult problems against time. . . .

The old Tripos and its vices are dead horses; I will not flog them. I do not claim to have suffered high-souled frustration. I took things as they came; the game we were playing came easily to me, and I even felt a sort of satisfaction in successful craftsmanship." In 1905 Littlewood was Senior Wrangler, bracketed with J. Mercer (Trinity), who had graduated at Manchester before coming to Cambridge.

Part II of the Mathematical Tripos, in which Littlewood was placed in Class I, division 1 in 1906, dealt in genuine mathematics of a formal and traditional kind (e.g. the special functions of analysis, conformal representation, differential geometry).

Littlewood started research in the Long Vacation of 1906. His tutor E. W. Barnes suggested the subject of integral functions of zero order. The pupil recalls that he "rather luckily struck oil at once (by switching to more elementary methods)." . . . Barnes suggested as the next problem—prove the Riemann hypothesis. As Sir Peter Swinnerton-Dyer said in his address at the memorial service to Littlewood (1977): "It is an amazing illustration of the isolation and insularity of British mathematics at that time that Barnes should have thought it suitable for even the most brilliant research student, and that Littlewood should have tackled it without demur."

Characteristically Littlewood is able to report: "As a matter of fact this heroic suggestion was not without result . . . there was a consolation prize." He would encourage his future pupils: "Try a hard problem. You may not solve it, but you will prove something else."

From October 1907 to June 1910 Littlewood was at Manchester University as Richardson lecturer (a special lectureship with a stipend of £250 instead of the usual £150 or £120). In retrospect he doubted the wisdom of the move, as he had the option of staying in Cambridge as a research scholar. Young mathematicians in later years, at such a parting of the ways, were emphatically warned: "Your days of leisure are going, never to return." The work-load at Manchester may have justified another recurrent observation: "Young men of today don't know what work is."

Littlewood won a Smith's Prize in 1908 and was elected a Fellow of Trinity in the same year, returning in 1910 to succeed A. N. Whitehead on the Trinity staff.

The last page of *A Mathematical Education* is a fascinating recollection of the discovery of the proof of the converse of Abel's theorem (1911).

> If $\Sigma a_n x^n \to s$ as $x \to 1$ from below and $a_n = O(1/n)$, then Σa_n converges

and concludes with "On looking back this time seems to me to mark my arrival at a reasonably assured judgement and taste, the end of my 'education.' I soon began my 35-year collaboration with Hardy."

The Hardy-Littlewood Collaboration

In 1911, Hardy and Littlewood had a number of common interests which each of them would be impelled to pursue. To name three, there were (1) summability *(C, k)*, Abelian and Tauberian theorems, (2) Diophantine approximation with its applications to function-theory, (3) the theory of numbers—the challenges to analysts underlined by Landau's inimitable Primzahlen (1909). If the two men were disposed to collaborate, the possibilities were boundless.

What did each of them bring to the partnership? Of both it is true that mathematical creation was the ruling passion of their lives. Their personalities were sketched by C. P. Snow, who in Hardy's later years was close to him. He pictures the two scrutinizing the manuscript sent to Hardy in January 1913 by Ramanujan to decide whether or not it was the work of a man of genius. There sat Hardy "with his combination of remorseless clarity and intellectual panache . . . : Littlewood, imaginative, powerful, humorous, . . . , a good deal more *homme moyen sensuel.*" People who knew both men believed that Hardy's deepest satisfaction came from the beauty of the mathematical patterns that he created or contemplated, whereas Littlewood found fulfilment in solving resistant problems. In 1943 he was awarded the Sylvester medal of the Royal Society with the citation "Littlewood, on Hardy's own estimate, is the finest mathematician he has ever known. He was the man most likely to storm and smash a really deep and formidable problem; there was no one else who could command such a combination of insight, technique and power."

This memoir would be incomplete without the canonical version of the four axioms of the Hardy-Littlewood collaboration as recounted in a lecture by Harald Bohr on his sixtieth birthday (*Collected Works,* vol. 1, p. xxviii).

1. When one wrote to the other, it was completely indifferent whether what they wrote was right or wrong.
2. When one received a letter from the other, he was under no obligation to read it, let alone answer it.
3. Although it did not really matter if they both simultaneously thought about the same detail, still it was preferable that they should not do so.
4. It was quite indifferent if one of them had not contributed the least bit to the contents of a paper under their common name.

The axioms contemplate collaboration by letter, and there is no doubt that departures from this practice were rare.

Normally Littlewood would make the penultimate version of a paper, with a skeleton of all the essential mathematics, simplifying and abbreviating in notation clear to Hardy. Hardy would add what they called the "gas" and write the paper in the elegant prose of which he was a master. Littlewood's own style, in its clarity and brevity, was equally magisterial. . . .

1911–19

A glance at the bibliography shows the immediate and abundant success of the Hardy-Littlewood partnership. Many of their early papers cover Diophantine approximation with important applications to the theory of functions. Other papers give a systematic treatment of summability of series. Tauber, fortunate in his name forming a euphonious adjective, became immortal with Archimedes, Newton, Euler, Gauss, Abel. . . . Tauberian theorems quickly acquired interest by providing an elegant approach to the prime number theorem. The results of Hardy and Littlewood on special expansions such as power series, Dirichlet series, Lambert series sufficed for applications until a general setting of Tauberian theorems was promulgated by N. Wiener (1932), simplified and extended by S. Bochner and H. R. Pitt.

Everything that Hardy wrote with Littlewood is reproduced in his *Collected Papers*. The contents of those volumes are arranged by subject. The appropriate editor has written an introduction to the papers on each subject, followed by illuminating remarks on the separate papers. As all this is available, the policy in this memoir is to give most space to comment on researches by Littlewood alone or with mathematicians other than Hardy.

In 1912–13, H. Bohr stayed in Cambridge for some time, working intensively with Littlewood on deep properties of the zeta function. Bohr wrote:[1]

> I have to add, that although we succeeded in preparing the complete manuscript, we were so exhausted afterwards that we did not have the strength to send it to the printer, and so it was left for a number of years until, at a later date, when the theory had been developed so much further, we turned it over to two younger English mathematicians, Titchmarsh and Ingham, to use freely in the preparation of their two excellent booklets in the series *Cambridge Tracts* about the subjects in question.

From 1914 to 1918 Littlewood served as 2nd Lieutenant in the Royal Garrison Artillery. After a stint of gun drill in which the opportunities of riding were congenial, he was seconded early in 1916 for ballistic work under the Ordnance Committee at Woolwich.

The mathematical problems of gunnery were neither formidable nor aesthetic, and they would have repelled Hardy, but Littlewood tackled them cheerfully and was able to reduce the man-hours of calculations needed in the pre-electronic age. In the R. S. *Obituary Notice* of R. H. Fowler (vol. V, 1945–48, pp. 61–78) E. A. Milne records that Littlewood devised a rapid and powerful method for the calculation of high-angle trajectories avoiding the painfully long process of "small-arc" integration. At the subsequent anti-aircraft trials "to the astonishment and joy of all concerned the observed positions of the shellbursts fell exactly on Littlewood's trajectories, at the correct time-markings, within very small errors of observation."

The higher brass at Woolwich recognized that Littlewood should not be subjected to routine chores or petty restrictions. He could "live out" and make his home with friends living in London. If, in uniform, he carried and used an umbrella, it would not be seen.

From the extent of the bibliography of (say) 1915 to 1922, it is plain that ballistics did not fill all of Littlewood's working hours during the war years.

1919–28

In 1919 and 1920, undergraduates and graduates in Cambridge could hear lectures from Hardy, a course on partitions including his work with Ramanujan, and from Littlewood on real and complex analysis. Hardy, with his physical grace and finely cut features, wrote on the board in his stylish handwriting from a well prepared script. Littlewood was less dependent on systematic notes, and there were interpolations, with the occasional shock of a hiatus and consequent improvisation. The enthusiasm of both the masters is a lifelong memory with their pupils. Of course a high proportion of those aspiring to research were attracted to analysis. Some, starting as analysts, turned later to other fields. For example M. H. A. Newman went on to modernize the teaching of algebra in Cambridge and introduced topology.

The pupils and colleagues who have known Littlewood longest date from the 1920s. An effort must be made to convey an impression of the man who inspired admiration, awe and, above all, affection.

Littlewood was slightly below average in height, strongly built and agile. At school he had been one of the best gymnasts and a hard hitting batsman. Like Hardy he was a keen follower of ball games and watched cricket at Fenner's on summer afternoons. He had an intense interest in music (classical—particularly Bach, Beethoven and Mozart); he had taught himself as an adult to play the piano—later he had a stereo gramophone with a large collection of records. An athlete with a sense of rhythm, he and Marie Stopes were the best dancers at Manchester University.

He was best known to unmathematical undergraduates at Trinity for his skill in circling the seven yards of a pillar of the Library on the narrow ledge of its base and for his daily walk across the court to the baths, with a towel but no shirt. On most days he walked many miles in the country.

Littlewood could maintain a long term cordial association with a man (or, presumably, with a woman) on the implicit assumption that mutual understanding and harmony could coexist with complete independence. The reader will observe that it was just this assumption which was central to the Hardy-Littlewood axioms for collaboration in research.

He would dread the impending advent of an insensitive visitor (say a mathematician other than a personal friend) who might count on long stretches of his company. I acquired inordinate credit for having collected from his rooms on successive days a foreign mathematician and diverted him for hours with walks and films.

Littlewood belonged to a generation before Christian names supplanted surnames, and so *a fortiori* did Hardy. Of all Littlewood's relationships with men, that with Hardy must have been one of the closest. It is inconceivable that either would have used the Christian name of the other. Their letters would begin "Dear H" and "Dear L," if there was any allocution at all. A typical letter in 1920 began "Quite a rush of ideas this morning . . . ."

With the later widespread use of Christian names, perhaps four or five of more than a hundred Fellows of Trinity called him Jack (without a response in kind?); only the latest of his five Masters could speak *ex cathedra* of Jack Littlewood.

We turn now to the mathematics of the 1920s. Hardy left for Oxford in 1920, returning to Cambridge in 1931 having established a school of analysis in Oxford, with E. C. Titchmarsh as his first pupil. The only dislocation in the Hardy-Littlewood partnership was that their thoughts were conveyed by the post instead of by a college messenger. The decade 1920–30 saw the series of papers on "Partitio Numerorum." There was also a systematic attack on problems of convergence and summability of Fourier series; this entailed large-scale prolegomena on fractional integrals and a wide variety of inequalities.

In the 1920s Littlewood included among his lectures a course on the foundations of set-theory, following the general approach of Whitehead and Russell. He lectured from printed notes on cardinal and ordinal numbers, the multiplicative axiom, well ordered series and general order-types. When the treatment had stabilized, he made the notes into a book (1926).

Littlewood's independent work during the 1920s was largely concentrated on complex function theory, as will appear from Professor Hayman's account. Among the research students whom he formally supervised as candidates for the Ph.D. degree were R. Cooper, T. A. A. Broadbent, E. F. Collingwood, H. P. Mulholland and S. Verblunsky. Others, including A. E. Ingham, S. W. P. Steen, R. M. Gabriel, not Ph.D. candidates, must have had boundless stimulus from Littlewood by his encouraging remarks like "When you have an idea, you can generally crash through the obstacles."

As a fellow of Trinity, Littlewood had a stipend and the right to rooms in college and dinner in hall from his election in 1908. In 1926 he took the option of tenure under the old statutes instead of new statutes which then came into force. Thus he became entitled, after his retirement in 1950, to retain all the privileges of a Fellow for life.

An unmarried Fellow of Trinity can enjoy a civilized and care-free life. The professorial and official Fellows holding university and college offices are the nucleus of a society rejuvenated by the annual election of new Fellows including about five young research Fellows. Littlewood had before him the prospect of a régime which would give full scope to the magnificent powers of which he had evidence. The responsibilities of domesticity, whatever the rewards, would have curtailed his freedom. Though he enjoyed the society of intelligent and good-looking women, it would have surprised his friends if he had married. Even as a bachelor he had reservations about college life as an environment for creative work. He spoke of "the wrong sort of mental stimulus and the constant bright conversation of the clever." Littlewood lived in college during term, this being the natural way to carry out his teaching duties. In vacations he would normally go to the sea or to mountains. For some years he went to the Isle of Wight; later he spent vacations with friends of long standing on the Cornish coast. Away from both the duties and the routine of college life he could give himself to sustained periods of creative work. When mentally tired he could walk and scramble on the rocks in complete relaxation until ready for the next onslaught on resistant problems.

He took up skiing in Switzerland in 1924 and found it an enjoyable activity. A little later he decided to accept the invitation of a well known rock climber to teach him the techniques. Returning from that instructional holiday, he said: "I found that I'd been rock climbing all my life" (mainly in Cornwall). His muscular strength, balance and agility made for success in these pursuits and he spent many holidays in Wales, the Lakes, Skye and Switzerland.

1928–50: Rouse Ball Professor

In 1928 Littlewood was elected as the first occupant of a chair of mathematics founded by a benefaction from W. W. Rouse Ball, author of *Mathematical Recreations* (who had been his tutor in his undergraduate days). Thenceforward he was free from college teaching and could give university lectures on topics of his own choice. His reputation as one of the strongest analysts of the world was established and there seemed to be no limits to what he might accomplish. There was, however, a cloud.

Notwithstanding his well ordered routine of work and recreation he had, even at school, endured periods of depression. As he approached middle age he spoke of his anxiety that they might continue indefinitely. His creativity seemed unimpaired, as did his physical health and strength. He dined regularly in hall and always appeared lively and sociable. Besicovitch's opinion "It is difficult to believe that Littlewood is not O.K." was shared by others, in Trinity and elsewhere. Whether from inner compulsion or deliberate decision he tended increasingly to withdraw from engagements and commitments. Invited to dine in another college he would accept with apparent pleasure; as the day drew near he would excuse himself: "I would rather dine in Trinity; it is a strain meeting people I don't know." He could say late in life: "I have never been chairman or secretary of any body." He carried out duties (e.g. relating to appointments or promotions) which could not be delegated to a deputy. Asked to exert a beneficent influence in some impending issue, he would reply in two or three days: "Action taken and I am not without hope."

Meanwhile the quality of his work was as high as ever, though he may have been right in deploring some reduction in output. No falling-off in excellence was apparent in the Hardy-Littlewood publications. Many of these deal with inequalities.

In 1928 Hardy gave his Presidential Address to the London Mathematical Society on "Prolegomena to a chapter on inequalities." In it he wrote: ". . . Littlewood, Pólya and myself have undertaken to contribute a tract on inequalities to the Cambridge series, and I am sure that we shall deserve the thanks of the mathematical world, even if we do not do it particularly well." The outcome (1934) was two or three times the size of a tract, a book of 300 pages, and the authors did deserve the thanks of the mathematical world. For sustained and inexhaustible interest it would rank high as a desert island book.

Of the Hardy-Littlewood papers, two which are conspicuous for originality are (1926) on rearrangements of sequences and (1930) on a maximal theorem. These papers were the foundations of later research by many mathematicians.

The paper (1931) announces a number of deep theorems on Fourier series and power series by Littlewood and R. E. A. C. Paley. Paley, born in 1907, was, in Hardy's words (*Journal of the London Mathematical Society*, 9 (1934), 76–80), "one of the two or three best English mathematicians who have made their reputations since the war" and "Littlewood's influence dominates nearly all his earliest work." Paley had written 25 papers, alone or in collaboration with Littlewood, Zygmund, Wiener or H. D. Ursell before, at the age of 26, he died in an avalanche while skiing in Canada.

Hardy, from his return to Cambridge in 1931, took the larger part in supervising the work of research students. Among Littlewood's pupils were J. Cossar, H. Davenport, S. Skewes and D. C. Spencer. Since 1928 he had held a weekly conversation class for advanced students. In December 1931, Hardy and Littlewood announced weekly meetings of a joint seminar to start in January 1932. . . .

Towards the end of the 1930s, Littlewood collaborated with A. C. Offord in investigations of zeros of the family of functions $\Sigma\epsilon_n a_n z^n$, random in the sense that the ϵ_n are $\pm$ 1.

Also in the late 1930s, under the shadow of probable war, the Department of Scientific and Industrial Research sought the interest of pure mathematicians in properties of nonlinear differential equations which were needed by radio engineers and scientists. The problems appealed to Littlewood and to Cartwright and they worked independently and together during the next 20 years. . . .

In contrast with World War I, in 1939–45 courses, particularly in scientific subjects, were maintained though often shortened. Some very able undergraduates had two years of tripos mathematics before being called up for work of national importance. Littlewood was lecturing in 1941–43 to a select class including F. J. Dyson and M. J. Lighthill. His help was sought from time to time when a mathematical puzzle emerged from the service ministries, but he had no systematic commitment to them.

Research in pure mathematics was at a discount in war-time, as irrelevant to national needs. Among the men whose research Littlewood supervised from 1947 onwards were A. O. L. Atkin, T. M. Flett, A. C. Allen, N. du Plessis, E. J. Watson, S. R. Tims, P. S. Bullen, F. R. Keogh, H. P. F. Swinnerton-Dyer and C. Obi (during Cartwright's absence in the U.S.A.).

1950–77: Emeritus Professor

In 1950, at the statutory age of 65, Littlewood became Emeritus Professor. He continued voluntarily to lecture until 1954 on nonlinear differential equations and the theory of functions.

Throughout his long life he lost little of his willpower or of his mental force and clarity. Moreover, from 1957, he was freed from his earlier periods of depression. A perspicacious psychiatrist, having traced their origin to a fault in the functioning of the central nervous system rather than to any adverse circumstances, surveyed the recent advances in knowledge of anti-depressant drugs and successfully prescribed treatment giving protection from recurrence of the symptoms.

This release from depression encouraged Littlewood to accept, during the 10 years from 1957, many invitations to the U.S.A. which had repeatedly been pressed on him. He was welcomed by friends of long standing, notably A. Zygmund at Chicago and L. C. Young at the University of Wisconsin's Mathematical Research Center. He stayed also three times at the University of California at Berkeley as a Visiting Professor acting as Consultant on Contracts with the Office of Naval Research. These visits gave pleasure to him and to those who met him. He gave lectures, seminars and was available for discussion. The main topics were the theory of functions and nonlinear differential equations.

[Littlewood authored] a steady flow of papers well into the ninth decade of his life. Many of the later papers are on differential equations and many show his interest in astronomy, physics and probability as well as in the problems of pure analysis which filled most of his life. His power at the age of 85 is shown [when] he solved a problem which "raised difficulties which defeated me for some time. I have now overcome them." The paper (1969) arose out of a long series of "card-guessing" trials in which he investigated the extent of perception between the participants.

Two products of this later period (1950–77) are characteristic of the author. *A Mathematician's Miscellany* (1953) is high-class entertainment. Some sections are for the professional only, but most of the book is within the range of the amateur in mathematics, and a number of pages (e.g. cross-purposes, misprints etc.) are independent of mathematics.

The second item of general interest is *The Mathematician's Art of Work* (1967).

From time to time Littlewood had experimented with a reordering of his life style towards greater creativity. In his twenties he had substituted morning work for the then fashionable 10 P.M. to 2 A.M. What and when to eat and drink is a theme with many variants. In the 1920s he gave up breakfast and ate lunch and dinner. He even temporarily reduced his meals to eight in the week, seven dinners and the lunch on Fridays with which the College Council celebrated their weekly meeting; he took generous helpings. Henry Head, whom he had consulted in 1918 after falling on his head climbing, later converted him from the practice of filling a row of corn-cob pipes, which would sustain him through a morning's work, to abstinence from tobacco except for a festive cigar.

The *Art of Work*, given as a lecture at Rockefeller University, is a vademecum to the theory and practice of creation (in mathematics, *par excellence*). There are four phases: preparation, incubation, illumination and verification. Preparation needs roots in an intense curiosity; the essential problem, stripped of its accidentals, must be brought and kept before the mind. The resulting drive is communicated to the subconscious "which does all the real work and would seem to be always on duty." "Incubation is the work of the subconscious during the waiting time, which may be several years. Illumination, which can happen in a fraction of a second, is the emergence of the creative idea into the conscious."

Direction of Research

The *Art of Work* reflects the experience of an elder who has guided younger men. It is of interest to complement this by quoting Swinnerton-Dyer whose career started at the receiving end.[2]

> Most research supervisors make use of their research students; they have their own research programme, and they split off the duller bits of it to give to their students. Littlewood was not like this at all. He had a list of twenty or thirty problems, gradually renewed. Some of them looked difficult, and were; others looked easy, and were not. A student could try any he chose, and if he failed on one, he could go on to another. It was not

until years later that I discovered how the list was put together, and there must be many of Littlewood's pupils who never discovered. Each problem was one which a mathematician whom Littlewood respected had seriously attempted and had failed to solve. Thus if a student solved one of the problems, Littlewood could with confidence write him a strong testimonial; and if not, there would be another research student coming along in due course. You have to be tough to survive this treatment, but research is tough; and Littlewood felt that he himself had come to no harm by being set to prove the Riemann Hypothesis. Besides, there was a second list of easier problems, which the more despondent pupils were allowed to see at the end of their second year; for Littlewood used to say that a Ph.D. was a degree which you had to take if you failed to get a Research Fellowship, just to show that you had been at Cambridge.

S. Skewes recalls his experiences of the tough treatment in Littlewood's "habit of summoning me to wherever he happened to be—Treen in Cornwall, Bideford in Devon, Seascale in Cumberland. There we walked miles at great speed, he talking mathematics all the time. The last-named place was a trial to me. Coping with the scree on the hill-sides was a full-time occupation, both mental and physical, and answering the questions he fired at me was often more than difficult—it was impossible. He was, as ever, as sure-footed and tireless as a stag."

Longevity

The *Art of Work* has this apt paragraph.

> Mathematics is very hard work, and dons tend to be above the average in health and vigor. Below a certain threshold a man cracks up, but above it hard mental work *makes* for health and vigor (also—on much historical evidence throughout the ages—for longevity).

In retirement Littlewood lived for most of the year in college in the set in Nevile's Court of two large communicating rooms and bedroom, without mod. cons., which he had occupied since 1913. From Christmas until the Commemoration Feast in mid-March he would go to Davos, skiing until an age at which his friends felt constrained to protest that a fall could (and did) mean a fracture. His contemporary and friend for 70 years, Professor H. A. Hollond, wrote in 1972:

> Old age has treated J. E. Littlewood kindly. The only falling off which there has been is that he has lost his capacity to take long walks and confines his exercise to the College grounds. Mathematics continue to be of the essence of his life. In his eighty-seventh year he is still working long hours at a stretch, writing papers for publication and helping mathematicians who send their problems to him. He reads voraciously both serious literature and science fiction: he has a large collection of records of classical music and plays them on a superb gramophone. He plays a much valued part in the life of the college, dines in Hall every evening and attends what is called in Cambridge the Combination Room for dessert and wine. His range of conversation is unequalled in our society: there is no gathering in which he does not make his contribution.

In 1972 his activity was impaired by two falls, the first near to his fire, burning his shoulder and entailing painful skin-grafts; in the second fall he sustained serious concussion and a fracture of the right arm, permanently restricting its movement.

The year 1965 was both the hundredth anniversary of the foundation of the London Mathematical Society and the eightieth of Littlewood's birth. Volume XIV A of the *Proceedings* was a Festschrift of papers by friends and colleagues, many of whom were former pupils, "offered as a tribute to a great mathematician who was also an inspiring teacher."

On Littlewood's ninetieth birthday, 9 June, 1975, the Institute of Mathematics and its Applications held a joint symposium with the London Mathematical Society on "Excitement in Mathematics" with lectures by M. F. Atiyah, M. L. Cartwright, W. K. Hayman, D. G. Kendall and E. C. Zeeman. In Trinity, in the second after-dinner speech of his life (ten years after the first), Littlewood replied to the Master's toast.

In August 1977 he fell out of bed and was not found until the morning. This made it clear that he must move from college to a nursing home, where he died suddenly on 6 September, 1977. A memorial service was held in Trinity College on 26 November with an address by Sir Peter Swinnerton-Dyer.

Honors

Littlewood was elected F.R.S. in 1916 and received the Royal (1929), Sylvester (1943) and Copley (1958) medals.

He was an honorary D.Sc. of Liverpool (1928) and honorary LL.D. of St. Andrews (1936). In 1965 he received the distinguished compliment of the honorary Sc.D. of his own university.

He was a foreign member of the Swedish (1948), Danish (1948) and Dutch (1950) Academies, and since 1925 a corresponding member of the Akademie der Wissenschaften in Göttingen. He was elected on 26 November, 1957 a corresponding member of the Paris Académie des Sciences, replacing Maurice Fréchet in the section of Geometry.

The London Mathematical Society awarded him in 1938 the de Morgan medal and in 1960 the Senior Berwick Prize for two papers on celestial mechanics. He was President from 1941 to 1943. (On the occasion of his election his only duty was to declare the meeting closed. At later meetings (few and short in the war years) a Vice-President took the Chair.)

Notes

1. Bohr, H. *Collected works,* Vol. 1, p. xxviii.
2. Swinnerton-Dyer, Sir Peter, Address at Memorial Service in Trinity College, *Cambridge Review,* Vol. C (1978), No. 2242, p. 74.

Hardy's *A Mathematician's Apology*

L. J. Mordell

L. J. Mordell was born in Philadelphia in 1888. He received his graduate training at Cambridge and then taught at the University of Manchester, Birkbeck College, and Cambridge. His research has been in the theory of numbers. He has traveled widely, with appointments at twelve universities outside of Great Britain and lectures at over 190 universities throughout the world.

One would hardly expect to find personal accounts by mathematicians on the New York Times *best-seller list. One such work, however, has received a remarkably long and surprisingly broad public acceptance. The work in question is* A Mathematicians's Apology *by G. H. Hardy.*

Hardy was quite probably the best-known figure of twentieth-century British mathematics. Through both the significance of his mathematics works and the force of his personality, he provided a nucleus of excellence about which revolved a large part of British mathematics in this century. Consequently, Hardy became a widely known and much respected mathematics folk hero. His Apology *was warmly and uncritically received and has done nothing but enhance his already substantial reputation. The following article, while in no way detracting from Hardy's well-deserved position, examines his autobiography with a bit more scholarly detachment. Mordell suggests that famous mathematicians, like the rest of humanity, are occasionally guilty of selective recollection of the past.*

A REPRINT OF this most interesting book appeared in 1967 with a foreword by C. P. Snow, Hardy's friend of long standing.

It has often been reviewed and highly praised, but there are, however, some opinions expressed by Hardy which, perhaps, have not been adequately dealt with by other reviewers. Furthermore, Snow's foreword calls for some comment, especially his references to Ramanujan (1887–1920). He writes that after Hardy and Littlewood read the manuscript sent by Ramanujan to Hardy (probably Jan. 16, 1913), "they knew, and knew for certain" that he "was a man of genius. . . . It was only later that Hardy decided that Ramanujan was, in terms of *natural* mathematical genius, in the class of Gauss and Euler; but that he could not expect, because of the defects of his education, and because he had come on the scene too late in the line of mathematical history, to make a contribution on the same scale." While one would readily accept that Ramanujan was a man of genius, the comparison with Gauss and Euler is very farfetched. I have some difficulty in believing that Hardy made such a statement, or at any rate made it in this form. What does natural mathematical genius mean? Undoubtedly Ramanujan was outstanding in some aspects of mathematics and had great potentialities. But this is not enough. What really matters is what he did, and one cannot accept such a comparison with Euler and Gauss, whose many-sided contributions were of fundamental importance and changed the face of mathematics. In fact in 1940, in the book on Ramanujan, Hardy said "I cannot imagine any-

Source: L. J. Mordell, "Hardy's *A Mathematician's Apology*," *American Mathematical Monthly* 77 (1970): 831–836.

body saying with any confidence, even now, just how great a mathematician he was and still less how great a mathematician he might have been."

Snow says that Ramanujan, as is commonly believed, was the first Indian to be elected (2 May, 1918) a fellow of the Royal Society. He was the second. The first was Ardaseer Cursetjee (1808–1877), shipbuilder and engineer, F.R.S. 27 May, 1841. Snow notes that Ramanujan was elected a fellow of Trinity four years after his arrival in England and continues, "it was a triumph of academic uprightness that they should have elected Hardy's protégé Ramanujan at a time when Hardy was only just on speaking terms with some of the electors and not at all with others." It is well that the merits of a fellowship candidate are judged by the quality of his original work and not by the political views of his sponsors.

Let us examine some of the views expressed by Hardy. They are sometimes stated too categorically, regardless of exceptions and limitations. A number of them had their origin in what he says, most gloomily, in the very first section of the Apology: "It is a melancholy experience for a professional mathematician to find himself writing about mathematics. The function of a mathematician is to do something, to prove new theorems, to add to mathematics, and not to talk about what he or other mathematicians have done."

His practice many years ago does not conform with this statement. He recalls in Section 6 that he did talk about mathematics in his 1920 Oxford inaugural lecture, which actually contains an apology for mathematics. Further in 1921, he gave an address on Goldbach's theorem to the Mathematical Society of Copenhagen. In this, he did talk about what he and other mathematicians had done. Such talks render a real service to mathematics and many have found great pleasure and inspiration in listening to or reading such expositions. Hardy had followed the practice of many eminent mathematicians in giving them. These have contributed to the richness and vividness of mathematics and make it a living entity. Without them, mathematics would be much the poorer.

No mathematician can always be producing new results. There must inevitably be fallow periods during which he may study and perhaps gather ideas and energy for new work. In the interval, there is no reason why he should not occupy himself with various aspects of mathematical activity, and every reason why he should. The real function of a mathematician is the advancement of mathematics. Undoubtedly the production of new results is the most important thing he can do, but there are many other activities which he can initiate or participate in. Hardy had his full share of these. He took a leading part in the reform of the mathematical Tripos some sixty years ago. Before then, it was looked upon as a sporting event, reminding one of the Derby, and was out of touch with continental mathematics. A mathematician can engage in the many administrative aspects of mathematics. Hardy was twice secretary and president of the London Mathematical Society and, while so occupied, must have done an enormous amount of unproductive work. He served on many committees dealing with mathematics and mathematicians. He wrote a great many obituary notices. He was well aware that a professor of mathematics is a representative of his subject in his University. This entails many duties which cannot be called doing mathematics.

His reference to a melancholy experience shows how much he took to heart and suffered from the loss of his creative powers. The result is, as Snow says, that the *Apology* is a book of haunting sadness.

Further in this first section, he says despairingly, "If then I find myself writing not mathematics but 'about' mathematics, it is a confession of weakness, for which I may rightly be scorned or pitied by younger and more vigorous mathematicians. I write about mathematics because, like any other mathematician who has passed sixty, I have no longer the freshness of mind, the energy, or the patience to carry on effectively with my proper job." He had been for many years a most active mathematician and his collected works now being published will consist of seven volumes. It seems almost nonsense to say that anyone would scorn or pity him, and the use of the term "rightly" is even more nonsensical.

We all know only too well that with advancing age we are no longer in our prime, and that our powers are dimmed and are not what they once were. Most of us, but not Hardy, accept the inevitable. There are still many consolations. We can perhaps find pleasure in thinking about some of our past work. We can read what others are doing, but this may not be easy since many new techniques have been evolved, sometimes completely changing the exposition of classical mathematics. Various reviews, however, may give one some idea of what has been done. (We can still be of service to younger mathematicians.)

His statement about a mathematician who has passed sixty is far too sweeping and any number of instances to the contrary can be mentioned, even among much older people. One need only note some recent Cambridge and Oxford professors. Great activity among octogenarians is shown by Littlewood, his lifelong collaborator, Sydney Chapman, his former pupil and collaborator, and myself. There is also Besicovitch in the seventies. Davenport, who had passed sixty, was as active and creative as ever, and his recent death is a very great loss to mathematics since he could have been expected to continue to produce beautiful and important work.

The question of age was ever present in Hardy's mind. In Section 4, he says, "No mathematician should ever allow himself to forget that mathematics, more than any other art or science, is a young man's game." It seems that he could not reconcile himself to growing old. For further on, he says, "I do not know an instance of a major mathematical advance initiated by a man past fifty." This may be so, but much depends on the definition of the *advance*. But there is no need to be troubled about it. Much important work has been done by men after the age of fifty.

A number of Hardy's statements must be qualified. In Section 2, he says, that "good work is not done by 'humble' men. It is one of the first duties of a professor, for example, in any subject, to exaggerate a little both the importance of his subject and his own importance in it. A man who is always asking, 'Is what I do worthwhile?' and 'Am I the right person to do it?' will always be ineffective himself and a discouragement to others."

Though one may naturally have a better opinion of one's work than others have, there are many exceptions to his statement. I never knew Davenport to exaggerate or emphasize the importance of his work, but he was a most effective mathematician and a very successful supervisor of research. Prof. Frechet told me a few years ago, that when Norbert Wiener was working with him a long time ago, he was always asking, "Is my work worthwhile?" "Am I slipping?" etc. S. Chowla is as modest and humble a mathematician as I know of, but he inspires many research students.

We comment on some more of Hardy's statements about mathematics. One of the most surprising is in Section 29, "I do not remember having felt, as a boy, any *passion* for mathematics, and such notions as I may have had of the career of a mathematician were far from noble. I thought of mathematics in terms of examinations and scholarships; I wanted to beat other boys, and this seemed to me to be the way in which I could do so most decisively."

It has often been said that mathematicians are born and not made. Most great mathematicians developed their keenness for mathematics in their school days. Their ability revealed itself by comparison with the performances of their schoolmates. Their ambition was to continue the study of mathematics and to take up a mathematical career. Probably no other motive played any part in the decision of most of them.

In Section 3, he considers the case of a man who sets out to justify his existence and his activities. I see no need for justification any more than a poet or painter or sculptor does. As Trevelyan says, disinterested intellectual curiosity is the life blood of real civilization. It is curiosity that makes a mathematician tick. When Fourier reproached Jacobi for trifling with pure mathematics, Jacobi replied that a scientist of Fourier's caliber should know that the end of mathematics is the great glory of the human mind. Most mathematicians do mathematics for the very good reason that they like and enjoy doing it. Davenport told me that he found it "exciting" to do mathematics.

Hardy says that the justifier has to distinguish two different questions. The first is whether the work which he does is worth doing and the second is why he does it, whatever its value may be. He says to the first question: The answer of most people, if they are honest, will usually take one or the other of two forms; and the second form is merely a humbler version of the first, which we need to consider seriously. "I do what I do because it is the one and only thing I can do at all well." It suffices to say that the mathematician felt no need to do anything else.

Hardy is very appreciative of the beauty and aesthetic appeal of mathematics. "A mathematician," he says in Section 10, "like a painter or poet, is a maker of patterns . . . ," and these ". . . must be beautiful. The ideas . . . must fit together in a harmonious way. Beauty is the first test: there is no permanent place in the world for ugly mathematics. . . . It may be very hard to *define* mathematical beauty . . ." but one can recognize it. He discusses the aesthetic appeal of theorems by Pythagoras on the irrationality of $\sqrt{2}$, and Euclid on the existence of an infinity of prime numbers. He says in Section 18, "There is a very high degree of *unexpectedness,* combined with *inevitability* and *economy*. The arguments take so odd and surprising a form; the weapons used seem so childishly simple when compared with the far-reaching results."

I might suggest among other attributes of beauty, first of all, simplicity of enunciation. The meaning of the result and its significance should be grasped immediately by the reader, and these in themselves may make one think, what a pretty result this is. It is, however, the proof which counts. This should preferably be short, involve little detail and a minimum of calculations. It leaves the reader impressed with a sense of elegance and wondering how it is possible that so much can be done with so little.

Somehow, I do not think that Hardy's work is characterized by beauty. It is distinguished more by his insight, his generality, and the power he displays in carrying out his ideas. Many of the results that he obtains are very important indeed, but the proofs are often long and require concentrated attention, and this may blunt one's feelings even if the ideas are beautiful.

Hardy does not define ugly mathematics. Among such, I would mention those involving considerable calculations to produce results of no particular interest or importance; those involving such a multiplicity of variables, constants, and indices, upper, lower, right, and left, making it very difficult to gather the import of the result; and undue generalization apparently for its own sake and producing results with little novelty. I might also mention work which places a heavy burden on the reader in the way of comprehension and verification unless the results are of great importance.

Hardy had previously said that he could "quote any number of fine theorems from the theory of numbers whose meaning anyone can understand, but whose proofs, though not difficult, may be found tedious." It often happens that there are significant results apparently of some depth, the proof of which can be grasped by those with a minimum of mathematical knowledge. Perhaps I may be pardoned if I give one of my own. The theorem of Pythagoras suggests the problem of finding the integer solutions of the equation $x^2 + y^2 = z^2$. This was done some 1000 years ago and is not difficult. But suppose we consider the more general equation $ax^2 + by^2 = cz^2$. This is a real problem in the theory of numbers. Legendre at the end of the eighteenth century gave necessary and sufficient conditions for its solvability. Then when the equation is taken in the normal form, i.e. abc is square-free and $a > 0, b > 0, c > 0$, Holzer showed in 1953 that a solution existed with $|z| < \sqrt{ab}$, from which it follows that $|x| \leqq \sqrt{bc}$, $|y| \leqq \sqrt{ca}$. I recently found a proof of this result that no one would call tedious by showing that if a solution (x_1, y_1, z_1) existed with $|z_1| > \sqrt{ab}$, then there was another with $|z_2| < |z_1|$. This arose by taking an appropriate line through the point (x_1, y_1, z_1) to meet the conic $ax^2 + by^2 = cz^2$ in the point (x_2, y_2, z_2). I call this a schoolboy proof, because the only advanced result required is that the equation $lx + my = n$ has an integer solution if l and m are coprime. A proof of the theorem could have been found by a schoolboy.

We conclude by examining Hardy's views about the utility or usefulness of mathematics. He seems to denigrate the usefulness of "real" mathematics. In Section 21, he says, "The 'real' mathematics of the 'real' mathematicians, the mathematics of Fermat and Euler and Gauss and Abel and Riemann is almost wholly 'useless.' " This statement is easily refuted. A ton of ore contains an almost infinitesimal amount of gold, yet its extraction proves worthwhile. So if only a microscopic part of pure mathematics proves useful, its production would be justified. Any number of instances of this come to mind, starting with the investigation of the properties of the conic sections by the Greeks and their application many years later to the orbits of the planets. Gauss' investigations in number theory led him to the study of complex numbers. This is the beginning of abstract algebra, which has proved so useful for theoretical physics and applied mathematics. Riemann's work on differential geometry proved of invaluable service to Einstein for his relativity theory. Fourier's work on Fourier series has been most useful in physical investigations. Finally one of the most useful and striking applications of pure mathematics is to wireless telegraphy which had its origin in Maxwell's solution of a differential equation. Many new disciplines are making use of more and more pure mathematics, e.g., the biological sciences, economics, game theory, and communication theory, which requires the solution of some difficult Diophantine equations. It has been truly said that advances in science are most rapid when their problems are expressed in mathematical form. These in time may lead to advances in pure mathematics.

These remarks may serve as a reply to Hardy's statement that the great bulk of higher mathematics is useless.

It is suggested that one purpose mathematics may serve in war is that a mathematician may find in mathematics an incomparable anodyne. Bertrand Russell says that in mathematics, "one at least of our nobler impulses can best escape from the dreary exile of the actual world." Hardy's comment on this reveals his depressed spirits. "It is a pity," he says, "that it should be necessary to make one very serious reservation—he must not be too old. Mathematics is not a contemplative but a creative subject; no one can draw much consolation from it when he has lost the power or desire to create; and that is apt to happen to mathematicians rather soon." What does he mean when he says mathematics is not a contemplative subject? Many people can derive a great deal of pleasure from the contemplation of mathematics, e.g., from the beauty of its proofs, the importance of its results, and the history of its development. But alas, apparently not Hardy.

Some Mathematicians I Have Known

George Pólya

George Pólya was born and educated in Hungary. He obtained a Ph.D. in mathematics from Budapest and taught in Switzerland and at Brown, Smith, and Stanford Universities in the United States. He is the recipient of numerous honors and awards in mathematics, has taught and lectured in virtually every civilized country, and in his ninety-seventh year is currently professor emeritus of mathematics at Stanford University.

One of the most remarkable figures of twentieth-century mathematics is George Pólya. Still actively engaged in his profession at the age of ninety-six when this volume was being published, he has had a great influence on mathematics and on the public's understanding of mathematics for more than seventy years. With some 230 publications on an amazing variety of topics his most recent efforts have been in effective communications of mathematical ideas. This autobiographical account is taken from an address given to the American Mathematical Society in 1969.

Anecdotes can teach some things that biographies can't. Those readers who gain additional insight into mathematics through Pólya's anecdotal form should consider reading How to Solve It *(Doubleday Anchor Books, New York, 1957) in which Pólya uses the anecdotal style to train people to solve problems.*

I AM VERY old, my days of invention are over. The little mathematical remarks I have made lately are too little and too few to make a speech about them, and I cannot tell you very well about my former work on which I spent almost sixty years, because too many of you would find it quite unfashionable.

So what shall I do? Make an after dinner speech? Well, it will be a before lunch speech: I shall tell you a few anecdotes about mathematicians I have known. These stories are not printed and perhaps they should not be printed. They are part of an oral tradition—you may find an occasion to tell them to friends or students.

When non-mathematicians discuss mathematicians (when faculty wives discuss their husbands and their husbands' friends) they often ask the same questions: What is particular about mathematicians? How do mathematicians differ from other people? And I often heard the same answers: "Mathematicians are absent-minded" or "Mathematicians are eccentric." Are mathematicians really absent-minded or eccentric? I don't know, but there are infinitely many stories purporting that they are, and I shall quote a few. Probably you will know several of them, but perhaps not all of them.

First, about absent-mindedness. Many such stories are told about Hilbert. Are they true? I doubt it, but some are quite good. Here is one of the very well known ones: There is a party in Hilbert's house and Frau (I mean Mrs.) Hilbert sud-

Source: Excerpts from George Pólya, "Some Mathematicians I Have Known," *American Mathematical Monthly* 76 (1969): 746–53. Address given at the meeting of the Northern California Section of the Mathematical Association of America at the University of Santa Clara, February 8, 1969.

denly notices that her husband forgot to put on a fresh shirt. "David," she says sternly, "go upstairs and put on another shirt." David, as it befits a long married man, meekly obeys and goes upstairs. Yet he does not come back. Five minutes pass, ten minutes pass, yet David fails to appear and so Frau Hilbert goes up to the bedroom and there is Hilbert in his bed. You see, it was the natural sequence of things: He took off his coat, then his tie, then his shirt, and so on, and went to sleep.

There is another story which I like even more because it reminds me of the Göttingen I knew where I studied more than half a century ago. Yes, that old-time Göttingen was rather formal. A new member of the faculty was supposed to introduce himself formally to his colleagues. He put on a black coat and a top hat, took a taxi, and made the round of the faculty houses. The taxi stopped in front of each, and the new colleague presented his visiting card at the door. Sometimes he got the answer that the Herr Professor is not at home, but when the Herr Professor was at home the new colleague was supposed to go in and chat for a few minutes. Once such a new colleague came to Hilbert's house and Hilbert decided (or Frau Hilbert decided for him) that he was at home. So the new colleague came in, sat down, put his top hat on the floor, and started talking. This was the proper thing to do, but he did not stop talking. And Hilbert—the visit probably interrupted some mathematical meditation—became more and more impatient. And what did he do finally? He stood up, took the top hat from the floor, put it on his head, touched the arm of his wife, and said: "I think, my dear, we have delayed the Herr Kollege long enough"—and walked out of his own house.

There are some authentic stories about absent-minded mathematicians, for instance, about Newton who, working intensively at his problems, often forgot to eat his lunch, or, when he ate it, forgot that he had eaten it. Yet other stories are less authentic.

I heard the following from Theodore von Kármán himself. Still, I would not swear that it actually happened; he liked good stories too much, and the best stories do not happen, they are invented. At that time he had a double position: He was professor at Aachen in Germany and also lectured at Cal Tech in Pasadena. As an important aeronautical engineer, he was consultant to several airlines, and so he got free transportation whenever he found an unoccupied seat on a plane of one of these lines. So he commuted more or less regularly between Aachen and Pasadena. He gave similar lectures at both places. Once he was somewhat tired when he arrived in Pasadena, but started lecturing. That was not so difficult: He had his notes which he also used in Aachen. He talked, but as he looked around he had the impression that the faces in the audience looked even more blank than usual. And then he caught himself: He was speaking in German! He became quite upset. "You should have told me—why did you not tell me?" The students were silent, but finally one spoke up: "Don't get upset, Professor. You may speak German, you may speak English, we will understand just as much."

Yet the most beautiful story of my collection is about Norbie—I mean Norbert—Wiener. (The name "Norbie" comes from a conversation I overheard between Wiener and a friend. "Confess," said Wiener, "that you call me Wienie behind my back." "No," said the friend, "we call you Norbie.") Now, here is the story which was widely told, but is hardly true. It is about a student who had a great admiration for Wiener, but never had an opportunity to talk to him. The student walked into a post office one morning. There was Wiener, and in front of Wiener a sheet of paper on the desk at which he looked with tremendous concentration. Suddenly Wiener ran away from, and then back to, the paper, facing it again with tremendous concentration. The student was deeply impressed by the prodigious mental effort mirrored in Wiener's face. He had just one doubt: Should he speak to Wiener or not? Then suddenly there was no doubt, because Wiener, running away from the paper, ran directly into the student who then had to say, "Good morning, Professor Wiener." Wiener stopped, stared, slapped his forehead and said: "Wiener—that's the word."

I cannot tell you a better story on this subject, so let us pass on to the other question: Are mathematicians eccentric? Are they odd, singular, out of the ordinary?

In a way they are, of course. To be really a

mathematician, to spend your best effort not in making money, not in working for power, but just in thinking about mathematics is singular behavior. Therefore, the question should be put so: Are mathematicians eccentric beyond this point, also in other respects?

Well, I don't know. When I think of the mathematicians I have known not quite superficially, I am inclined to refrain from any general statement. Let me tell you about three mathematicians I have known fairly well: Leopold (Lipót) Fejér (1880–1959), Adolf Hurwitz (1859–1919), and Godfrey Harold Hardy (1877–1947). I knew each of them for several years, I had the privilege of working with, and I am deeply grateful to, all three. Let me tell you a little about their career, their personality, the style of their work, and (this is a before lunch talk) a few characteristic stories.

Lipót Fejér was born in Hungary. He was about twenty years old when he discovered "Fejér's theorem" (on the arithmetic means of the Fourier series—but I should not enter into mathematical details today). His dissertation (he passed his Ph.D. at the age of twenty-two) deals with that theorem and he came back again and again to his initial discovery: he found sharper formulations, analogies, applications, extensions, he followed up the underlying idea into adjacent domains. Although he also found good remarks on other subjects, his initial discovery remained the center of his work.

His papers are particularly well written, they are very easy to read. This is due to his style of work: When he found an idea, he tended it with loving care; he tried to perfect it, simplify it, free it from unessentials; he worked on it carefully and minutely until the idea became transparently clear. He eventually produced a work of art, not of too large dimensions, but highly finished.

He had artistic talents besides mathematics. He loved music and played the piano. He had a special gift for telling stories, he was a "raconteur." In telling his stories, he acted the part of the persons he was telling about, and underlined the points with little gestures. He liked to talk about his teacher, who was in a rather indirect way responsible for his first discovery, Hermann Amandus Schwarz. When he told about the little misadventures of this great mathematician, he was irresistible, you could not help laughing.

This variety of talents has a bearing on a question which I have often heard: Why did Hungary produce so many mathematicians? Hungary was a small country (it is even smaller today) not much industrialized, and it produced a disproportionately large number of mathematicians, several of whom were active in this country. Why was that so? There is no complete answer, I think—Hungary produced not only mathematicians, but also many musicians and some physicists. Yet, I think, as far as mathematicians are concerned a good part of the answer can be found in Fejér's personality: He attracted many people to mathematics by the success of his own work and by his personal charm. He sat in a coffee house with young people who could not help loving him and trying to imitate him as he wrote formulas on the menus and alternately spoke about mathematics and told stories about mathematicians. In fact, almost all Hungarian mathematicians who were his contemporaries or somewhat younger were personally influenced by him, and several started their mathematical career by working on his problems.

To round out the picture I must quote some of his witty remarks I heard myself.

It happened at a meeting in Germany. At that time I was a "Privatdozent." I cannot completely explain what that is: A financially shaky position, somewhat similar to, but not quite, an Assistant Professor—thank goodness, this institution of Privatdozents has started to disappear nowadays. I was married, and my wife took photographs of the mathematicians. She also stopped Fejér in the company of three or four others, in front of the university on the street car tracks, took a picture and was about to take a second one as Fejér spoke up. "What a good wife! She puts all these full professors on the tracks of the street car so that they may be run over and then her husband will get a job!"

At another meeting (that was several years later) Fejér was very angry (and had some reason to be angry) at a Hungarian mathematician, a topologist whose name I shall not tell you. I walked up and down a long time with Fejér who could not stop talking about the target of his anger and wound up by saying: "And what he says is a topological map of the truth." You must realize how distorted a topological map may be.

Oh yes, let us not forget the question: Was Fejér eccentric? After all these stories, if you could see him in his rather Bohemian attire (which was, I suspect, carefully chosen) you would find him very eccentric. Yet he would not appear so in his natural habitat, in a certain section of Budapest middle class society, many members of which had the same manners, if not quite the same mannerisms, as Fejér—there he would appear about half eccentric.

Adolf Hurwitz was very much like Fejér in one respect, in the style of his work. Felix Klein, in his *History of Mathematics in the Nineteenth Century,* calls Hurwitz an "aphoristician." An aphorism is a concise weighty saying. The aphorism is short, but its author may have worked a long time to make it so short. Also Hurwitz tended his ideas with loving care, until he arrived at the simplest attainable expression, devoid of superfluous ornament or ballast and transparently clear. He was not unlike Fejér in another respect: He preferred not too large problems which are more amenable to perfect clarity. But his range was much wider than that of Fejér. He mastered the whole width of mathematical knowledge of his time as far as that was possible at the beginning of this century. And he learnt much of what he knew at the source: number theory and algebra from Kummer and Kronecker, the Riemannian aspect of complex variables from Felix Klein, the Weierstrassian aspect from Weierstrass himself. His mastery of wide domains of mathematics is described much better than I could, and with infinitely more authority, by Hilbert in the necrology prefixed to Hurwitz's collected works.

Yes, Hurwitz's papers are like aphorisms: In the wide range of his mathematical knowledge he spotted well circumscribed weighty problems capable of a surprisingly simple solution and presented the solution in perfect form. If you wish to have an easily accessible sample, read two pages in his collected works: the proof for the transcendence of the number *e*.

There was another point of resemblance with Fejér: Music played an important role in Hurwitz's life and he was an excellent pianist. In fact, as a young man he hesitated: should he become a mathematician or a pianist—fortunately he decided against the piano.

Yet here the similarity ends. As a personality, Hurwitz was very different from Fejér. First of all, nothing was farther from Hurwitz than to appear Bohemian or eccentric. He was always correct, reserved, inconspicuous, exceedingly modest, lifting his hat to the servants of the neighbors. A stranger could not suspect that there was more behind this unassuming exterior than middle class respectability. Only those who read his writings or attended his classes could suspect, and only those who knew him better could begin to understand, his strong sense of duty and his deep devotion to truth and clarity.

I never heard Hurwitz utter a sharp sentence in public. Yet in the circle of his family or with good friends he could find a sharp and witty word. I must preface a little what I wish to quote. In discharging conscientiously his duties as a professor, he took care of many Ph.D. candidates, treating them with much consideration and patience. Among so many there were some who needed a lot of help, and even the patient Hurwitz was once led to say: "A Ph.D. dissertation is a paper of the professor written under aggravating circumstances."

G. H. Hardy was very much like Fejér in one respect: Fejér developed mathematics in Hungary by his example, his personal charm, and his personal drive; and Hardy did very much the same in England. Yet the similarity ends there: In other respects, the circumstances and the personalities are very different.

England had a great tradition in applied mathematics, starting with Newton, but did not contribute comparably to pure mathematics which was developed mainly in France and Germany. Hardy insisted on pure mathematics and his insistence changed the trend of mathematical work in England. (That he occasionally misjudged, and was unjust to, applied mathematics is of comparatively little importance.)

Hardy wrote very well and with great facility, but his papers, especially some of his joint papers with Littlewood, make no easy reading: The problems are very hard and the methods unavoidably very complex. He valued clarity, yet what he valued most in mathematics was not clarity but power, surmounting great obstacles that others abandoned in despair. He himself had very great power, and he was fascinated by the Riemann hypothesis.

(Relatively concrete problems, such as the proof of the Riemann hypothesis, are less in vogue nowadays, for reasons partly good and partly bad—"Mostly bad" Littlewood would interject if he were present.)

Yet things were different a few decades ago, and I cannot resist telling here a story, although it needs a long preface.

There is a German legend about Barbarossa, the emperor Frederick I. The common people of Germany liked him and as he died in a crusade and was buried in a far away grave, the legend sprang up that he was still alive, asleep in a cavern of the Kyffhäuser mountain, but would awake and come out, even after hundreds of years, when Germany needed him.

Somebody allegedly asked Hilbert, "If you would revive, like Barbarossa, after five hundred years, what would you do?" "I would ask," said Hilbert, "Has somebody proved the Riemann hypothesis?"

Yet I must come back to Hardy—there are too many stories to tell. He must have seemed eccentric in almost any company, perhaps even in the colleges of Oxford and Cambridge where at the time eccentricity was tolerated, even encouraged. He was strikingly good looking, and very elegant when he put on a dinner jacket (tuxedo). Yet often, especially when travelling outside England, he was sloppily dressed—it was an artistic sloppiness and he was still very good looking.

He had very personal and very definite views about all sorts of things: He liked cats, but could not stand dogs, he loved cricket, but despised rowing. (He was a Cambridge man, but for some time professor in Oxford. It happened there that somebody, who was unaware of Hardy's idiosyncrasies, asked him: "For which university are you in sports?" "It depends," said Hardy. "In cricket I am for Cambridge, in rowing I am for Oxford.")

Hardy liked to shock people mildly by stating unconventional views, and he liked to defend such views just for the sake of good argument, because he liked arguing. And he loved jokes.

There is an inexhaustible multitude of stories about Hardy's jokes, but I must control myself and must not delay your lunch too long. Yet I must tell you one about myself. In working with Hardy, I once had an idea of which he approved. But afterwards I did not work sufficiently hard to carry out that idea, and Hardy disapproved. He did not tell me so, of course, yet it came out when he visited a zoological garden in Sweden with Marcel Riesz. In a cage there was a bear. The cage had a gate, and on the gate there was a lock. The bear sniffed at the lock, hit it with his paw, then he growled a little, turned around and walked away. "He is like Pólya," said Hardy. "He has excellent ideas, but does not carry them out."

Hardy loved sunshine, but there is not much sunshine in England. Therefore, in the summer vacation he went regularly to the continent (Europe) as soon as the cricket season was over and visited with friends. His principal friend was Harold Bohr. They had a set routine. First they sat down and talked, and then they went for a walk. As they sat down, they made up and wrote down an agenda. The first point of the agenda was always the same: "Prove the Riemann hypothesis." As you are probably aware, this point was never carried out. Still, Hardy insisted that it should be written down each time.

The picture would be incomplete if I were not to mention Hardy's principal standing joke: God was his personal enemy. You understand: God has nothing more urgent to do than to annoy Hardy. The multitude of stories belonging in this chapter is inexhaustible. Here is the best known.

Hardy stayed in Denmark with Bohr until the very end of the summer vacation, and when he was obliged to return to England to start his lectures there was only a very small boat available (there was no airplane traffic at that time). The North Sea can be pretty rough and the probability that such a small boat would sink was not exactly zero. Still, Hardy took the boat, but sent a postcard to Bohr: "I proved the Riemann hypothesis. G. H. Hardy." You are not laughing? It is because you don't yet see the underlying theory: If the boat sinks and Hardy drowns, everybody must believe that he has proved the Riemann hypothesis. Yet God would not let Hardy have such a great honor and so he will not let the boat sink.

I believe this story, because almost the same thing happened in my presence. Another summer Hardy stayed in Engelberg, an Alpine valley in Switzerland where we had a chalet. He liked the sunshine, but it rained all the time, and as there was nothing else to do, we played bridge: Hardy, who was quite a good bridge player, my wife, myself, and a friend of mine, F. Gonseth, mathematician and philosopher. Yet after a while Gonseth had to leave, he had to catch a train. I was present as Hardy said to Gonseth: "Please, when the train starts you open the window, you stick your head through the window, look up to the sky, and say in a loud voice: 'I am Hardy.' " Now, some of you are laughing. You have understood the underlying theory: When God thinks that Hardy has left, he will make good weather just to annoy Hardy.

Prof. Pólya received his Univ. Budapest degree in 1912 and holds honorary degrees from the E. T. H. Zürich, Univ. Alberta, and Univ. Wisconsin. He taught at the E. T. H. until 1940 and has been at Stanford Univ. since then. . . . The scientific contributions of George Pólya include over 230 research papers and six books. . . . Prof. Pólya's personal influence on three generations of mathematicians has been enormous. Perhaps no book in existence has influenced the direction of thinking of young mathematicians more than his two volume masterpiece with G. Szegö, *Aufgaben und Lehrsätze aus der Analysis*. . . .

Women Mathematicians

Marie-Louise Dubreil-Jacotin

Marie-Louise Dubreil-Jacotin is professor of mathematics and on the faculty of science at the University at Poitiers in France.

In the nineteenth century women were not allowed to enroll in many universities. At one school, which will remain nameless, a talented woman had persuaded the administration to let her audit a mathematics course. However, when the professor walked in and found four men and one woman in the classroom, he let her know in no uncertain terms that despite the administration's weakness and willingness to allow her in the class, he was not changing his attitude. Forthwith he began to lecture and lecture he did—with a vengeance. Week by week the lectures took their toll and by midterm only the woman and one struggling male student remained. The professor did not reduce the pace, and within another week the woman student found that she was the only person in the classroom when the professor walked in to begin a lecture. Glancing around, the professor announced, "Since there are no students here, I will not give a lecture," and walked out of the room.

This essay presents a brief overview of women in mathematics. Dubreil-Jacotin surveys the lives of several of the most historically notable women mathematicians and suggests why they are, relatively, so few. This survey is somewhat depressing in its implication that unknown genius has been wasted or suppressed by societies that have not recognized mathematics as an appropriate activity for women. However, the publication and dissemination of this article suggest there is hope for the future. We now live in an era at least superficially devoted to equality of opportunity. Surely any future catalogue of women mathematicians will be longer. Indeed, one hopes that in the future such a list will be irrelevant.

WOMEN WHO HAVE left behind a name in mathematics are very few in number—three or four, perhaps five. Does this say, as a common prejudice would tend to persuade us, that mathematics, so very abstract, is not congenial to the feminine disposition? This would be to ignore the true character of mathematics, to forget, as Henri Poincaré said, "the feeling of mathematical beauty, of the harmony of numbers and of forms, of geometric elegance. It is a genuinely esthetic feeling, which all mathematicians know. And this is sensitivity." And should not this very sensitivity, contrary to the prejudice, make mathematics a feminine domain? Moreover, if we consider the other sciences, the great inventions whose applications have staggered the world, we do not find any more women's names associated with these areas. Must it then be said, as Maurice d'Ocagne seems to conclude in his *Etudes sur les Femmes de Science et sur les Mathématiciennes*, that woman is generally destitute of inventive spirit and creative genius? The growth of female education, the overthrow of prejudices, the profound changes in the kind of life and in the role assigned to woman during the last few years will doubtless bring about a revision of her position in science. Then

Source: Marie-Louise Dubreil-Jacotin, "Women Mathematicians," in F. Le Lionnais, *Great Currents of Mathematical Thought* (New York: Dover Publications, Inc., 1971). (References omitted.)

we shall see in what measure she can, as the equal of man, emerge from the role of the excellent pupil or the perfect collaborator, and join those of our scientists whose work has opened new paths and bears the mark of genius.

Yet, it is as early as Greek antiquity that the first woman who can be considered as a mathematician makes her appearance: Hypatia, born at Alexandria in the year 370 of our era. In a school which she opened in her native city, she expounded at one and the same time Plato, Aristotle, the works of Diophantus and the conic sections of Apollonius of Perga. It is also believed that a commentary on the tables of Ptolemy, which has come down to us under Theon's name, is due to her. Hypatia acquired a great reputation as much for her science as for her eloquence and her beauty. She died tragically at the age of 45; she was thrown from her carriage and stoned by a crowd inflamed by monks at the instigation, it appears, of the patriarch St. Cyril.

Then, from the 16th century on, while the lords devoted themselves to politics and war, the ladies of high society gave themselves to the joys of the mind, and mathematics, by its very abstraction and beauty, had a preferential place there which did nothing but expand in the two following centuries, when a certain snobbishness seems, however, to have been mingled with the charm found by certain women in algebra and geometry. And it is perhaps this very epoch that saw the fullest appearance of those truly feminine qualities which make some great lady not a scientist, but the best of pupils or the fine and clear-sighted collaborator of some great scientist. There was Vieta, the creator of algebra, who had as his best pupil Catherine de Parthenay, Princess of Rohan-Soubise. Then Descartes, who had as his disciple Christine, Queen of Sweden, and Elizabeth of Bohemia, Palatine Princess; and Leibniz with Sophie, Electress of Hanover, and her daughter Sophie-Charlotte, Queen of Prussia and mother of the great Frederick. Euler wrote his *Lettres à une princesse d'Allemagne* to teach science to the Princess Anhalt-Dessau. Newton was studied and understood by Caroline of Brandenburg Anspach, and Emilie de Breteuil, Marquise du Châtelet, noted for her liaison with Voltaire, translated his principles and added certain personal notes to her translation—notes with which it may be Clairaut was not entirely unfamiliar. The latter, moreover, used the long calculations made for him by Mme. Lepaute, the charming wife of the celebrated watchmaker of Louis XV, without even mentioning her, "to oblige," say Lalande, "a women jealous of the attainments of Mme. Lepaute and who had pretensions without any kind of knowledge. She succeeded in making a wise but weak scientist, whom she had captivated, commit this injustice."

But let us abandon these brilliant women, more or less amateurs, and the admirable astronomers' assistants such as Mme. Yvon Villarceau, Mme. Lefrançais de Lalande, and others, in order to come to those who, if they cannot all be set among the greatest names, can at least cut an honorable figure among the great mathematicians.

First we have the Italian Maria Gaetana Agnesi (1718–1799), the first woman professor of mathematics on a faculty; indeed, she was appointed professor at the University of Bologna by Pope Benedict XIV. Maria Gaetana Agnesi was a very gifted and precocious child. Her father, Don Pietro Agnesi, feudatory of Monteveglia, actively encouraged his daughter in her studies and, proud of her, exhibited her in the kinds of academic meetings fashionable at that time. But, of a self-effacing nature, she went into the background again after the death of her father and finally went into retreat among the nuns of the "Azure" order, where she is said to have given of herself unsparingly and to have been an object of great edification. She left behind, under the name of *Instituzioni analitiche,* a "remarkable account of ordinary algebra, with the solution of several solved and unsolved geometric problems"; a second volume, entirely devoted to infinitesimal analysis, a science then quite new, was declared "the most complete and the best done in this field" by the commissioners of the Academy of Sciences of Paris, who were assigned to examine this work at their meeting of December 6, 1749. This book was translated into French and published "under the license of the Royal Academy of Sciences" by a decision of August 30, 1775, signed by the permanent secretary of the Academy, upon the advice of the commission composed of d'Alembert, the Marquis of Condorcet and Vandermonde.

The contribution of Agnesi to the study of curves of the third degree was great enough that one of them still bears her name: "the Witch of Agnesi."

Far from being encouraged by their families, it was only by main force against their parents that the two following women succeeded, almost simultaneously, in becoming brilliant mathematicians. First, Sophie Germain, born in Paris in 1776, the daughter of a rich middle-class silk merchant. As a child, Sophie received a very careful education and upbringing. At the age of 13, she found by chance in her father's library Montucla's history of mathematics. She thus read that Archimedes was killed by Roman soldiers because, entirely absorbed in the study of a problem, he was unaware of the capture of Syracuse and did not answer their questions. That one could be so thoroughly absorbed by a mathematical question as to forget everything, even the threat of death, filled her with such admiration that she wished at any cost to plunge herself into the study of this science. It was then that she encountered paternal hostility; nothing daunted her—she worked at night wrapped in a blanket, because they had taken her clothing away to keep her from getting up; they took away her heat, her light—all this only hardened her resolve and increased her fervor, so that her father finally gave in and she was at last able to devote herself to mathematics.

Her knowledge was already quite extensive in 1794 when the Ecole Polytechnique was founded; she was then able to obtain the notes of Lagrange's course and studied them with profit. It was from their study that her first personal observations took form. She wished to submit them to her master Lagrange and decided to write him; but, as she confessed much later, fearing the "ridicule attached to the name of *femme savante*," she signed her letter "Le Blanc, pupil at the Ecole Polytechnique." But Lagrange wanted to meet the Polytechnic student with such interesting observations. Astonished and charmed, Lagrange became a valued adviser for Sophie Germain and introduced her to all the French scientists of the time. She was quickly appreciated in scientific circles, as much for her learning as for the charm of her conversation. Nevertheless, later on, when she wanted to write Gauss, after the publication of his *Disquisitiones arithmeticae* in 1801, to discuss with him results she had obtained in the theory of numbers, she once more concealed herself under the pseudonym of Le Blanc, Polytechnic student. But Gauss, too, learned the true identity of Le Blanc. It was during the time of the German campaign and French troops were entering Brunswick, Gauss's city; Sophie Germain, haunted by the memory of Archimedes' death, began to fear for the scholar and wrote to a friend of her father, General Pernety, who was at the very moment in Brunswick, to commend her master to him and to entreat him to watch out for his safety. The latter hastened to reassure Sophie Germain and . . . to show her letter to the party concerned, who naturally continued with the young feminine mathematician the epistolary contact begun with the self-styled male Polytechnic student.

Sophie Germain died at the age of 55, after two years of terrible suffering which she bore with an admirable courage and stoicism. Her moral worth was a match for her beautiful intelligence; she loved virtue, it was said, like a geometric truth. She had the great honor in 1816 of receiving from the Paris Academy of Sciences the Grand Prize of the Mathematical Sciences, for a paper on the vibrations of thin elastic plates, a question put up for competition since 1811. It was in the year of her death that her important work on the curvature of surfaces appeared, in which she introduced for the first time the notion—classic today—of mean curvature. Her arithmetic work is no less important. Attacking the proof of Fermat's last theorem with the help of Legendre's formulas, she supplied an important theorem and its application to the proof of Fermat's theorem up to the hundredth degree. Finally, aside from her mathematical work, she left a certain number of articles on the history and philosophy of the sciences, articles of a genuine value, which Auguste Comte quoted with praise in his course on positive philosophy.

Her contemporary, Mary Fairfax, born in 1780, met from her father, a Scotch admiral, the same hostility toward her mathematical studies; and, despite her precocious propensities, it was only after a short widowhood and then remarriage to her cousin Somerville, that she succeeded in asserting herself as a mathematician. A long life—

she died in Naples at 92—made it possible for her to leave a respectable output under the name of Mary Somerville, as well as the memory of a devoted wife and a good mother. Mary Somerville's principal work consisted of translating and thus making known to her contemporaries the celestial mechanics of Laplace and of adding to it personal notes of real value. Mary Somerville also left a goodly number of papers in mathematics and physics; she was pensioned by Queen Victoria for her scientific work.

Mary Somerville had also had the honor of introducing to mathematics the only daughter of Lord Byron, Ada Byron—Countess Lovelace—born in 1815, died in 1852. Raised far from all parental influence, Ada Byron was early attracted to mathematics, and distinguished herself therein; she left original works which she signed A. L. L., a pseudonym whose true meaning was disclosed only thirty years later by General Menabrea, a correspondent of the Paris Academy of Sciences, and Italian ambassador to France.

Here finally are the last two, the two most striking—the Russian Sophie Kovalevski (1850–1891) and the German Emmy Noether (1882–1936)—so different from one another.

The first, Sophie Kovalevski, née Korvina-Krukovski, was a direct descendant of Mathias Corvinus, King of Hungary; her grandfather, by his marriage to a Gypsy, had lost his hereditary title of prince; on her maternal side, her grandfather was a stern scholar and to these contrary heritages Sophie attached great importance, explaining by them the sudden changes of her passionate character, so essentially feminine. "This morning," she wrote to her friend, Mme. Anne-Charlotte Leffler, when, as professor at Stockholm, she was spending a few days of vacation at Berlin, "I awakened with the greatest desire of having fun; all of a sudden who came in but my maternal grandfather, the German pedant, that is, the astronomer. He looked at the scholarly dissertations which I had promised myself to study during the Easter recess, and reproached me most seriously for unworthily wasting my time. His severe words put to flight my poor Gypsy grandmother."

It is impossible to recount here the childhood of Sophie in Russia, her life in the great family house, that winter spent in Saint Petersburg, where at thirteen she was infatuated with Dostoevski and jealous of her sister, whom he came to see frequently; one should read all this recounted by herself with so much talent in her *Recollections of Childhood,* which has merited this appreciation: "The Russian and Scandinavian literary critics have been unanimous in declaring that Sophie Kovalevski was the equal of the best writers of Russian literature, in style as well as in subject matter." Her premature death prevented her from realizing all her literary projects; in particular she had wanted to write "The Razhevski Sisters during the Commune," a reminiscence of her trip to Paris in 1871. She loved enormously to write and found in literature an escape and a respite from the excessive labor to which she applied herself during the periods of her life when mathematical research engrossed her. She wrote, moreover, in her memoirs: "At twelve years old I was thoroughly convinced I was born a poet." But at the same age, she listened eagerly to her uncle, who had just bought several books on mathematics, speaking to her about the quadrature of the circle, and, she says: "If the meaning of his words was unintelligible to me, they struck my imagination and inspired me with a kind of veneration for mathematics, as for a superior, mysterious science, opening to its initiates a new and marvelous world, inaccessible to the ordinary mortal." As to her subsequent ease in understanding analysis, she attributed that to the deep impression left in her mind by the hours she spent as a child contemplating the unusual wall covering in her room, in the large country mansion; not having any wallpaper, her father had used the lithographed pages of Ostrogradsky's course on integral calculus!

But let us come back to the life of Sophie Kovalevski. At seventeen she shared the thirst for science of all the intelligent Russian youth of the period, its desire for freedom and its political aspirations. Being a girl, she did not have the necessary independence to pursue her studies, so she decided with her sister and a girlfriend that one of them would make a fictitious marriage which would permit the three of them to attend a foreign university. Kovalevski, a very gifted young student, consented to help them, but on the condition that Sophie marry him. The young couple were installed at Heidelberg and spent a year wholly devoted to work—he studied geology, she mathematics. According to the report of her contemporaries, she seemed perfectly happy, with very great good fortune, of which she nevertheless later spoke with a certain bitterness. Small, slender, her hair short and curly, she captivated everybody with an unconscious charm. The coming into her home of her sister and her friend put an end to this intellectual intimacy with her husband; the latter left for Jena and then Munich, and if he came to see her frequently, Sophie nonetheless suffered from this separation and from the impression that her husband, totally absorbed in his studies, suffered from it less than she did. "He loves me only when near me," she said, at the same time refusing to end their irregular state of affairs. She always had, it seemed, a certain predilection for strained relations; she was a tortured being; she needed to be encouraged and admired. Too impassioned, she was not satisfied with abstract research. She had to be understood, encouraged with each new idea which came to her; a woman's weakness, Mittag-Leffler said deservedly—but how complex and winning a personality.

In 1870 she decided to go to Berlin to see Weierstrass, who, at the age of 55, was then in all his glory, and asked him for his advice. "To conceal her emotion, Sophie had put on a large hat with a broad brim, so that Weierstrass could see nothing of those marvelous eyes whose eloquence no one could resist when she wanted something." Weierstrass was won over nevertheless, and not being able to have Sophie admitted to his courses at the faculty—the university then being closed to women—he decided to give her private lessons. Sophie went to see him every Sunday afternoon, and Weierstrass came to see her one day a week. For four years, Sophie worked thus furiously, having no diversion, entirely absorbed in her studies, which were directly inspired by Weierstrass' work, of which they were either appendages or developments.

In 1874 she received from the University of Göttingen the title of doctor *in absentia* for her thesis *On the Theory of Partial Difference Equations*. But she was then completely exhausted by this excessive labor. She also lived in lamentable material circumstances, for, thoroughly unpractical and totally absorbed in mathematics, she was unaware of the insufficiency of her diet, the lack of comfort in her room and even, so they said, of whether her clothing was torn. Excessively worn out, surfeited with science to the point of no longer taking pleasure in research or discovery, she returned to Russia; she enjoyed herself, read novels, played cards and, upon the death of her father, feeling herself all alone, finally allowed herself to be loved by her husband and became his wife. The couple settled down in Saint Petersburg and led a brilliant life. The joy of living carried her away in a whirl of pleasures and parties; she was surrounded, admired, flattered. She forgot mathematics and her teacher Weierstrass; she no longer wrote to him and did not answer his letters, not even the one in which he urgently asked her to deny the rumor current in Berlin that she had become a society woman and had abandoned mathematics.

In 1878, Sophie brought into the world a little girl, Foufie; the enforced rest gave her back some desire for work and she wrote Weierstrass for a research project. The latter, forgetting her former ingratitude, always ready to help her, answered her; but it was only two years later that Sophie Kovalevski truly returned to mathematics. The immediate cause was a financial disaster: the brilliant life of the young couple required an income greater than theirs, and they had gambled on speculations which had good success only in the beginning. In this misfortune, Sophie found herself again, the born mathematician, able to live only for science. Her husband, on the contrary, could not decide to return to the simple life of the scholar and persisted in continuing his collaboration with the speculator who had involved them; it was the break-up of their marriage. Sophie went away again to work abroad. She stopped off first at Berlin, and upon the advice of Weierstrass, attacked the problem of the propagation of light in a crystalline medium.

It was at Paris in 1883 that she learned of her husband's death, and it was a terrible shock for her. Several months later, Sophie Kovalevski became Privatdocent of Professor Mittag-Leffler at Stockholm; moreover, she was to become professor for life on this same faculty. Weierstrass had finally succeeded in finding for his pupil, despite the difficulties caused by her sex, a position worthy of her. Sophie, completely happy, blossomed in the pleasant life of Stockholm. However, she quickly became weary of it—her Gypsy grandmother, she said—and soon considered Stockholm as an exile and went to Berlin or Paris as often as possible. The sentimental element then played an important part in her life. Indeed, she had met a man who finally seemed bound to give her perfect happiness. She was prey to a terrible struggle between her feminine aspirations and her scholarly ambitions. She wrote to Mme. Leffler from Paris, in the midst of her work: "Yesterday was a rough day for me, for big M . . . left in the evening. If he had stayed here I do not know how I would have been able to work. He is so tall, so powerfully built, that he manages to take up a great deal of room, not only on a sofa, but also in my thoughts and I would never have been able in his presence to think of anything but him." It was exactly at this moment of her life that she produced her master work. Her reputation as a mathematician was also at stake; all her scientific friends knew that she was competing for the Bordin Prize. She succeeded in presenting in time her paper entitled *On the Rotation of a Solid Body about a Fixed Point,* and was awarded the prize.

The equations for the problem had been set up in definitive form by Euler, who had himself given an integrable case of it (a case named after Euler and Poinsot); later Lagrange and Poisson discovered a second one, and Jacobi gave the general solution of the problem in terms of elliptic functions. But in her paper, Sophie Kovalevski found a new integrable case, that is to say a new case in which it is possible to find for the equations of the notion of a heavy solid moving about a fixed point a third algebraic "first integral" distinct from that of kinetic energy and from the integral for areas with arbitrary initial conditions. Husson, moreover, proved later that these three cases are the only ones. Sophie Kovalevski's paper was of such exceptional worth that the Commission of the Academy of Sciences, to attest its importance, increased the value of the prize from 3000 to 5000 francs. Her teacher Weierstrass expressed his joy to her in these terms: "I do not need tell you how much your success has rejoiced my heart and those of my sisters, as well as all your friends here. I have particularly experienced a real satisfaction; competent judges have now given their verdict that my faithful pupil, my 'weakness,' is not a frivolous marionette."

Sophie experienced a real triumph in Paris on Christmas Eve, 1888, at the time of the conferring of her prize. Feted, invited everywhere, she was in full glory, she seemed completely happy. Her friend had come to rejoin her. But this happiness was of short duration; with her demands and her tyrannical and jealous love, she asked too much of him; she fancied that she found in him an admiration rather than a love comparable to hers. Besides, she was unwilling to give up her career and become simply the wife of this man whom she loved and admired, and so it was that they separated often, full of bitterness, becoming reconciled only to part violently once more. Unable to do without him any more than to live with him, exhausted, torn by this incessant strife, she became ill and died in 1891 at the age of 41 after a brief attack of influenza. Her death was universally mourned; every journal published articles in her praise. Her funeral was magnificent; from every country people sent a profusion of flowers and the women of Russia had a monument erected to her.

Emmy Noether was born in Germany in 1882 in the little university town of Erlangen where her father, Max Noether, had been appointed professor in 1875 and where he remained until his death in 1921. Max Noether, a celebrated mathematician who played a considerable part in the development of the theory of algebraic functions, had two children: Emmy and her brother Fritz, two years younger than she. Both inherited mathematical ability from their father. Fritz was attracted to applied mathematics—Emmy, on the contrary, "thought only in concepts," calculation and intuition being equally strange to her! Emmy Noether developed, then, in a mathematical environment, with her father and his friendly colleagues, Gordan in particular. She received a carefully supervised education and a solid foundation, but nothing seemed to foreshadow the great mathematician she would become. Nothing seemed to impel her particularly toward research. She was a girl of good family, cultivated, taking part in household duties and in social activities. Admittance to German universities and to scientific careers becoming possible for women, she naturally directed her steps in this direction, always ready to accept life as she found it. Hers was not a tormented nature, still less rebellious. Devoid of feminine charms, she had neither women's trickeries nor their false-heartedness. Hers was a simple and good soul, without ambition, full of courage and life, a faithful friend. "There remained about her to the very end," said Weyl, "something child-like, as if an entire part of her being, overwhelmed by her mathematical genius, had not been developed." How far she was from Sophie Kovalevski, a prey to the torments of her ambition struggling against her love; a less complex personality, she was without doubt also happier. Quite naturally then she first worked with Gordan and did her thesis, *On Complete Systems of Invariants for Ternary Biquadratic Forms,* with him—she upheld it in 1907. But she quickly turned away from Gordan's influence, which so ill suited her nature—Gordan, who turned out pages of calculation with hardly a word of text, and who said one day about an abstract proof of one of his results, given by Hilbert simultaneously with a broad generalization, "This is no longer mathematics, but theology." Yet it was probably the influence of the Erlangen environment which made an algebraist of her. She learned much from Fisher before settling at Göttingen under the influence of Klein and Hilbert. There Emmy Noether found a long period of work, and results which were interesting, but in which her personality had not yet appeared. She was still somewhat dependent upon her teachers; her creative power, so original, even brilliant, revealed itself only later. Weyl made it begin in 1920 with a work published in collaboration with Schmeidler on modules of non-commutative domains. It was starting from this moment that she began to change the aspect of algebra. She did it in a series of papers, but also in her teaching and by her effect on her pupils. She had, in fact, been "qualified" at Göttingen in 1919, thanks to the new regulations put into force by the German Republic. During the war Hilbert had already tried to obtain her qualification, but in vain; he ran afoul principally of the hostility of the philologists and historians. There was an anecdote often told at Göttingen that to defend her he thought it clever to say to the council of the university: "I do not see why the sex of the candidate should be an argument against her appointment as Privatdocent; after all, we are not a bath-house . . ." Meanwhile she was authorized to give lectures at Göttingen, then in 1922 appointed "nichtbeamteter ausserordentlicher Professor" [unofficial professor extraordinary], a simple title with no obligations and no salary. She was entrusted, nevertheless, with an algebra course with modest pay—a mediocre position which she kept until 1933 and which was unworthy of her personal ability and the wide influence of her classes, which were attended by many foreigners attracted by her methods. She was nevertheless not a good lecturer; she even seemed devoid of pedagogical qualities. Van der Waerden relates that "the touching trouble she took to explain her statements, even before she had completely uttered them, by corollaries rattled off at top speed, had rather the opposite effect." And what work to follow her! But what profit for a small group of those who succeeded in doing so; each lesson was a syllabus! She did not teach completed, polished things, but rather theories in the process of becoming, distributing her ideas generously to her pupils, happy when they profited by them and pursued her investigation. Round about her—I was going to say "him," for they more often called her *der* Noether, not so much because of her face with its masculine features and her corpulence as for the unfeminine authority which her mathematical superiority conferred upon her—they formed a little family. She loved them, was interested in their personal affairs, was kind and motherly to them, always ready to help them but also always an implacable judge. Of those on whom she had the greatest influence, let us mention among the best-known: Krull, Grell, Koethe, Deuring, Fitting and F.-K. Schmidt. Van der Waerden came from Holland to follow her courses, and learned from her the notions which enabled him to formulate his own ideas and solve the problems which he had set himself. A part of his remarkable book *Modern Algebra* is only the work of Emmy Noether, but clarified and set in order by him. She also usefully collaborated with Hasse and Richard Brauer. With Alexandroff she was linked by a deep friendship; moreover, she spent a semester at Moscow around 1930 and Alexandroff often came to Göttingen as a guest.

It was in the midst of her scientific activity and influence that the racial laws introduced by the revolution of 1933 interrupted her career at Göttingen and led finally to her departure along with Born, Courant, Landau, Neugebauer and many others. In the tragic months which preceded their departure she was a comfort to all with her courage and her indifference to her own fate. After a short stay at the Institute for Advanced Studies at Princeton, she became a professor at the nearby women's college Bryn Mawr. With her happy nature, she adapted herself there admirably and "her Bryn Mawr girls became as dear to her as her Göttingen boys." There and at nearby Princeton she once more formed about her a nucleus of researchers. In America as in Germany she pursued her research, the source of her greatest joy. Emmy Noether was at full maturity, she had just published her major works on non-commutative algebras—the principal subject of her research, once she had completed her axiomatic theory of ideals—when her accidental death interrupted this beautiful, hard-working life. She had just undergone an operation which seemed to have been completely successful, when in a few hours she died, on April 14, 1935, of an unforeseen complication.

She left a considerable body of work, published both alone and in collaboration. . . .

Dedekind had introduced ideals in algebraic number fields and established the decomposition of an ideal into the product of prime ideals. Lasker had proven that for ideals of polynomials there is in general only one decomposition into the lowest common multiple of primary ideals. In her first paper, Emmy Noether introduced her famous axiom of divisor chains, which permitted her to establish the theorem of the decomposition of an ideal into l.c.m. of primary ideals in every ring satisfying this axiom, and in particular in polynominal rings according to Hilbert's finite base theorem. In the second paper she exhibited the axioms which allowed the final result of decomposition into the product of prime ideals.

Her works on hypercomplex systems, the theory of representation and, more generally, noncommutative algebra are especially characterized by the important role played in them by the notions of module, ideal and automorphism, and by the fact that her results are valid whatever the fundamental field, whereas in the works of Frobenius and his immediate successors, this field was that of complex numbers or that of real numbers. By theorems such as that of the "cross product," developed by herself alone or in collaboration with Hasse and Brauer, Emmy Noether attained results in algebra and arithmetic of great depth, where hypercomplex methods are brilliantly applied to difficult problems in the theory of class fields.

This work not only puts Emmy Noether in the first rank of women mathematicians, but places her among the very great mathematicians.

Reminiscences of a Mathematical Immigrant in the United States

Solomon Lefschetz

Solomon Lefschetz was born in Moscow, Russia, in 1884. Educated in the United States he was awarded the Ph.D. in mathematics from Clark University. The author of numerous powerful results in mathematics, he was awarded the National Medal of Science by the United States, the Bocher Prize of the American Mathematical Society, and other international honors. He died in October 1972.

As a nation of immigrants, the United States in the twentieth century has been the beneficiary of one of the most remarkable transfers of intellectual ability in history. It is hardly an exaggeration to state that until 1900 the total contribution of native-born Americans to mathematics would not fill one page in a multivolume history. Yet, by 1950 the United States was universally recognized as the world's principal center of mathematics research. Many of the most significant mathematical results of this century, solutions of problems hundreds of years old, have been achieved by Americans and have been done since 1950.

The foundation for this remarkable transformation was laid in the 1920s and 1930s, accelerating as the Nazi horrors swirled toward their ghastly climax. A seemingly endless stream of the best mathematical minds of continental Europe emigrated to the United States. At first, most of the immigrants were seeking mere survival. In time, many took root to form an American mathematical community amazingly parallel to the nation as a whole—diverse, somewhat nomadic, multilingual, temperamental, yet amazingly fruitful. One mathematical immigrant, Solomon Lefschetz, relates his experiences in this article.

MY CAREER AS mathematical immigrant began in 1911 upon my receiving the Ph.D. degree from Clark University (Worcester, Mass.). While small, Clark had as its President G. Stanley Hall, an outstanding psychologist, and several distinguished professors. The mathematical faculty consisted of three members: W. E. Story, senior professor (higher plane curves, invariant theory); Henry Taber (complex analysis, hypercomplex number systems); De Perrott (number theory).

There were great advantages for me at Clark. I graduated from the *École Centrale* (Paris) (one of the French *"Grandes Écoles"*) in 1905, and for six years was an engineer. I soon realized that my true path was not engineering but mathematics. At the *École Centrale* there were two Professors of Mathematics: Émile Picard and Paul Appel, both world authorities. Each had written a three-volume treatise: *Analysis* (Picard) and *Analytical Mechanics* (Appel). I plunged into these and gave myself a self-taught graduate course. What with a strong French training in the equivalent of an undergraduate course, I was all set.

To return to Clark, I soon obtained a research topic from Professor Story: to find information

Source: Solomon Lefschetz, "Reminiscences of a Mathematical Immigrant in the United States," *American Mathematical Monthly* 77, (1970): 344–350.

about the largest number of cusps that a plane curve of given degree may possess. An original contribution which I made secured my Ph.D. thesis and my doctorate in 1911.

At Clark there was fortunately a first rate librarian, Dr. L. N. Wilson, and a well-kept mathematical library. Just two of us enjoyed it—my fellow graduate student in mathematics and future wife, and myself. I took advantage of the library to learn about a number of highly interesting new fields, notably about the superb Italian school of algebraic geometry.

My first position was an assistantship at the University of Nebraska (Lincoln), soon transformed into a regular instructorship. This meant my first contact with a regular midwestern American institution and I enjoyed it to the full. I owed it mainly to the very pleasant and attractive head of the department, Dean Davis of the College. The teaching load, while heavy, did not overwhelm me since it was confined to freshman and sophomore work.

Not too many weeks after my arrival, the Dean got me to speak before a group of teachers in Omaha on "Solutions of algebraic equations of higher degree." And then and there I learned an all-important lesson. For I spoke three quarters of an hour—three times my allotted time! When I found this out some weeks later from the Dean, my horror knew no bound. I decided "never again," to which I have most strictly adhered ever since.

A second lesson was of another nature. I utilized my considerable spare time in reading Hilbert's recent papers on integral equations. At Clark I had also read Fredholm's Acta paper on the same topic and my enthusiasm for integral equations was very great. I offered to lecture on Hilbert's work in my fourth term and this was accepted. Consequence: a very heavy teaching load for two students who I fear were quite bewildered. One of them, Oliver Gish, a graduate student in physics (later a distinguished geophysicist) remained my lifelong friend. I also formed a close friendship with his mentor and a capable mathematician, Professor L. B. Tuckerman (later of the Bureau of Standards).

The course taught me a valuable lesson: the experience generally absorbs too much energy. I have since expressed this opinion to many a recent doctor, but I fear that few heeded it.

My two years in Nebraska made me realize a widespread feature of American institutions of higher learning which were State institutions. By general state rule they had to accept any graduate from an accredited high school. Consequence: in the freshman year a flood of very poorly prepared students and a large number of sections, especially in the first term. By the end of the first year the entrance flood was reduced to half; the sophomore sections—in mathematics at least—were in much smaller number, more readily handled and better taught. This went on down to the last year, with the flood in mathematics reduced to 10–15 or so (mostly girls) and the total number of graduates much smaller than at entrance.

Lincoln, the capital of the State (population about 50,000) was a very pleasant city, with a distinct urban flavor. It was not too far from Omaha, the major city of the State. Most family houses were surrounded by a small garden and the whole made a very good impression. The University was at one end of the town; the Agricultural College, part of the University (pet of the very rural Board of Regents), at the other end. There were a couple of small colleges situated in Lincoln.

At the end of two years (1913) a larger offer, plus my approaching marriage to my Clark fellow student, made me accept an instructorship at the University of Kansas in Lawrence. The teaching conditions there were the same as in Lincoln, but with a slightly smaller load. At the University of Kansas the department was divided into two groups: college plus graduate work and engineering. I was assigned to the latter. While the students were somewhat more purposeful, the preparation was equally weak in both parts.

Lawrence (population 12,000) had a rather severe New England tradition. Except for the University with about 3000 students, it was really a most pleasant rural community. The University was on top of quite a hill, with well-constructed and mostly recent buildings. The view from the top was exceptionally attractive.

The major city nearby was Kansas City. Lawrence was about 25 miles from Topeka (the capital), while Kansas City was 50 miles away. This was all before the automobile age and my friends and I indulged in many country walks.

The general entrance preparation in Lawrence and Lincoln was so feeble that early teaching could only be technical and deprived of theory. As the freshman flood eroded, this situation improved somewhat.

The rule in Lawrence for beginning faculty members was three years in each position and it was rather rigidly enforced. The situation did not seem perfect—far from it. However, I discovered in myself first a total lack of desire to "reform" coupled with a large adaptive capacity. At Lawrence I only cooperated with a colleague in driving out several unattractive texts, notably Granville's Calculus, for which my taste was $< \epsilon$.

Years later I inquired of Professor Lusin (Moscow) why the Soviet mathematicians translated Granville. Reply: "We only took his excellent collection of problems, but provided our own theory." This may explain our efforts to move this book out of Kansas.

At this place I was prepared to indulge in extensive criticism, at least of the midwestern system. The fact is, however, that in both Nebraska and Kansas I found good and well-kept mathematical libraries, ample at least for my own purposes. Moreover, I came to realize the enormous advantage over the European system: it provided uncountably many opportunities for younger research men with ideas to grow and develop their powers, as instructors for example, with ample leisure. For the teaching loads, while considerable, were not really intolerable. Moreover, they generally went with colleagues who had other interests, mathematical or administrative, but not intent upon imposing on one uncongenial mathematical interests. At all events, in my case, it turned out to be of great value. Needless to say, special research favors were rare indeed.

In spite of the general level, I had in Lawrence three or four excellent students. One of them, Warren Mason, went to work for Bell Laboratories in New York (later near Elizabeth, N.J.), took his Ph.D. in physics at Columbia, and at Bell became a top specialist in the theory of sound and its applications. I am very proud of him. Still another strong student, Clarence Lynn, joined forces with Westinghouse in Pittsburgh (electrical department) and was most successful there.

I have found that in freshman courses in mathematics, and less so in the next year, hardly one third of the students care for and are not totally bored by mathematics. Hence at that early level a teacher must be exceptionally lively and have a sympathetic understanding of the students. Needless to say this must be coupled with a complete grasp of the topic taught.

Here are a few very radical suggestions for later years. From the junior year on through graduate work they should be merged into a professional school, with teaching, at least in mathematics, of seminar type plus abundant but easy contact with faculty on an individual basis. In other words "baby talk" should end with sophomore years.

The guidelines in my research were: Picard-Simart: *Fonctions algébriques de deux variables* (two volumes, mostly Picard); Poincaré's papers on topology (= analysis situs) and on algebraic surfaces; Severi's two papers on the theory of the base; Scorza's major paper (dated 1915) in *Circole di Palermo* on Riemann matrices.

Around 1915 and for a long time, a certain result of Picard baffled me. Let H be a hyperelliptic surface. Direct calculation yielded: the Betti number $R_2(H) = 6$. Picard, however, appeared to give its value as 5. The discovery of the missing link played a major role for me. Namely, Picard only wanted R_2 for the *finite* part of H, neglecting the curve C *at infinity. Hence* C *was a 2-cycle,* and so was any algebraic curve! This launched me into Poincaré-type topology, the 1919 Bordin Prize of the Paris Academy and in 1924 Princeton! (The translated prize paper appeared in the *Trans. Amer. Math. Soc.*, vol. 22, 1921.)

The immediate effect of the Prize was the Kansas promotion (January 1920) to Associate Professor plus a schedule reduction. Also (1923) there came a promotion to a Full Professorship. I spent the year 1920–21 in Europe, half in Paris, half in Rome. I gathered little mathematical profit in Europe; some from the summer of 1921 which I spent in Chicago.

About Paris I particularly remember an interview with Émile Borel lasting five minutes in which I offered to write for his series my future monograph *L'Analysis Situs et la Géométrie Algébrique*. He accepted at once! (In such matters our "speedy" country knew no such speed.) Proof sheets, etc., were dealt with rapidly and not a syllable was changed.

I come now to my Princeton period. In 1923 an invitation came from Dean Fine, the Chairman of the Department of Mathematics and Dean of the Faculty at Princeton, to spend the following year there as Visiting Professor of Mathematics. Dean Fine was the long-time head of the department and the true founder of what became an outstanding department of the University. With reason, upon the construction of the mathematical building it was called "Fine Hall." (Dean Fine was killed in an automobile accident just before Christmas 1928 and his lifelong friend, Mr. Thomas D. Jones, immediately granted $600,000 as a memorial to Dean Fine for a new mathematical building.)

Well, upon receiving Dean Fine's invitation, I accepted. For the following year I received a permanent offer to stay at Princeton as Associate Professor. This was changed 18 months later (January 1927) to a Full Professorship and January 1932 to a Research Professorship (Fine professorship) as successor to Oswald Veblen. In this position I had no assigned duties whatever.

At Princeton I found myself in a world-renowned University and in one of its outstanding Departments. Among the great mathematical Professors there were: Eisenhart, Veblen, Wedderburn, Alexander, Hille. I was in closest contact with Alexander—a top authority in topology.

My joining the Princeton faculty coincided with a definite change of direction in my research from the applications of nascent topology to algebraic geometry (*vide* my prize paper) to a pure topological problem: coincidences and fixed points of transformations. For this problem I invented a completely new method of attack, which by 1925 culminated in a well-known fixed point expression $\phi(f)$, f a mapping of a manifold into itself, that said: $\phi \neq 0$ implies that f has fixed points; if f has none, $\phi = 0$. The preparation and extensions required occupied me for several years. One of my early graduate students, A. W. Tucker, an outstanding Princeton mathematician, found the way to a far simpler method than my early one, which I have accepted *in toto*.

Much of my Princeton teaching, until 1930, was still freshman-sophomore. However the students, selected with care at entrance, were much better prepared than in the midwest. The contrast of the systems was very great.

Princeton system: A strictly private school, with limited funds and space, could not accept all comers. Hence it had, unavoidably, to fix the number of admissions, utilize a strict selection, and keep the admitted men practically through the four collegiate years. The same system, in some form, was also applied to admission to the Graduate School.

Midwestern system: As I already stated, they had to admit all duly certified high school students. The freshman entrance flood resulted in teaching mostly by graduate students, many of uncertain quality.

The Princeton system had two important consequences. First, it enabled one to organize preferred sections even before entrance. Second, courses could be initiated at a more advanced stage and proceed more speedily. Thus algebra and trigonometry were done each in two weeks, analytical geometry in five weeks, calculus started in the second freshman semester (in Kansas-Nebraska in the sophomore year).

Some years later, good students from strong preparatory schools or high grade secondary schools (where they already had these subjects) were allowed to skip, even the whole first year. Moreover, such A-1 men (not many) were soon treated like graduate students, allowed to participate in advanced seminars and thus to become well acquainted with the members of the mathematical faculty.

The Princeton aim was decidedly different from the Nebraska-Kansas aim. The latter had to provide for a considerable number of teachers in their states, to form moderate level technicians of all kinds, sending a very few of the best for better training to major eastern institutions. Princeton on the contrary was planned to form the top echelons, notably in the sciences. This meant aiming first for the doctorate. In mathematics it soon became customary to retain the best men for at least one year after the Ph.D. on some fellowship, or in some teaching position with very light duties. A number of the men so developed occupy today major posts in outstanding institutions.

In 1932 a major change took place through the establishment at Princeton of the Institute for Advanced Study, with mathematics as its first and strongest group. This resulted in the migration of three of our major members: Veblen, Alexander and von Neumann.

The basic effect on me was regaining the mathematical calm of Nebraska-Kansas, which I had so enjoyed without realizing it. Our mathematics chairman, Dean L. P. Eisenhart, with the unstated motto "live and let live" had much to do with this return of calm. During this period my mathematical work progressed. My first Topology treatise (1930) appeared and was many times approved by friendly colleagues. A second Algebraic Topology appeared in 1942, rather less satisfactory, because it was too algebraic. Other books came. I was editor of the *Annals of Mathematics,* which grew to occupy an A-1 place in mathematics, but did not overwhelm me with work. Then came World War II and I turned my attention to Differential Equations. With Office of Naval Research backing (1946–1955) I conducted a seminar on the subject from which there emanated a number of really capable fellows, also a book: *Differential Equations, Geometric Theory* (1957).

When Dean Eisenhart retired (1945) I succeeded him as Chairman, until my own retirement in 1953.

In 1944 I joined as a part-time connection the *Instituto de Mathematicas* at the National University of Mexico. This continued until 1966. At the *Instituto* I was as free as under my Princeton professorship. I conducted seminars in topology and differential equations, gave a couple of times a "volunteer" course on "general mathematical concepts" directed at beginners and, thanks to a good working library, was able to continue research. Conditions were of course quite different from ours, but as I became rapidly fluent in Spanish, it gave me many advantages. Through the years I found quite a number of capable young men, several of whom I directed to Princeton for further advanced training up to the doctorate and later. Among them I may mention Dr. José Adem, Chairman of the Department of Mathematics of the newly founded *Centro de Estudios Avanzados* in Mexico City.

My long connection with Mexico has been the occasion of many side trips (especially in connection with meetings of the Mexican Mathematical Society), so that I have a fair acquaintance with that wonderful country.

In 1964 the rarely awarded order of the Aztec Eagle was conferred upon me by the government of Mexico.

My work as Russian reviewer for differential equations had made me aware of our lag relative to the Soviets in this all important field in all sorts of applications. The arrival of Sputnik in 1957 convinced me that this lag had to be remedied. As I attributed it to our scattered efforts, I came to the conviction that the only remedy was to establish a Center for study and research in differential equations.

From Dr. Robert Bass, formerly a member of my project, I learned of the formation in Baltimore, as a division of the Martin Aircraft Company, of a new Research Institute for Advanced Study (RIAS) under the direction of Welcome Bender, a graduate of MIT and long time Martin engineer. When I approached him with my (modest) plans he was enthusiastic. In a few days I was entrusted with the formation of a group of say five top men and about ten younger associates, with myself as director. Suffice it to say that I had considerable success. I first was able to obtain the cooperation of Prof. Lamberto Cesari of Purdue, one of the major specialists anywhere; also of Notre Dame, Prof. J. P Lasalle as my second in command (my best appointment) and complete the group with Dr. J. K. Hale of Purdue (Cesari's best student there) and Dr. Rudolph Kalman of Columbia (an electrical engineer coupled with good mathematics). My strong basic group was thus complete.

I demanded (and obtained) from Mr. Bender that my group operate under standard university conditions.

Very shortly we became known. A considerable number of the good differential-equationists visited us, and some few were invited for a year or so.

After some six years it was necessary to transfer our Center elsewhere. This operation, carried by Lasalle, resulted in our becoming part of the Division of Applied Mathematics at Brown University as "Center for dynamical systems" with Lasalle as Director and myself as (once weekly) Visiting Professor. At Brown our general relationship has been excellent. A year or so ago the Director of the Division died and was succeeded by Lasalle whose general performance could not be excelled.

In conclusion I must recognize a budget of debts which I may never succeed in liquidating to the full.

The first is my enormous debt to my wife Alice, my Clark companion. Without her constant and unfailing encouragement through 59 years, 56 as my wife, I would have long since ceased to operate. . . .

Second major debt: to the United States, which through their (however imperfectly organized) universities made it possible for me to follow my deep bent for mathematics. I should also include here the contribution of the National University of Mexico from 1944 to 1965—years after my Princeton retirement, and also of RIAS and Brown.

In this long and agreeable route of 57 years I encountered so many *simpáticos amigos* that to name them all would be impossible. May they one and all accept my fervent *gracias* for my debt to them. I hope that they have felt that it was not incurred in vain.

Prof. Lefschetz [had] an astonishingly productive career. His profound influence in the development of topology and of algebraic geometry is expounded at length in articles by W. V. D. Hodge and Norman E. Steenrod in the Princeton Symposium volume in honor of S. Lefschetz, *Algebraic Geometry and Topology* (1957) edited by R. H. Fox, D. C. Spencer, and A. W. Tucker. . . .

Graduate Student at Chicago in the Twenties

W. L. Duren, Jr.

William Larkin Duren, Jr., born in Mississippi in 1905, received his doctorate from the University of Chicago. He was chairman of the University of Tulane from 1947 to 1955 and then became dean of the mathematics department at the College of Arts and Sciences, University of Virginia, from 1955 to 1962. Duren has also been president of the Mathematical Association of America.

The University of Chicago was and continues to be one of the few truly outstanding American universities. Its reputation is international and it has sheltered a truly distinguished assembly of scholars. In the 1920s it was one of a relatively few centers of excellence not on the eastern seaboard. The following is a personal account of what it was like to be a student there at the time that American mathematics was beginning its amazing mid-twentieth century efflorescence.

Graduate students sometimes think that what they are going through is truly a unique experience in the world's history. In fact, we humans often tend to think that the period we are going through is The *pivotal point in history. It is refreshing and sobering to read about the graduate experience of sixty years ago and realize that except for the prices, everything seems vaguely familiar.*

In reading these biographies one recognizes that there is little new in the human experience. Through such recognition we may each gain a measure of comfort that we are part of a community of people who are all experiencing or have experienced similar human problems.

AS AN UNDERGRADUATE at Tulane in New Orleans, 1922–'26, I was programmed to go to the University of Chicago and study celestial mechanics with F. R. Moulton. My teacher, H. E. Buchanan, had been a student of Moulton. That was an example of the great strength of the University of Chicago. Its Ph.D. graduates made up a large part of the faculties of universities throughout the Mississippi Valley, Midwest and Southwest. So they sent their good students back to Chicago for graduate work. I went there first in the summer of 1926 and came to stay in 1928. In the interim I studied Moulton's *Celestial Mechanics* and some of his papers in orbit theory. I met Moulton at a sectional meeting of the MAA where he was the invited speaker. He was a man of great charm and energy and was most encouraging to me. But by the time I got to Chicago in 1928 Moulton had resigned. I was told that he felt it was an ethical requirement, since he and his wife were getting a divorce. On the advice of T. F. Cope, another former student of Buchanan, who was working with Bliss, I turned to Bliss as an advisor in the calculus of variations.

It was a down cycle for mathematics at Chicago. All the great schools have their downs as well as ups, partly because great men retire, partly because their lines of investigation dry up. At Chicago at that time a young student could see the holdovers of the great period, 1892–1920, in Eliakim Hastings Moore, officially retired, Leonard E. Dickson, rounding out his work in algebra, Gilbert A. Bliss, busy with administration and planning for the projected Eckhart Hall. Also

Source: W. L. Duren, Jr., "Graduate Student at Chicago in the Twenties," *American Mathematical Monthly* 83 (1976): 243–248.

there was Herbert E. Slaught, teacher and doer, one of the original organizers of the Mathematical Association of America and its *Monthly*, even if he played only a supporting role in mathematics itself. He had an extrovert, friendly personality that reached out and got hold of you, whether he was organizing a department social or the Mathematical Association of America. He was the teacher of teachers and key figure in Chicago's hold on education in the midwest and south. Every graduate department needs a man like Slaught if it is fortunate enough to find one. He was being succeeded by Ralph G. Sanger, a student of Bliss, an outstanding undergraduate teacher, though not the organizer Slaught was.

The University of Chicago was founded in 1892 with substantial financial support from John D. Rockefeller. William Rainey Harper, the first president, had bold educational ideas, one of which was that the United States was ready for a primarily graduate university, not just a college with graduate school attached. Harper brought E. H. Moore from Yale to establish his department of mathematics. Moore's graduate teaching was done in a research laboratory setting. That is, students read and presented papers from journals, usually German, and tried to develop new theorems based on them. The general subject of these seminars was a pre-Banach form of geometric analysis that Moore called "general analysis." It was itself not altogether successful. But even if general analysis did not succeed, Moore's seminars on it generated a surprising number of new results in general topology, among them the Moore theorem on iterated limits and Moore-Smith convergence. Moore's seminars also produced some outstanding mathematicians. His earlier students had included G. D. Birkhoff, Oswald Veblen, T. H. Hildebrant and R. L. Moore, who took off in different mathematical directions. R. L. Moore developed the teaching method into an intensive research training regimen of his own, which was very successful in producing research mathematicians at the University of Texas.

I studied general analysis with other members of the faculty including R. W. Barnard, whom Moore had designated as his successor and whose notes record the second form of the theory [Am. Philosophical Soc., *Memoirs,* v. 1, Philadelphia, 1935]. Instead of taking the general analysis courses, my old friend E. J. McShane, from New Orleans, worked in Moore's small seminar on the foundations of mathematics. Although he was officially a student of Bliss, I think he was in a sense Moore's last student.

Moore himself was meticulous in manners and dress. He would stop you in the hall, gently remove a pen from an outside pocket and suggest that you keep it in the inside pocket of your jacket. Nobody thought of not wearing a jacket. But Moore was less gentle if you used your left hand as an eraser, and he displayed towering anger at intellectual dishonesty. To understand him and his times one must read his retiring address as President of the Society [*Science,* March 1903]. In those days the Society accepted responsibility for teaching mathematics and Moore's address was largely devoted to the organization of teaching, the curriculum, and the ideas of some of the great teachers of the time, Boltzman, Klein, Poincaré, and, in this country, J. W. A. Young and John Dewey, whose ideas Moore supported by proposing a mathematics laboratory. This address was adopted as a sort of charter by the National Council of Teachers of Mathematics and republished in its first Yearbook (1925). By the time I got to Chicago the Association had been formed to relieve the Society of concern for college education, and NCTM to relieve it of responsibility for the school curriculum and training teachers. In the top universities only research brought prestige, even if a few, like Slaught, upheld the importance of teaching.

L. E. Dickson's students tended to identify themselves strongly as number theorists or algebraists. I felt this particularly in Adrian A. Albert, Gordon Pall and Arnold Ross. All his life Albert strongly identified himself, first as an algebraist, later with mathematics as an institution and certainly with the University of Chicago. I remember him as an advanced graduate student walking into Dickson's class in number theory that he was visiting, smiling and self confident. He knew where he was going. Dickson was teaching from the galley sheets of his new *Introduction to the Theory of Numbers* [University of Chicago Press, 1929] with its novel emphasis on the representation of integers by quadratic forms. I think he requested Albert to sit in for his comments on this aspect. He was tremendously proud of Albert. I remember A^3 too with his beautiful young wife, Frieda, at the perennial department bridge parties. He had superb mental powers; he could read a page at a glance. One could see even then that as heir apparent to Dickson he would do his own mathematics rather than a continuation of Dickson's, however much he admired Dickson.

In the conventional sense Dickson was not much of a teacher. I think his students learned from him by emulating him as a research mathematician more than being taught by him. Moreover, he took them to the frontier of research, for the subject matter of his courses was usually new mathematics in the making. As Antionette Huston said, "He made you want to be with him intellectually. When you are young, reaching for the stars, that is what it is all about." He was good to his students, kept his promises to them and backed them up. Yet he could be a terror. He would sometimes fly into a rage at the department bridge games, which he appeared to take seriously. And he was relentless when he smelled blood in the oral examination of some hapless, cringing victim. He was an indefatigable worker and in public a great showman, with the flair of a rough and ready Texan. An enduring bit in the legend is his blurt: "Thank God that number theory is unsullied by any application." He liked to repeat it himself as well as his account of his and his wife's honeymoon, which he said was a success, except that he got only two papers written.

The theme of beauty for its own sake was expressed more surprisingly by another Texan who worked in mechanics and potential theory, W. D. MacMillan. According to the story he had come to Chicago as a mature man, without a college education, to sell his cattle. Having sold them, he went to Chicago's Yerkes Observatory to see the Texas stars through the telescope. He was so fascinated that he stayed on to get his degrees in rapid succession, all *summa cum laude.* Then he remained as a member of the faculty. One day in his course on potential theory he wrote some important partial differential equations on the board with obvious pleasure, drawing the partial derivative signs with a flourish. Standing back to admire these equations, he said: "That is just beautiful. People who ask, 'What's it good for?', they make me tired! Like when you show a man the Grand Canyon for the first time and you stand there as you do, saying nothing for a while." And we could see that old Mac was really looking at the Grand Canyon. "Then he turns to you and asks, 'What's it good for?' What would you do? Why, you would kick him off the cliff!" And old Mac kicked a chair halfway across the room. He was a prodigy, a good lecturer, an absolutely fascinating personality with a twinkling wit. Some of his work was outstanding, yet he had few doctoral students.

Celestial mechanics was being carried on by the young Walter Bartky, who was, I think, Moulton's last student. But celestial mechanics had gone into a barren period and Bartky with his superb talents turned to other applications of differential equations, to statistics and to administration.

Lawrence M. Graves was the principal hope of the department for carrying on the calculus of variations, which he did in the spirit of functional analysis. He was my favorite professor because he knew a lot of mathematics, knew it well, and in an unassuming way was glad to share it with you. Although he taught Moore's general analysis, he pointed out the difficulties in it to me. His own brand of functional analysis was more oriented towards the use of the Frêchet differential in Banach space.

Research in geometry at Chicago was a continuation of Wylczinski's projective differential geometry. There was no topology, though we heard that Veblen's students studied something called *analysis situs* at Princeton. I knew so little about the subject that years later when I wanted to prepare for Morse theory I spent months studying Kuratowski's point set topology before it dawned on me that what I wanted was algebraic topology. E. P. Lane and his students carried on the study of projective differential geometry using rather crude analytical methods, that is, expansions in which one neglected higher order terms. We who were not Lane's students tended to look on it with disdain as being non-rigorous. But the structure of the theory was beautiful, I thought. Lane was honest about the shortcomings of the methods, though he did not know how to overcome them.

Lane was a very fine man. I had come to Chicago in 1926 to run the high hurdles in the National Intercollegiate Track and Field Meet at Soldiers Field. I placed in the finals and some members of the U.S. Olympic Committee urged me to keep working for the 1928 Olympics. So I worked on the Stagg Field track until an accident set off a series of leg infections. I was very sick in Billings Hospital in the days before antibiotics and it was Lane who came to the hospital to see me and make sure that I got the best available care. The only way I was ever able to express my thanks to him was to do a similar service to some of my own students in later years. I guess that is the only way we ever thank our teachers.

Bliss was an outstanding master of the lecture-discussion. He could come into a class in calculus of variations obviously unprepared, because of the demands of his chairmanship, and still deliver an elegant lecture, drawing the students into each deduction or calculation, as he looked at us quizzically and waited for us to tell him what to write. His students learned their calculus of variations very thoroughly. Yet we did not work together, except in so far as we presented class assignments. Each research student reported to Bliss by appointment. The subject itself had come to be too narrowly defined as the study of local, interior minimum points for certain prescribed functionals given by integrals of a special form. Generalization came only at the cost of excessive notational and analytic complications. It was like defining the ordinary calculus to consist exclusively of the chapter on maxima and minima. A sure sign of the decadence of the subject was Bliss's project to produce a history of it, like Dickson's *History of the Theory of Numbers*. The history reached publication only in the form of certain theses imbedded in *Contributions to the Calculus of Variations*, 4 vols, 1930–1944, University of Chicago Press.

It is perhaps surprising that this narrowly prescribed regimen turned out men who did important work in entirely different areas as, for example, A. S. Householder did in biomathematics and numerical analysis, and Herman Goldstine did in computer theory. Among all of us Magnus Hestenes has been most faithful to the spirit of Bliss's teaching in carrying on research in the calculus of variations. Yet when Pontryagin's Optimal control papers revived interest in the subject many years later, students of Bliss were easily able to get into it. Optimal control theory really contained relatively little that was correct and not in the calculus of variations. In fact, optimal control was anticipated by the thesis of Carl H. Denbow, *loc. cit.*

Quantum mechanics was breaking wide open in the twenties. Bliss himself got into it with his students by studying Max Born's elegant canonical variable treatment of the Bohr theory. While that was going on, Sommerfield's *Wellenmechanische Ergänzungsband* to his *Atombau und Spektrallinien* [Vieweg, Braunschweig, 1929] came out. It was the first connected treatment of the new wave mechanics formulation of quantum mechanics due to de Broglie and Schrödinger. We dropped everything to study wave mechanics. Bliss was a remarkably knowledgeable mathematical physicist and quite expert in the boundary value problems of partial differential equations. That was not so remarkable in a mathematician of his generation. The narrowing of the definition of a mathematician and withdrawal into abstract specializations was just beginning. In fact Bliss had been chief of mathematical ballistics for the U.S. Government in World War I, and later was commissioned to do a mathematical study of proportionate representation for purposes of reassigning Congressional districts. Bliss did not follow up his move into quantum mechanics but returned to the classical calculus of variations.

There were always more students in summers with all the teachers who came. Visiting professors like Warren Weaver, E. T. Bell, C. C. Mac Duffee and Dunham Jackson came to teach. And there was the memorable visit of G. H. Hardy which was supposed to provide a uniting of Hardy's analytic approach to Waring's theorem with Dickson's algebraic approach. Even with this infusion of talent, the offerings of the department were rather narrow. Besides having no topology as such, more surprisingly, there was little in complex function theory. And I do not recall being in a seminar, either a research or journal seminar. Essentially all teaching was done in lectures. Yet the only one of the abler students who I remember taking the initiative to go elsewhere was Saunders MacLane, when he did not find at Chicago what he was looking for.

I once asked Edwin B. Wilson, a famed universalist among mathematicians, how he came to switch from analysis to statistics at Yale. With a humorous twinkle he said: "An immutable law of academia is that the course must go on, no matter if all the substance and spirit has gone out of it with the passing of the original teacher. So when (Josiah Willard) Gibbs retired, his courses had to go on. And the department said: 'Wilson, you are it.' " A graduate student at Chicago in the late twenties could see this immutable academic law in effect. In each line of study of the, then passing, old Chicago department, a younger Chicago Ph.D. had been designated to carry on the work. If, in one's immaturity, this was not apparent, the point was made loud and clear in a blast from Dickson during a colloquium with graduate students present. Dickson charged the chairman with permitting the department to slide into second rate status. It was true that the spirit of original investigation had given way to diligent exposition in some of these fields. In some cases the fields themselves had gone sterile.

It was the lot of Bliss to preside over this ebb cycle of the department. He did an impressive best possible with what he had, with high mathematical standards, firmly, kindly and quietly. Most of the difficulties he had inherited. Bliss was able to appoint some outstanding young men but, if he had asked for the massive financial outlay to bring in established leading mathematicians to make a new start like the original one under President Harper, the support would not have been forthcoming, even with a mathematician, Max Mason, as president and certainly not with the young Robert M. Hutchins, bent primarily on establishing his new college. It took the Manhattan Project, the first nuclear pile under the Stagg Field bleachers and Enrico Fermi to convince Hutchins of the importance of physical science and mathematics and to throw massive resources into the reorganization of the department near the end of World World II. Such reorganizations are necessary from time to time in every graduate department. They can be effective only when the time is right. It is the mark of a great university to recognize the necessity to break the immutable law of academia, and the opportunity, and to do it when the time is right. However, there were deep hurts, symbolized by Bliss's refusal ever again to set foot in Eckhart Hall to his death. But this is really getting ahead of my story.

It was no ebb cycle for the University of Chicago as a whole in the twenties. There was intellectual excitement in many places in the university. I attended the physics colloquia where the great innovators of the day came to talk. With Mr. Bliss's grudging consent, I took Arthur Compton's course in X-rays. He already had the Nobel Prize for his work on the phenomena of X-rays colliding with electrons. Yet he seemed so naïvely simple minded to me, far less expert and mentally profound than other physicists in the department. Somewhere in here Einstein came for a brief visit. He permitted himself to be escorted by the physics graduate students for a tour of their experiments. To one he offered a suggestion. The brash young man explained immediately why it could not work. Einstein shook his head sadly. "My ideas are never good," he said.

Michelson, another Nobel Prizeman, was around, though retired. So was the great geologist, Chamberlin, with his planetesimal hypothesis in cosmology. In biology and biochemistry the great breakthroughs on the chemical nature of the steroid hormones and their effects on growth and development were excitingly unfolding. Young Sewall Wright was attracting students to his mathematical genetics. Economics promised a real breakthrough, though as it turned out, it was slow in coming. Linguistics was burgeoning. Anthropology and archeology were still actively following up the results of digs in Egypt, Turkey and Mesopotamia. The great debates over the truth of theories of relativity and quantum mechanics were raging. What was later to be planet Pluto had been observed as "Planet X" but heated arguments persisted on what it really was. On Sundays the University Chapel produced a succession of the leading Christian and Jewish spokesmen of the day. The textbook, *The Nature of the World and of Man,* H. H. Newman ed., University of Chicago Press, 1926, by illustrious Chicago faculty members was the best survey of physical and biological knowledge for college students that I have ever seen, though now dated, of course.

And outside the university the dangerous and ugly city of Chicago nevertheless had its charms, cultural and otherwise, that could take up all the time (and money) of a country boy. One could hear Mary Garden or Rosa Raisa at the Chicago Opera by getting a job as usher or super, or attend a fiesta in honor of the patron saint of some Halstead Street community that maintained its identity with the home village in the old country. One could drink wine at Alexander's clandestine speakeasy. For recall that it was Prohibition and the height of the bootlegging days of Al Capone and rival gangs. The famous Valentine Day massacre was just one of the lurid stories in the Chicago Tribune. We students formed an informal protective association to promulgate rules to optimize safety for oneself and date. One old boy from Georgia, a graduate student in history, was so impressed by our admonition never to approach a car asking him to get in, that, when a police car challenged him with order to stop, he just took off in a blaze of speed. Caught later, out of breath, his one phone call brought some of us to police court to testify to his character. The officer who had made the arrest moved to dismiss the charges on the condition that "the defendant appear at Soldiers Field next Saturday and run for our company in the policemen's track meet." But it was grim business. Police, armed with machine guns, in such a car once arrested me on suspicion of rape on the Midway (not guilty!). Other students were mugged, raped, robbed and even killed.

Like today it was a time of inflation and most of us were poor. I had a full fellowship of $410, of which $210 had to be returned in tuition for three quarters. A dormitory room cost $135 out of what was left. We could get cheap meals at the Commons, and on Sundays one could go to the Merit Cafeteria and splurge on a plate-sized slab of roast beef. It cost 28¢ but it was worth it. We all looked forward to a teaching job, I think. Those jobs required 15 hours of teaching for about $2700. Soon the depression hit and, if we were lucky, we kept our jobs with salary cut to $2400. Some beginning salaries for Chicago Ph.D.'s were as low as $1800 in the early thirties.

Before closing these recollections I must write something about women as graduate students in those times, not long after the victory of women's suffrage. Only years later did I learn that it was considered unladylike to study mathematics. Many of the graduate students in mathematics were women. In fact there were 26 women Ph.D.'s in mathematics at Chicago between 1920 and 1935. I shall mention only a few by name. Mayme I. Logsdon (1921) was in the faculty of the department. Mina Rees (1931) was already showing the kind of ability that led her to a distinguished administrative career at Hunter College and CUNY. She did more than any other person to gain federal support for mathematics through her position as chief, Mathematics Branch ONR, when the National Science Foundation was established. Others included Abba Newton (1933), chairman at Vassar, and Frances Baker (1934) also of Vassar, Julia Wells Bower (1933), chairman at Connecticut College, Marie Litzinger (1934), chairman, Mt. Holyoke, Lois Griffiths (1927) Northwestern, Beatrice Hagen (1930) Penn State, and Gweneth Humphreys (1935) Randolph Macon. Graduate students married graduate students, though of necessity only after the man had his degree. In the department Virginia Haun married E. J. McShane. Emily Chandler, student of Dickson, married Henry Pixley and continued her publishing and teaching career at the University of Detroit. Antoinette Killen married Ralph Huston. They both later taught at Rensselaer Polytech. Aline Huke married a non-Chicago mathematician, Orrin Frink, and continued her teaching at Penn State. Jewel Hughes Bushey was in the department of Hunter College. These, and a number of others, were able to continue their professional work in spite of family obligations. Even intermarriage between departments was permitted! My wife to be, Mary Hardesty, was in zoology. We got our Ph.D. degrees in the same commencement.

Looking back on those days, I wonder if the current women's liberation has even yet succeeded in pushing the professional status of women to the level already reached in the twenties. Maybe this time women can hold their gains in universities.

Helmut Hasse in 1934

S. L. Segal

Sanford Leonard Segal received his Ph.D. from the University of Colorado in 1963. He has been associated with the University of Rochester since 1964. He was a Fulbright fellow to Austria in 1965 and a visiting lecturer at the University of Nottingham in 1972.

The mathematician as superman is a common theme in biographical works by mathematicians about mathematicians. Always wiser, more truly moral, kinder, more broad-minded, less petty, transported above the niggling concerns of mortals on the wings of pure reason, the mathematician is both hero and savant. Anyone who wants to really understand the essential nature of almost anything should ask a mathematician. No mathematician ever fell for a crude political scheme, or rallied behind a demagogic rabble-rouser, or unwittingly bought swampland in Florida. The mind that grasps, generalizes, and extends calculus as an evening's entertainment can surely discern the nonsense and sham so readily accepted by the common folk. That is too often the way mathematicians prefer to view themselves. Reality, however, tends to be rather less charitable.

Consider, as exhibit A, the eighteenth-century French geometer who liquidated a substantial portion of his considerable fortune to purchase antiquities such as an original letter, in French no less, from the Apostle John to the Virgin Mary. Or exhibit B, the contemporary mathematician who, when convinced of the imminence of atomic warfare, spent a whole week in his backyard with pick and shovel constructing an earth-covered fallout shelter. The structure was well reinforced with timbers and covered with three feet of earth and lacked nothing but a way to get inside it. There is a virtually endless supply of such eccentric anecdotes, but why bother? No one but mathematicians believe the first description anyway. Beyond such eccentricity lies the rather gritty and much less charming reality that mathematicians are no more noble or virtuous than the rest of us. And in the Germany of 1934 nobility was a commodity in increasingly short supply as this account demonstrates.

IN 1934 HELMUT HASSE became Professor of Mathematics at Göttingen. Hasse's attitudes and behavior during the Nazi period were representative of the ambiguous position of much of the mathematical community at that time. Various aspects of Hasse's situation make him appear to be almost the ideal example of the apolitical conservative and ideologically naive German academic in extremis. The purpose of this article is to present him as such an example.

In 1934 Helmut Hasse left Marburg to become, at age 36, a Professor at Göttingen. Hasse, who is still alive and rightly much honored among mathematicians, was a conservative nationalist; in 1937 he applied for membership in the Nazi party. Hasse's attitudes and behavior during the period of Nazi control of the Universities were representative of the ambiguous position of much of the mathematical community at that time. His case

Source: S. L. Segal, "Helmut Hasse in 1934," *Historia Mathematica* 7 (1980): 46–56. Reprinted by permission of Academic Press. (Some notes omitted.)

has surfaced in Constance Reid's recent book about the Göttingen mathematical community, *Courant in Göttingen and New York*. However, there is unpublished material which apparently escaped Reid's attention, and which sheds considerable light upon the relationship of Hasse and mathematicians in general to the Nazi period [Papers of Oswald Veblen, Library of Congress Manuscript Collection]. Courant (himself a forced émigré), on meeting Hasse during a return visit to Germany in 1947, noted in his Journal: "Met Hasse. Mixed feelings" [Reid 1976, 263]. Indeed Hasse sought no alibi for his actions during the Nazi period; he was no vicar of Bray, nor, to use a marvelous German expression, a "Konjunktur-Karrierist."

In fact, it is hard not to have mixed feelings about Hasse, a political conservative and brilliant scientist who apparently wished not only to survive but to do the best for his discipline in difficult times.

In a taped conversation with Constance Reid around 1975, Hasse expressed his political creed as follows:

> My political feelings have never been National-Socialistic but rather "national" in the sense of the Deutschenationale Partei, which succeeded the Conservative Party of the Second Empire [under Wilhelm II]. I had strong feelings for Germany as it was created by Bismark in 1871. When this was heavily damaged by the Treaty of Versailles in 1919, I resented that very much. I approved with all my heart and soul of Hitler's endeavors to remove the injustices done to Germany in that treaty. It was from this truly national standpoint that I reacted when the Faculty [at the first meeting after the war's end] more or less suggested that such a view was not permissible in one of its members. It was also the background for my remarks to the Americans [at the same time]. They were talking about reeducating Germany, and I said some strong things against this. It irked me that everything against Hitler was desirable, and everything that he had done was wrong. I continued to be a national German, and I resented Germany being trampled under the feet of foreign nations [Reid 1976, 250–251].

Hasse is referring to the DNVP (Deutsche Nationale Volks Partei) with which the majority of professors apparently sympathized [Ringer 1969, 201 and Chap. 4, *passim*. See also Mosse 1964, Chap. 13]. The DNVP played upon nostalgia (it advocated loyalty to the Hohenzollern as late as 1928); it was popularly referred to as the "Conservatives," and in standing for the primacy of Church, State, and private property, it was the party of the orthodox right. At the same time, from its inception the DNVP was influenced by *Völkisch* and anti-Semitic ideologies, both of which became increasingly influential in the party. Eventually the party collaborated with and then capitulated to Hitler. Especially after its takeover by Hugenberg in 1928, it was the party of the elitist conservatives who thought they could use Hitler to bring the masses into line, and later found that they had been used instead to make Hitler respectable. The DNVP members appeared to be quite respectable; they were solid pillars of the community, neither fanatics nor hooligans. Yet, despite the anti-Semitism which was a feature of the party's program even in 1919, and with increasing prominence afterwards (in 1929 the party was closed to Jews), it continued to have Jewish and half-Jewish adherents. It may well be that this anomaly can be explained by an inherent conservatism of established assimilated Jews who regarded themselves as German, especially in contrast to the East European immigrants. Men as diverse as Rathenau and Kapp had made this distinction. The same sort of inward retreat might be said to characterize the orthodox majority of professors like Hasse, who as members of the DNVP, continued in Fritz Ringer's phrase "to exploit the antidemocratic and antimodernistic implications of the mandarin tradition to the fullest possible extent" [Ringer 1969, 214]. Devoid of inherent republican sentiments and blinded by traditional prerogative, status, and political attitude, they were even unable to be *Vernunftrepublikaner* [republicans because "moderation" was the only sane and practical course to follow].

On April 26, 1933, the *Göttingen Tageblatt* announced that six professors (among them Richard Courant) had, as Jews, forcibly been placed on leave. Courant had been head of the world-famous Göttingen Mathematical Institute. Otto Neugebauer was appointed as Courant's successor, but served for only a day. He refused to take the oath of fealty to the new state and hence was suspended as *untragbar*. Herman Weyl, one of the most famous mathematicians of his generation, became acting director of the institute and attempted to reverse some of the governmental decisions. Weyl's position was at best extremely uneasy, not only because of his philosophical opposition towards the régime, but because his wife was Jewish. The previous year Weyl had turned down an offer from Abraham Flexner to join the Institute for Advanced Study in Princeton. Negotiations were reopened, and by Christmas he was in the United States. Before leaving, Weyl urged Hasse to accept a call to Göttingen. Both Weyl and Courant believed that Hasse, a political conservative and a first-class mathecian with no immediate Jewish relations, was exactly the sort of person who might be acceptable to the Nazi régime and expected to uphold the Göttingen mathematical tradition.

By Easter of 1934 Hasse had indeed been called to Göttingen and was negotiating his acceptance. On Easter Tuesday, 1934, the prominent German mathematician Ludwig Bieberbach, who also became a prominent pro-Nazi force in German University life, gave a speech at the Technische Hochschule in Berlin, entitled (when published) "Persönlichkeitsstruktur und Mathematisches Schaffen" [The Structure of Personality and Mathematical Creation] in which, to quote the newspaper report (of April 8) in *Deutsche Zukunft*, "he seems to show that the teaching of Blood and Race also applies here, and places the most abstract of sciences beneath the total state."[1] In the same year Bieberbach [1934] also published another article with similar aim in the *Proceedings of the Berlin Academy*, entitled "Stilarten Mathematischen Schaffens" [Styles of Mathematical Creation]. Bieberbach's point of departure was the SS-organized boycott of lectures by the German Jewish mathematician Edmund Landau the preceding fall. Here is Courant's description of the event in a letter to Abraham Flexner:

> The following event is characteristic of the course of things. Professor Landau . . . went to start his lectures last week. In front of his lecture hall were some seventy students, partly in S.S. uniforms, but inside not a soul. Every student who wanted to enter was prevented from doing so by the commander of the boycott [Werner Weber who had once been Landau's Assistant]. Landau went to his office and received a call from a representative of the Nazi students, who told him that Aryan students want Aryan mathematics and not Jewish mathematics and requested him to refrain from giving lectures. . . . The speaker for the students is a very young, scientifically gifted man, but completely muddled and notoriously crazy. It seems certain that in the background there are much more authoritative people who rather openly favor the destruction of Göttingen mathematics and science [Reid 1976, 155–156].

It is perhaps of interest that Landau was socially prominent as well; he is listed in the *Reichshandbuch der Deutschen Gesellschaft (1930)*; his father was a well-to-do Berlin gynecologist; his father-in-law the famous microbiologist Paul Ehrlich. Essentially the same version of these events was communicated personally to me by the late Hans Heilbronn who had just obtained his degree under Landau. The "speaker for the students" was probably the brilliant young mathematician Oswald Teichmüller who made fundamental mathematical contributions before disappearing in 1943 [Reid 1976, 155–156]. Bieberbach's lectures, which leaned on the racial-typological theories of the Marburg Nazi psychologist E. R. Jaensch, caused a stir abroad. Distinguished foreign mathematicians such as Harold Bohr (in Copenhagen), Oswald Veblen (in Princeton), and G. H. Hardy (in Cambridge) publicly denounced the ideas expressed. Bohr's protest in particular led to an incident which became a focus of the early attempt to Nazify German mathematics.

On May 16, Bohr initiated a correspondence with Hasse on the subject. Most of this correspondence is contained in the unpublished papers of Oswald Veblen in the Library of Congress [Veblen papers, Bohr folder. The papers are otherwise uncatalogued. Dates of all letters cited are in the text]. Bohr apparently hoped that Hasse, who was taking over the Institute at Göttingen where the Landau incident had occurred, would take a strong stand against Bieberbach. Hasse replied from Marburg on June 6 with apologies for not answering sooner, blaming "difficulties in connection with my taking over the position in Göttingen, which have affected me both outwardly and inwardly" [letters between Hasse and Bohr were written in German, between Bohr and Veblen in English; all English direct quotations from a German original (here and elsewhere) are my translations]. He then suggested that the pedagogical suitability of Landau's rather distinctive mathematical style, which had been attacked by Bieberbach, was a matter on which individuals might well differ. However, Hasse rejected Bieberbach's preoccupation with "blood and race" and instead sought to explain individual stylistic differences in terms of "environment and education." Hasse went on to say that the press reports of Bieberbach's lecture were inaccurate (a claim also made by Bieberbach, but refuted by the published lecture). In the last sentence of the letter he declared that he was unable to take a personal stand on the issues raised by Bieberbach's lecture.

One of the difficulties which Hasse had to face at the time was Bieberbach's wish to attack Bohr's protest with a denunciation in the official journal of the German Mathematical Society. The editors were Bieberbach, Hasse, and Konrad Knopp, and although Hasse and Knopp protested that this was an inappropriate use of the journal, Bieberbach's letter, dated May 21, was finally published.

On June 9, Bohr replied to Hasse's letter (of June 6) in which the latter attempted to trivialize the matter of Bieberbach's lecture by regarding it as a discussion about style. Bohr pointed out that it is one thing to discuss stylistic differences, and another to draw the sort of consequences therefrom which Bieberbach did (such as the necessary rejection of Landau). Bohr called Hasse's stylistic discussion a red herring, expressed his astonishment that Hasse was unable to take a personal stand and enclosed a copy of Veblen's "open letter" of protest against Bieberbach's ideas (written May 19).

Hasse replied on June 11. He asked Bohr "most heartfelty" not to take ill his refusal to take a personal stand, adding "I can explain the grounds which move me thereto only very badly in a letter. They are connected in any case with my present situation and also with the events in Göttingen at the end of May, of which you perhaps have heard." He agreed that matters of pedagogical style were not in fact the central issue, and suggested that Bieberbach's statements were meant "mainly to raise questions." Hasse then earnestly asked Bohr "because of the [unspecified] ground indicated above" to "in no way make use of this, my letter, publicly and above all to leave my name out of the whole matter." In an additional remark, probably referring to the editorial conflict over Bieberbach's as yet unpublished "open letter" to Bohr, Hasse noted that he had tried to make Bieberbach aware of the impression his remarks had made on many foreign and German mathematicians, although Bieberbach's business was ultimately his own. The letter then drew enigmatically to a close: "I believe an oral discussion with you would contribute much to clarification. Perhaps I shall make a lecture trip in the autumn, and could come to Copenhagen, either on my way to or from Helsinki. I would be very happy to meet you there and to speak then in more detail."

What Hasse feared was unclear; his behavior may simply have reflected a desire to do nothing that might jeopardize his appointment. Such a desire would not necessarily have been motivated solely by personal ambition; it was important to preserve what one could of a famous mathematical tradition. This was precisely the reason Courant and Weyl had thought the conservative Hasse the very man for the job. The verdict of the Göttingen student magazine *Politikon* in 1965 may well be true:

> Many professors explain that they entered the [Nazi] Party because they wished to save what could be saved. By 1935 there was nothing more to save at the University of Göttingen, except perhaps that remnant of personally decent behavior which allowed some professors . . . to retain their positions out of fear that even more NS-philistines (NS-Banausen) would replace them [*Politikon* 1965, 25].

However, in 1934, this would have been far from obvious to a man of Hasse's political beliefs. It may be that Hasse also feared the possibility of a personal investigation which would reveal that he had a Jewish ancestor, thereby jeopardizing more than just the Göttingen position.

As to the unnamed disturbances in Göttingen that May, Bohr asked Veblen to respect Hasse's wish for confidentiality, and added that:

> the situation in Göttingen is really extremely absurd, as the assistants in the Mathematical Institute declined to give the keys to Hasse when he came as the official new professor sent from the government. This is characteristic for the Führerprinzip [Veblen papers, Library of Congress, Bohr folder].

In fact, the mathematician Erhard Tornier, long a secret Nazi, had been appointed to fill Landau's chair and to be interim Institute director until Hasse's arrival in Göttingen. Tornier had earlier worked closely with Willy Feller, the promising young Yugoslavian probabilist whose brilliant career was to be largely as an émigré, first in Sweden, and after 1939 in the United States. Tornier had followed Feller to Kiel by appealing to the then *Ordinarius* A. A. Fraenkel (who was Jewish) that it would facilitate their continued mathematical collaboration. After the Nazi take-over of the government, Tornier revealed his true political sympathies. Apparently he played a major role in determining Feller's "non-Aryan" origin, thus driving the latter away from Kiel. Moreover, in a letter to Fraenkel, Tornier seems to have admitted that he only went to Kiel in order to inherit Fraenkel's chair at the politically appropriate time, which he felt was coming [Fraenkel 1967, 154–155]. However, a far more prestigious position than Kiel was to be Tornier's reward. Certainly Weyl and Courant were right in thinking that a mathematician who happened to be a political conservative, like Hasse, should take over in Göttingen. The alternative to Hasse was a political fanatic like Tornier who was only incidentally interested in mathematics. One cannot stress too strongly the difference between someone like Hasse who felt a duty to his profession and country to save what he could, perhaps hoping even to mitigate ideological interference, and opportunistic adventurers like Tornier. Tornier and Teichmüller, then a student leader, kept Göttingen mathematics in turmoil in the period preceding Hasse's take-over. In July, when Hasse came to Göttingen in connection with his future appointment, he was greeted by a demonstration of pro-Nazi students [Reid 1976, 162].

However frightened Hasse may have been before the demonstration, he would become even more so. On the "Night of the Long Knives," June 30, 1934, Tornier was to threaten Hasse directly. Here is the story as related by Bohr in his own English to Veblen on August 11, 1934:

> . . . Together with Courant, F. K. Schmidt, a collaborator and friend of Hasse and an extremely bold and honest man, came to see me here [Fynshav] and told the whole history of Göttingen since the revolution in all details. Being "Stellvertreter" there from the very beginning, he has played a prominent part in the whole development. He did his best to defend Hasse's whole behavior and in fact I think that he is right, that Hasse is simply extremely scared and not able to act as an ordinary human being in the present circumstances. Only to give an instance of the diabolic way in which Tornier (Bernstein's successor and "Vertrauensmann" in mathematical questions for the Nazi party and the government) deals with Hasse I may tell you the following incredible detail. In the night of the famous 30, June, Tornier wrote an express-letter to Hasse, who was in Marburg, telling him that Tornier thought Hasse's life was in danger (since his so-called friendly attitude towards people like Courant and Emmy Noether made him suspectible) and advised him strongly to take the train immediately to Göttingen and come to Tornier's hotel so that he could protect him—what Hasse did. Tornier's purpose seemed to Schmidt to be to destroy Hasse's personality but to keep him as a mathematical sign for Göttingen. Tornier's next idea is to get Brouwer to Göttingen. Schmidt told us that Brouwer was extremely pleased about the idea and was very much inclined to come; in this moment he discusses the financial terms with the Nazi-government. All we who have had the privilege of coming

in nearer contact with Hilbert feel the idea of bringing Brouwer to Göttingen perhaps the most dirty trick, especially as the people do not really believe in Brouwer's ideas of a new foundation of mathematics, but simply wish to liquidate completely the epoch of Hilbert and his school. From Courant you can hear about the struggle in Göttingen in all details.[2]

There is very little to add in describing the events surrounding Hasse's appointment to Göttingen in 1934. If nothing else these events speak eloquently to the difficult position of serious academics at that time, whatever their political beliefs. In 1935 the British mathematician Harold Davenport wrote Courant ". . . I gather that Teichmüller, who was the ring leader of the opposition to Hasse among the students, has become reconciled to him. Hasse thinks he is a quite good mathematician but I am unable to judge . . ." [Reid 1976, 178–179]. At the autumn meeting of the *German Mathematical Society* in 1934, Hasse was one of those (in the majority) who opposed the attempt by Bieberbach and Tornier to introduce the *Führerprinzip* into the German mathematical community. By 1936 both Teichmüller and Tornier had moved to Berlin, which became a center for those devoted to a racially conditioned mathematics. In 1937, Hasse applied for membership in the Nazi party. As he explained in an interview with Constance Reid:

My endeavor at that time was to keep up Göttingen's mathematical glory. For this I needed the consensus of party functionaries at the university whenever I wanted to get some distinguished mathematician to fill a vacancy in Göttingen. Among these functionaries I had one close friend and one who was leaning towards helping me. They asked me to join the party so that they could help me better. It is true that I gave in and applied for membership. But on my application I put that there was a Jewish branch in my father's family. I was almost sure that this would lead to my application being declined. And so it was. The answer, which I received only after the outbreak of the war, was that the application was not going to be acted upon until the war was over. In the meantime, however, I had been able to help several mathematicians who were having political difficulties at other universities, by offering them positions in Göttingen [Reid 1976, 203].

A. A. Fraenkel's memoirs, *Lebenskreise,* tell a somewhat different, albeit second-hand story. Fraenkel says he was told by G. H. Hardy that Hasse was threatened by an unfriendly colleague at Göttingen with the revelation that he had a Jewish great-grandfather. Hasse then applied successfully for that peculiar distinction of the Third Reich: honorary Aryan, following which he entered the Nazi Party [Fraenkel 1967, 153–154]. In 1937 Hasse seems to have advised Emil Artin (whose wife was half-Jewish and who was to emigrate in October of that year) that he could have his children, only a quarter-Jewish, declared honorary Aryans [Reid 1976, 203]. Furthermore, Hasse was one of eighteen Göttingen faculty who, according to the Göttingen *Vorlesungsverzeichnis* for 1938–1939 (winter semester), was a member of the National Socialist Academy of Sciences [*Politikon,* 1965, 21].

On March 22, 1939, Carl Ludwig Siegel wrote Courant:

After the November pogrom, when I returned to Frankfurt from a trip, full of nausea and anger at the bestialities in the name of the higher honor of Germany, I saw Hasse for the first time wearing Nazi-party insignia! It is incomprehensible to me how an intelligent and conscientious man can do such a thing. I then learned that the foreign policy occurrences of recent years had made Hasse into a convinced follower of Hitler. He really believes that these acts of violence will result in a blessing of the German people [Veblen papers, Library of Congress, Siegel folder].

While Hasse displayed open adherence to the Nazi party, he apparently never involved himself in ideological extremes like those found in *Deutsche Mathematik,* the journal which was devoted to the publication of "Aryan mathematics." Also, as we have seen above, he was considered a "liberal" in his rejection of the *Führerprinzip* as a governing principle for scientific societies and in his opposition to Bieberbach's desire to use an established official mathematical journal for essentially political purposes. Siegel's letter emphasizes what Hasse later indicated in his interview with Constance Reid: his primary political orientation was one of intense nationalism. Hasse never seemed to have realized the qualitative difference between bourgeois nationalism and the Nazi program. He was not the only such professor, nor, as Helmut Kuhn has remarked, was the confusion that surprising: the conservative nationalism of the educated bourgeoisie had acquired a touch of resentment-filled radicalism which narrowed the distance between it and the Nazi movement. This, coupled with the condescension of the educated classes for the "plebeian" Nazis, and the politics of the "Harzburger Front," made it easy for a highly educated conservative nationalist who was not greatly concerned with political distinctions to view the Nazis as just a little bit further to the right [Kuhn 1966].

In 1945, Hasse was removed from his position at Göttingen, according to Fraenkel because of his party membership. G. H. Hardy apparently wrote a letter in June 1946 to the British occupation authorities asking Hasse's reinstatement [Fraenkel 1967, 153–154]. Hasse eventually went to Hamburg. We know that Courant had "mixed feelings" when he met Hasse. Hasse was perhaps a typical example of the distinguished, conservatively apolitical and ideologically naive German professor, who, thrust into the ideologically charged situation of Nazi Germany, coped as best he could for the preservation of his discipline and himself.

Notes

1. Bieberbach's lecture was published in 1934 in the series *Unterrichtsblätter für Mathematik und Naturwissenschaften.* The report in *Deutsche Zukunft,* 8 April 1934, Nos. 14, 15, entitled "Neue Mathematik," is generally accurate and somewhat critical of Bieberbach's not going far enough in the reviewer's opinion. The review is signed "P.S."
2. The very distinguished Dutch mathematician and logician L. E. J. Brouwer was an ardent Germanophile and long-time friend of the mathematician Theodor Vahlen, an "alter Kämpfer" who became a Nazi functionary. Brouwer and Hilbert had strong differences on issues affecting the foundations of mathematics which spilled over into personal animosities. No mention of this Tornier-Hasse incident is in Constance Reid's book; she speaks only of the July demonstration—"in the late summer of 1934, at the urging of other German mathematicians, who were disturbed by what had happened to mathematics in Göttingen, Hasse had agreed to go back to that university; but his life was still being made miserable by Tornier and the pro-Nazi students led by Teichmüller" [Reid 1976, 164].

References

Bieberbach, Ludwig. 1934. Stilarten Mathematischen Schaffens, *Sitzungsberichte der Akademie Berlin phys.-math. Kl.* (Sitzung of 5 July), 351–360.

Fraenkel, A. A. 1967. *Lebenskreise.* Stuttgart: Deutsche Verlags-Anstalt.

Kuhn, Helmut. 1966. *Die Deutsche Universität am Vorabend der Machtergreifung,* in *Die Deutsche Universität im Dritten Reich.* München: Piper.

Mosse, George L. 1964. *The Crisis of German Ideology.* New York: Grosset and Dunlap.

Politikon. 1965. *Göttingen Studenten-Zeitschrift für Niedersachser* (Nummer 9, Januar 1965). A copy is in the Wiener Library, 4, Devonshire St., London W. I.

Rathenau, Walter. 1965. *Schriften.* Berlin: Berlin-Verlag.

Reid, Constance, 1976. *Courant in Göttingen and New York.* New York: Springer-Verlag.

Ringer, Fritz. 1969. *The Decline of the German Mandarins.* Cambridge, Mass.: Harvard University Press.

Veblen, Oswald. Unpublished papers in the Library of Congress Manuscript Collection.

Adventures of a Mathematician

Stanislaw M. Ulam

Stanislaw M. Ulam was born and educated in Poland, receiving the Doctorate of Science in mathematics from the Polytechnic Institute. He subsequently emigrated from Poland to the United States where he worked at Los Alamos during World War II: after the war he taught at the University of Colorado. A winner of the Polish Millennium Prize he is currently professor emeritus of mathematics at the University of Colorado.

Stanislaw M. Ulam was a very bright Polish boy. In time he became a very bright young mathematician, a Polish expatriot and American immigrant, and a well-known and respected man of modern mathematics. He has written an autobiography, one chapter of which is reprinted here. It provides one of the few glimpses in print into the emotions of a mathematician.

Most autobiographical material by mathematicians is uniformly encyclopedic in recitation of facts: names, dates, and precedence of publication. Indeed, many mathematicians seem to define their existence in terms of mathematical achievements; like Hardy, they cease to exist when the muse departs. That they have loved, hated, feared, quarreled, we know from accounts by others; but to the principals, it seems the theorems are all that matter.

Fortunately for us, Stanislaw Ulam has a rather broader perspective. In this passage he offers a suggestion of what happens within a man whose very being lies in his intellect when he feels that intellect slowly skidding into a seemingly irreversible oblivion.

THE WAR WAS over and the world was emerging from the ashes. Many people left Los Alamos, either to return to their former universities like Hans Bethe, or to go to new academic positions like Weisskopf to MIT or Teller to Chicago. The government had not reached any decision yet about the fate of the wartime laboratory.

The University of Chicago took steps to start a great new center for nuclear physics, with Fermi, Teller and several others from the Manhattan District Project. Von Neumann, better than anyone else, it seems to me, argued that as a result of the role science had played in the winning of the war, the postwar academic world would not be recognizable in pre-1939 terms.

On the purely personal plane, I had no evidence that any member of my immediate family had survived (two cousins did reappear many years later, one in France, the other in Israel). Françoise had lost her mother in the concentration camp of Auschwitz. We were both American citizens now, the United States was our country, and the idea of returning to Europe never entered our heads. But the question of what job to return to from war work was very much on our minds.

I had some correspondence with Langer, who was then the chairman, about returning to Madison. He was very honest and open, and he told me with admirable frankness when I inquired about my chances for promotion and tenure: "No reason to beat around the bush, were you not a foreigner, it would be much easier and your career would develop faster." So it seemed that my chances in Wisconsin were not very good, and I looked elsewhere. Elsewhere came in the form of a letter from an old Madison friend, Donald Hyers, who had become a professor at the University of Southern California in Los Angeles. Hyers was well established there, and he asked whether I would be interested in joining the faculty as an associate professor at a salary somewhat higher than the one in Madison. The university was small, not very strong academically, and certainly not a very prestigious place, but the professors there, he said, were engaged in vigorous attempts at improving the academic standing of the institution. He invited me for a visit, and I flew to Los Angeles in August of 1945.

This was the first time I saw that city, and it gave me a very strange impression. It was a different world from any I had known, climatically, architecturally, and otherwise. I mentioned this job possibility to Johnny [von Neumann], and although he was rather surprised at my interest in this rather modest opportunity, he did not react negatively. His tendency was to go along. I did not see much sense in marking time in Los Alamos after the war, so I accepted the USC offer.

In early September of 1945, I went to Los Angeles to look for housing and to prepare our move from Los Alamos. In the immediate postwar period, the housing situation in Los Angeles was critical. Since we did not own a car, we were restricted to searching for a house in the vicinity of the University. I used to say that any two points in Los Angeles were at least an hour's drive apart, a "discrete" topological space. I managed to sublet for one semester a typical small Los Angeles house on a modest street lined with spindly palm trees. To me it seemed adequate, but it appeared rather miserable to Françoise. Nevertheless, we settled there temporarily for lack of anything better. I noticed that in our various moves from one habitat to the next all our material possessions, clothes, books, furnishings had a way of diminishing in transit. I used to say that they dwindled to $1/e$, in analogy to the energy losses of particles in transit through "one mean free path."

For the second semester of that academic year (1945–46), Hal and Hattie von Breton, good friends of the Hawkinses, invited us to stay in their summer cottage on Balboa Island across from Newport Beach. It was on the water, beautiful and comfortable—a wonderful change from the university neighborhood but a little too far for me to commute daily—so during the week I lived in a hotel near the campus and went home to the island on weekends. Françoise remained on Balboa with our baby daughter Claire, who had been born in Los Alamos the year before.

At USC I found the academic atmosphere somewhat restricted, rather anticlimactic after the intensity and high level of science at Los Alamos. Everyone was full of good will, even if not terribly interested in "research." The "teaching load" to which I was reluctantly returning was not too heavy. All in all, things looked promising had it not been for a violent illness which struck me suddenly. I had returned to Los Angeles from a mathematics meeting in Chicago with a miserable cold. It was a stormy day; on the walk from the bus to the house in Balboa the violent winds almost choked me. That same night I developed a fantastic headache. Never in my life had I experienced a headache of any kind; this was a new feeling altogether—the most severe pain I had ever endured, all-pervading and connected with a sensation of numbness creeping up from the breast bone to the chin. I remembered suddenly Plato's description of Socrates after he was given the hemlock in prison; the jailor made him walk and told him that when the feeling of numbness starting in the legs reached his head he would die.

Françoise had difficulty in finding a doctor who would come to the island in the middle of the night. The one who finally came could not find anything visibly wrong and gave me a shot of morphine to alleviate the excruciating pain. The next morning I felt almost normal but with a lingering feeling of lassitude and an inability to express myself clearly, which came and went. Nevertheless, I returned to Los Angeles and gave my lectures at the university. The following night the violent headache reappeared. When I tried to telephone Françoise from my hotel room, I noticed that my speech was confused, that I was barely able to form words. I tried to talk around the expressions which would not come out and form equivalent ones, but it was mostly a meaningless mumble—a most frightening experience.

Greatly alarmed by my incoherent phone call (I don't know how I managed to remember the phone number at home), Françoise called the von Bretons and asked them to send a doctor to see me. In fact, two doctors appeared. Perplexed by my symptoms that came and went, they took me to Cedars of Lebanon hospital. A severe attack of brain troubles began, which was to be one of the most shattering experiences of my life. By the way, many of the recollections of what preceded my operation are hazy. Thanks to what Françoise told me later I was able to put it together.

For several days I underwent various tests—encephalograms, spinal taps, and the like. The encephalogram was peculiar. The doctors suspected a tumor, which could be benign or malignant. Dr. Rainey, a neurosurgeon pupil of Cushing, was called in and an operation was planned for the following day. Of all this I knew nothing, of course. I remember only trying to distract the nurse's attention by telling her to look out of the window so I could read my chart. I saw there some alarming notation about C-3 which I suspected to mean the third convolution of the brain. Through all this I was overcome by an intense fear and began to think I was going to die. I considered my chances of surviving to be less than half. The aphasia was still present; much of the time when I tried to speak I uttered meaningless noises. I do not know why no one thought of ascertaining whether I could write instead of speak.

Françoise, alerted by the von Bretons, rushed all the way back from Balboa by taxi and arrived on the scene just as I was beginning to vomit bile, turning green and losing consciousness. She feared I was dying and made a frantic telephone call to the surgeon, who decided the operation should be performed immediately. This probably saved my life; the emergency operation relieved the severe pressure on my brain which was causing all the trouble. I remember that in my semiconscious state my head was being shaved by a barber (he happened to be a Pole) who said a few words in Polish, to which I tried to reply. I remember also returning to consciousness briefly in a pre-operating room and wondering whether I was already in the morgue. I also remember hearing the noise of a drill. This was a true sensation as it turned out, for the doctors drilled a hole in my skull to take some last-minute X-rays. The surgeon performed a trepanation not knowing exactly where or what to look for. He did not find a tumor, but did find an acute state of inflammation of the brain. He told Françoise that my brain was bright pink instead of the usual gray. These were the early days of penicillin, which they applied liberally. A "window" was left on the brain to relieve the pressure which was causing the alarming symptoms.

I remained in a post-operative coma for several days. When I finally woke up, I felt not only better, but positively euphoric. The doctors pronounced me saved, even though they told Françoise to observe me for any signs of changes of personality or recurrence of the troubles which would have spelled brain damage or the presence of a hidden growth. I underwent more tests and examinations, and the illness was tentatively diagnosed as a kind of virus encephalitis. But the disquietude about the state of my mental faculties remained with me for a long time, even though I recovered speech completely.

One morning the surgeon asked me what 13 plus 8 were. The fact that he asked such a question embarrassed me so much that I just shook my head. Then he asked what the square root of twenty was, and I replied: about 4.4. He kept silent, then I asked, "Isn't it?" I remember Dr. Rainey laughing, visibly relieved, and saying, "I don't know." Another time I was feeling my heavily bandaged head, and the doctor chided me saying the bacteria could infect the incision. I showed him I was touching a different place. Then I remembered the notion of a mean free path of neutrons and asked him if he knew what the mean free path of bacteria was. Instead of answering, he told me an unprintable joke about a man sitting on a country toilet and how the bacteria leaped from the splashing water. The nurses seemed to like me and offered all kinds of massages and back rubs and special diets, which helped my morale more than my physical condition (which was surprisingly good).

Many friends came to visit me. Jack Calkin, who was on leave on Catalina Island, appeared several times at the hospital. So did colleagues from the University. I remember the mathematician Aristotle Dimitrios Michael. He talked so agitatedly that I fell out of bed listening to him. This scared him very much. But I managed to scramble back even though I was still slightly numb on one side. Nick Metropolis came all the way from Los Alamos. His visit cheered me greatly. I found out that the security people in Los Alamos had been worried that in my unconscious or semi-conscious states I might have revealed some atomic secrets. There was also some question as to whether this illness (which was never properly diagnosed) might have been caused by atomic radiation. But in my case this was highly improbable, for I had never been close to radioactive material, having worked only with pencil and paper. University officials visited me, too. They seemed concerned about my ability to resume my teaching duties after I got well. People were acutely concerned about my mental faculties, wondering whether they would return in full. I worried myself a good deal about that, too; would my ability to think return in its entirety or would this illness leave me mentally impaired? Obviously in my profession, complete restoration of memory was of paramount importance. I was quite frightened, but in my self-analysis I noticed that I could imagine even greater states of panic. Logical thought processes are very much disturbed by fright. Perhaps it is nature's way of blocking the process in times of danger to allow instinct to take over. But it seems to me that mere instincts, which reside in nerves and in muscle "programming," are no longer sufficient to cope with the complicated situations facing modern man; some sort of reasoning ability is still needed in the face of most dangerous situations.

I regained my strength and faculties gradually and was allowed to leave the hospital after a few weeks. I obtained a leave of absence from the university.

I remember being discharged from the hospital. As I was preparing to leave, fully dressed for the first time, standing in the corridor with Françoise, Erdös appeared at the end of the hall. He did not expect to see me up, and he exclaimed: "Stan, I am so glad to see you are alive. I thought you were going to die and that I would have to write your obituary and our joint papers." I was very flattered by his pleasure at seeing me alive, but also very frightened to realize that my friends had been on the brink of giving me up for dead.

Erdös had a suitcase with him and was just leaving after a visit to Southern California. He had no immediate commitments ahead and said, "You are going home? Good, I can go with you." So we invited him to come with us to Balboa and stay awhile. The prospect of his company delighted me. Françoise was somewhat more dubious, fearing that it would tire me too much during the early part of my convalescence.

A mathematical colleague from USC drove us all back to the von Bretons' house on Balboa Island. Physically, I was still very weak and my head had not yet healed. I was wearing a skullcap to protect my incision until my hair grew back. I remember having difficulty walking around the block the first few days, but gradually my strength returned, and soon I was walking a mile each day on the beach.

In the car on the way home from the hospital, Erdös plunged immediately into a mathematical conversation. I made some remarks, he asked me about some problem, I made a comment, and he said: "Stan, you are just like before." These were reassuring words, for I was still examining my own mind trying to find out what I might have lost from my memory. Paradoxically, one can perhaps realize what topics one has forgotten. No sooner had we arrived than Erdös proposed a game of chess. Again I had mixed feelings: on one hand I wanted to try; on the other, I was afraid to in case I had forgotten the rules of the game and the moves of the pieces. We sat down to play. I had played a lot of chess in Poland and had more practice than he had, and I managed to win the game. But the feeling of elation that followed was immediately tempered by the thought that perhaps Paul had let me win on purpose. He proposed a second game. I agreed, although I felt tired, and won again. Whereupon it was Erdös who said, "Let us stop, I am tired." I realized from the way he said it that he had played in earnest.

In the days that followed we had more and more mathematical discussions and longer and longer walks on the beach. Once he stopped to caress a sweet little child and said in his special language: "Look, Stan! What a nice epsilon." A very beautiful young woman, obviously the child's mother, sat nearby, so I replied, "But look at the capital epsilon." This made him blush with embarrassment. In those days he was very fond of using expressions like *SF* (supreme fascist) for God, *Joe* (Stalin) for Russia, *Sam* (Uncle Sam) for the United States. These were for him objects of occasional scorn.

Gradually my self-confidence returned, but every time a new situation occurred in which I could test my returning powers of thought, I was beset by doubts and worries. For example, I received a letter from the Mathematical Society asking me if I would write for the *Bulletin* an obituary article on Banach, who had died in the fall of 1945. This again gave me reason to ponder. It seemed a little macabre after having barely escaped death myself to write about another's demise. But I did it from memory, not having a library around, and sent in my article with apprehension, wondering if what I had written was weak or even nonsensical. The editors replied that the article would appear in the next issue. Yet my satisfaction and relief were again followed by doubt for I knew that all kinds of articles were printed, and I did not have such a high opinion of many of them. I still felt unsure that my thinking process was unimpaired.

Normally primitive or "elementary" thoughts are reactions to or consequences of external stimuli. But when one starts thinking about thinking in a sequence, I believe the brain plays a game—some parts providing the stimuli, the others the reactions, and so on. It is really a multi-person game, but consciously the appearance is of a one-dimensional, purely temporal sequence. One is only consciously aware of something in the brain which acts as a summarizer or totalizer of the process going on and that probably consists of many parts acting simultaneously on each other. Clearly only the one-dimensional chain of syllogisms which constitutes thinking can be communicated verbally or written down. Poincaré (and later Pólya) tried to analyze the thought process. When I remember a mathematical proof, it seems to me that I remember only salient points, markers, as it were, of pleasure or difficulty. What is easy is easily passed over because it can be reconstituted logically with ease. If, on the other hand, I want to do something new or original, then it is no longer a question of syllogism chains. When I was a boy I felt that the role of rhyme in poetry was to compel one to find the unobvious because of the necessity of finding a word which rhymes. This forces novel associations and almost guarantees deviations from routine chains or trains of thought. It becomes paradoxically a sort of automatic mechanism of originality. I am pretty sure this "habit" of originality exists in mathematical research, and I can point to those who have it. This process of creation is, of course, not understood nor described well enough at present. What people think of as inspiration or illumination is really the result of much subconscious work and association through channels in the brain of which one is not aware at all.

It seems to me that good memory—at least for mathematicians and physicists—forms a large part of their talent. And what we call talent or perhaps genius itself depends to a large extent on the ability to use one's memory properly to find the analogies, past, present and future, which, as Banach said, are essential to the development of new ideas.

I continue to speculate on the nature of memory and how it is built and organized. Although one does not know much at present about its physiological or anatomical basis, what gives a partial hint is how one tries to remember things which one has temporarily forgotten. There are several theories about the physical aspects of memory. Some neurologists or biologists say that it consists perhaps of permanently renewed currents in the brain, much as the first computer memories were built with sound waves in a mercury tank. Others say that it resides in chemical changes of RNA molecules. But whatever its mechanism, an important thing is to understand the access to our memory.

Experiments seem to indicate that the memory is complete in the sense that everything we experience or think about is stored. It is only the conscious access to it that is partial and varies from person to person. Some experiments have shown that by touching a certain spot in the brain a subject will seem to recall or even "feel" a situation that happened in the past—such as being at a concert and actually hearing a certain melody.

How is memory gradually built up during one's conscious or even unconscious life and thought? My guess is that everything we experience is classified and registered on very many parallel channels in different locations, much as the visual impressions that are the result of many impulses on different cones and rods. All these pictures are transmitted together with connected impressions from other senses. Each such group is stored independently, probably in a great number of places under headings relevant to the various categories, so that in the visual brain there is a picture, and together with the picture something about the time, or the source, or the word, or the sound, in a branching tree which must have additionally a number of connecting loops. Otherwise one could not consciously try and sometimes succeed in remembering a forgotten name. In a computing machine, once the address of the position of an item in the memory is lost, there is no way to get at it. The fact that we succeed, at least on occasion, means that at least one member of the "search party" has hit a place where an element of the group is stored. Thus it is common to recall a last name once the first name has been recalled.

Then I thought, how about smell? Smell is something we sense; it is not related to any sound or picture. We do not know how to call it. It has no visual impact either. Does this contradict my guesses about simultaneous storage and connections? Then I remembered the famous incident related by Proust of the smell and taste of the "madeleine" (little cake). There are many descriptions in the literature of cases where a smell previously experienced and felt suddenly brings back a long-forgotten occasion when it was first associated with a place, or a person, many years before. So, perhaps on the contrary, this is another indication.

This feeling of analogy or association is necessary to place the set of impressions correctly on the suitable end points of a sequence of branches of a tree. And perhaps this is how people differ from each other in their memories. In some, more of these analogies are felt, stored, and better connected. Such analogies can be of an extremely abstract nature. I can conceive that a concrete picture, a visual sequence of dots and dashes, may bring back an abstract thought, which apparently in a mysterious coding had something in common with it. Some part of what is called mathematical talent may depend on the ability to see such analogies.

It is said that seventy-five percent of us have a dominant visual memory, twenty-five percent an auditory one. As for me, mine is quite visual. When I think about mathematical ideas, I see the abstract notions in symbolic pictures. They are visual assemblages, for example, a schematized picture of actual sets of points on a plane. In reading a statement like "an infinity of spheres or an infinity of sets," I imagine a picture with such almost real objects, getting smaller, vanishing on some horizon.

It is possible that human thought codes things not in terms of words or syllogisms or signs, for most people think pictorially, not verbally. There is a way of writing abstract ideas in a kind of shorthand which is almost orthogonal to the usual ways in which we communicate with each other by means of the spoken or written word. One may call this a "visual algorithm."

The process of logic itself working internally in the brain may be more analogous to a succession of operations with symbolic pictures, a sort of abstract analogue of the Chinese alphabet or some Mayan description of events—except that the elements are not merely words but more like sentences or whole stories with linkages between them forming a sort of meta- or super-logic with its own rules.

For me, some of the most interesting passages about the connections between the problem of time, as involved in the memory, and the physical or even mathematical meaning of it, whether it is classical or relativistic, were written, not by a physicist or a neurologist or a professional psychologist, but by Vladimir Nabokov in his book *Ada*. Some utterances by Einstein himself, as quoted in his biographies, show the great physicist's wonder at what living in time means, since we experience only the present. But, in reality, we consist of permanent and immutable world lines in four dimensions.

With such thoughts and worries about the thinking process, I was recovering my physical strength during this period of convalescence. What comforted me the most was the receipt of an invitation to attend a secret conference in Los Alamos in late April. This became for me a true sign of confidence in my mental recovery. I could not be told on the telephone or by letter what the conference was about. Secrecy was most intense at that time, but I guessed correctly that it would be devoted to the problems of thermonuclear bombs.

The conference lasted several days. Many friends were present. Some had been directly involved, like Fraenkel, Metropolis, Teller, and myself; others were consultants, like von Neumann. Fermi was absent. The discussions were active and inquisitive. They began with a presentation by Fraenkel of some calculations on the work initiated by Teller during the war. They were not detailed or complete enough and required work on computers (not the MANIACs but other machines in operation at the Aberdeen Proving Grounds). These were the first problems attacked that way.

The promising features of the plan were noticed and to some extent confirmed, but there remained great questions about the initiation of the process and, once initiated, about its successful continuation.

(All this was to have great importance in a later lawsuit between Sperry Rand and Honeywell over the validity of patents involving computers. The claim was that computers were already in the public domain then because the government of the United States used them and therefore the patents granted later were invalid. I was one of many who were called to testify on this in 1971.)

I participated in all the Los Alamos meetings. They lasted for hours, mornings and afternoons, and I noticed with pleasure that I was not unduly tired.

I remember telling Johnny about my illness. "I was given up for dead," I said, "and thought myself that I was already dead, except for a set of measure zero." This purely mathematical joke amused him. He laughed and asked, "What measure?"

Edward Teller and Johnny were often together, and I joined them in private talks.

In one conversation they discussed the possibility of influencing the weather. They had in mind global changes, while I proposed more local interventions. For example, I remember asking Johnny whether hurricanes could not be diverted, attenuated, or dispersed with nuclear explosions. I wasn't thinking of a point source, which is symmetrical, but several explosions in a line. I reasoned that the violence and enormous energy of a hurricane lies on top of a mass of air (the weather) which itself moves gently and slowly. I wondered if one could not, even ever so slightly, change its course in time and in trajectory on the slow-moving overall weather, thus making it avoid populated areas. There are, of course, many questions and objections about such an undertaking. One of the necessary conditions would be to make detailed computations on the course of the motion of the air masses, calculations which do not exist even now. Through the years Johnny and I ocassionally talked about this with experts in hydrodynamics and meteorology.

The conference over, I returned to Los Angeles. Upon alighting from the plane, two FBI agents approached me, showed their identification and asked for permission to search my luggage. A copy of the very secret Metropolis and Fraenkel report was missing, and they wondered if I might have taken it by mistake. We searched, but I did not have it. Later I learned that everybody who had attended the conference had been contacted. The authorities were very nervous, for this was potentially of grave consequence. The missing document reappeared much later among some of Teller's papers in a Los Alamos safe.

The time was rapidly approaching when I could resume teaching, but I was developing strongly negative feelings about Los Angeles. Rides through the streets where I had been driven in an ambulance reminded me of my recent illness. My feelings toward the University were colored by this, as well, and I was dissatisfied. I felt impatiently that it was not changing quickly enough from a glorified high school into a genuine institution of higher learning. I had disagreements with a dean about building up the academic level and increasing the staff. I was told he joked that he almost had a heart attack every time he saw me, even from a distance, so afraid was he that I was bringing him new proposals for expansion!

The best part of the University was the Hancock Library. It had an impressive building and some good books—but the building was better than the collection inside. The University had just acquired an old municipal library from Boston, and when I learned what it contained, I compared it to a priceless collection of hundred-year-old Sears Roebuck catalogs. This sarcastic remark probably did not enhance my popularity.

Even though I had friends like Donald Hyers, and some new acquaintances among mathematicians, physicists, and chemists, with this growing disenchantment I wanted to leave. The Los Angeles experience had not been satisfactory.

Just then I received a telegram inviting me to return to Los Alamos in a better position and at a higher salary. It was signed by Bob Richtmyer and Nick Metropolis. Richtmyer had become head of the theoretical division.

This offer to return to Los Alamos to work among physicists and live once again in the exhilarating climate of New Mexico was a great relief for me. I replied immediately that I was interested in principle. When the telegram arrived at the laboratory, it read that I was interested "in principal."

Adventures of a Mathematician: A Review

H. E. Robbins

H. E. Robbins was born in Pennsylvania and received a Ph.D. in mathematics from Harvard University in 1938. The recipient of a number of honors for scientific achievement he is professor of mathematical statistics at Columbia University and a member of the National Academy of Sciences.

A vast literature is available to readers of mathematics. Such readers are accustomed to journal articles, books, anthologies, biographies, extracts, and reprints. But there is one form of literature that people rarely use for learning a subject: the book review.

There is no substitute for actually going through each page of a long and sometimes boring book (not that Adventures of a Mathematician *is boring!). Nevertheless, a book review can be profitable to the reader. In this book review H. E. Robbins offers anecdotes that he feels epitomize the auther, Stanislaw Ulam. The reader may consider whether Robbins's choice of anecdotes tells more about Ulam or about Robbins.*

I.

Ulam is a magic name in modern mathematics. One thinks of Leonardo's letter to the Duke of Milan:

> Most Illustrious Lord;
>
> . . . Item: In case of need I will make big guns, mortars, and light ordnance of fine and useful forms, out of the common type.
>
> Item: I can carry out sculpture in marble, bronze, or clay, and also I can do in painting whatever may be done, as well as any other, be he who he may. . . .

And so he could.

In Ulam's writing, as in Leonardo's, scarcely a mention of mother and father. At eleven Ulam began to be known as a bright child who understood the special theory of relativity. He was an A student but did not study much, active in sports, played bridge, poker, and chess. At 15 he absorbed the calculus, number theory, and set theory. At 18, when he matriculated from gymnasium, the choice of profession presented difficulties. His father wanted him to join his successful law practice, while Ulam longed for a university career. But university positions in Poland were almost impossible to obtain if one's family, however wealthy and culturally assimilated, had a Jewish background. As a compromise, Ulam entered Lwów Polytechnic Institute to study engineering.

From the first, mathematics took complete possession of him. Kuratowski quickly recognized the young student's gifts and took special pains with him. The names of Mazur, Lomnicki, Borsuk, Kacmarz, Nikliborc, Tarski, Schauder, Averbach, Schreier, Steinhaus, and above all Banach dominated a euphoric period of feverish activity. At 23 Ulam was sufficiently well known to be an invited speaker at the Zürich congress. Meeting foreign mathematicians for the first time, he found them nervous and given to facial twitches, or short and old, like Hilbert; certainly less impressive than his fellow Poles. Returning to

Source: H. E. Robbins, Review of *Adventures of a Mathematician,* Bulletin of the American Mathematical Society 84 (1978): 107–110. Reprinted by permission of the American Mathematical Society.

Lwów, Ulam wrote a master's thesis which among other things outlined what is now category theory, and at 24 won his doctorate with a thesis in measure theory. But still there were no prospects of a university position for him in Poland.

Financed by his parents he visited Menger in Vienna, Hopf in Zürich, Cartan in Paris, and Hardy in Cambridge. Returning to Poland, he began a correspondence with von Neumann who invited him to visit the Institute at Princeton. In December 1935, Ulam sailed on the *Aquitania* for New York.

It was von Neumann whom Ulam came to admire above all others as a mathematician and kindred spirit. (The book was originally intended as a biography of von Neumann.) Things really began to happen when Ulam met G. D. Birkhoff at von Neumann's house and was in due course invited to Harvard as a Junior Fellow for three years. But soon after Ulam's return in 1939 from his customary three month visit to Poland, Hurewicz telephoned to say in somber tones "Warsaw has been bombed, the war has begun."

Next spring, when things looked darkest, it was Birkhoff who came to the rescue again by securing for Ulam an instructorship at Madison. This was no easy matter, for there were many emigrés by then and even modest positions were hard to find. At Madison he was promoted quickly to assistant professor, a position which held good hope for the future. He became an American citizen, married, and in 1943 rejoined von Neumann at Los Alamos, ignorant until he arrived of just what was going on there.

For Ulam, the transition from pure mathematics to applied physics was remarkably easy. (Not so for von Neumann, who had little physical intuition.) The physicist Otto Frisch on his first visit to Los Alamos from embattled Britain wrote "I also met Stan Ulam early on, a brilliant Polish topologist with a charming French wife. At once he told me that he was a pure mathematician who had sunk so low that his latest paper actually contained numbers with decimal points!"

II.

Although Ulam's three intellectual heros were Banach, von Neumann, and Fermi, none of them is portrayed so vividly in the book as Birkhoff. The Ulam-Birkhoff relationship seems to have been somewhat ambiguous on both sides.

"He liked the way I got almost furious when—in order to draw me out—he attacked his son Garrett's research on generalized algebras and more formal abstract studies of structures. I defended it violently. His smile told me that he was pleased that the worth and originality of his son's work was appreciated.

"In discussing the general job situation, he would often make skeptical remarks about foreigners. I think he was afraid that his position as the unquestioned leader of American mathematics would be weakened by the presence of such luminaries as Hermann Weyl, Jacques Hadamard, and others. He was also afraid that the explosion of refugees from Europe would fill the important academic positions, at least on the Eastern seaboard. He was quoted as having said, 'If American mathematicians don't watch out, they may become hewers of wood and carriers of water.'"*

Even after Birkhoff's death the American suspicion of foreigners—even those who as Ulam describes himself were "not unpresentable"—continued to cause trouble. When the war ended in 1945 and Ulam wanted to return to Madison, chairman R. E. Langer answered when Ulam inquired about his chances for promotion and tenure: "No reason to beat around the bush, were you not a foreigner, it would be much easier and your career would develop faster."

At the time Ulam was 36, by any standards an outstandingly creative mathematician, pleasant and courteous in manner, and well supplied by now with friends in high places. How is it that all this did not suffice to overcome the Wisconsin xenophobia, nor to secure for him then or later a position commensurate with his talents at some leading American university? Surely there is a mystery here.

Before 1945 mathematicians were about as numerous in the academic world as professors of French literature, and their importance in the military-industrial-intellectual complex about as great. During the next twenty years American

*Birkhoff's statement on the subject can be found in *American Mathematical Society Semicentennial Publications*, Vol. II, New York, 1938, pp. 276–277.

mathematics was a growth industry, since mathematicians had contributed essentially to making the weapons on which our safety now depended and would be needed in the future to keep ahead of possible rivals. Contrary to Birkhoff's fear, the refugees had created several jobs for American mathematicians for every one they occupied. Only the German rocket engineers imported after the war had a comparable effect.

III.

Turned down by Wisconsin, Ulam spent an unhappy year at U.S.C., interrupted by a mysterious illness which brought him close to death, and in 1946 returned to Los Alamos. There he proposed the Monte Carlo method in a conversation with von Neumann. "Little did we know in 1946 that computing would become a fifty-billion-dollar industry annually by 1970." Teller and von Neumann were emotionally committed to constructing an H bomb at all costs. Ulam was not so obsessed, but it was he who thought of a way to make it work. "Contrary to those people who were violently against the bomb on political, moral, or sociological grounds, I never had any questions about doing purely theoretical work. . . . I sincerely felt it was safer to keep these matters in the hands of scientists and people who are accustomed to objective judgments rather than in those of demagogues or jingoists, or even well-meaning but technically uninformed politicians."

In 1967 Ulam returned to university life at Boulder and became an elder statesman of government science.

IV.

Some readers will be put off by the frequent examples of mathematical humor characteristic of Ulam and his friends. Thus of Erdös: "Once he stopped to caress a sweet little child and said in his special language: 'Look, Stan! What a nice epsilon.' A very beautiful young woman, obviously the child's mother, sat nearby, so I replied, 'But look at the capital epsilon.' This made him blush with embarrassment." In fact, these episodes provide almost the only evidence of the humanity of the characters portrayed in this book. Erdös apart, they are preoccupied with seeking recognition of their precise rightful place in the official pecking order. It is a pity that this aspect of the world of mathematicians is so much emphasized in a book for the general reader; the more pity if indeed the emphasis is justified. The appearance of being thinking machines on the make, without discernible relation to parents, spouses, or children, and oblivious to the human concerns of our times, may be due in part to foreign systems of higher education that were devised to turn out idiot savants in the sciences as being more likely to be useful to the state. But if mathematical intelligence is strongly associated with emotional deprivation and social alienation, then even we earthy, super-honest, solid, and simple native Americans—the qualities that Ulam admires in us—are in for trouble.

The Legend of John von Neumann

P. R. Halmos

P. R. Halmos was born in Budapest, Hungary, and received a Ph.D. in mathematics from the University of Illinois. He has held professorial positions at the Universities of Chicago, Michigan, and Hawaii, and is currently professor of mathematics at the University of Indiana. He is the author of numerous books and research articles in pure mathematics.

What follows is a brief but very insightful biography of a most remarkable man. John von Neumann was one of the most influential figures in twentieth-century mathematics and must surely appear on any list of the ten most significant persons in the mathematics of this era. P. R. Halmos, a significant mathematical figure in his own right, assesses both von Neumann the man and his work.

No account of von Neumann's life can be considered complete without an understanding of the tragedy of his death. The reader should be sure to examine Steve J. Heims's article, found just after this one, which movingly describes the end of von Neumann's life.

JOHN VON NEUMANN was a brilliant mathematician who made important contributions to quantum physics, to logic, to meteorology, to war, to the theory and applications of high-speed computing machines, and, via the mathematical theory of games of strategy, to economics.

Youth

He was born December 28, 1903, in Budapest, Hungary. He was the eldest of three sons in a well-to-do Jewish family. His father was a banker who received a minor title of nobility from the Emperor Franz Josef; since the title was hereditary, von Neumann's full Hungarian name was Margittai Neumann János. (Hungarians put the family name first. Literally, but in reverse order, the name means John Neumann of Margitta. The "of," indicated by the final "i," is where the "von" comes from; the place name was dropped in the German translation. In ordinary social intercourse such titles were never used, and by the end of the first world war their use had gone out of fashion altogether. In Hungary von Neumann is and always was known as Neumann János and his works are alphabetized under N. Incidentally, his two brothers, when they settled in the U.S., solved the name problem differently. One of them reserves the title of nobility for ceremonial occasions only, but, in daily life, calls himself Neumann; the other makes it less conspicuous by amalgamating it with the family name and signs himself Vonneuman.)

Even in the city and in the time that produced Szilárd (1898), Wigner (1902), and Teller (1908), von Neumann's brilliance stood out, and the legends about him started accumulating in his childhood. Many of the legends tell about his memory. His love of history began early, and, since he re-

Source: P. R. Halmos, "The Legend of John von Neumann," *American Mathematical Monthly* 80, (1973): 382–394.

membered what he learned, he ultimately became an expert on Byzantine history, the details of the trial of Joan of Arc, and minute features of the battles of the American Civil War.

He could, it is said, memorize the names, addresses, and telephone numbers in a column of the telephone book on sight. Some of the later legends tell about his wit and his fondness for humor, including puns and off-color limericks. Speaking of the Manhattan telephone book he said once that he knew all the numbers in it—the only other thing he needed, to be able to dispense with the book altogether, was to know the names that the numbers belonged to.

Most of the legends, from childhood on, tell about his phenomenal speed in absorbing ideas and solving problems. At the age of 6 he could divide two eight-digit numbers in his head; by 8 he had mastered the calculus; by 12 he had read and understood Borel's *Théorie des Fonctions*.

These are some of the von Neumann stories in circulation. I'll report others, but I feel sure that I haven't heard them all. Many are undocumented and unverifiable, but I'll not insert a separate caveat for each one: let this do for them all. Even the purely fictional ones say something about him; the stories that men make up about a folk hero are, at the very least, a strong hint of what he was like.

In his early teens he had the guidance of an intelligent and dedicated high-school teacher, L. Rátz, and, not much later, he became a pupil of the young M. Fekete and the great L. Fejér, "the spiritual father of many Hungarian mathematicians". ("Fekete" means "Black", and "Fejér" is an archaic spelling, analogous to "Whyte.")

According to von Kármán, von Neumann's father asked him, when John von Neumann was 17, to dissuade the boy from becoming a mathematician, for financial reasons. As a compromise between father and son, the solution von Kármán proposed was chemistry. The compromise was adopted, and von Neumann studied chemistry in Berlin (1921–1923) and in Zürich (1923–1925). In 1926 he got both a Zürich diploma in chemical engineering and a Budapest Ph.D. in mathematics.

Early Work

His definition of ordinal numbers (published when he was 20) is the one that is now universally adopted. His Ph.D. dissertation was about set theory too; his axiomatization has left a permanent mark on the subject. He kept up his interest in set theory and logic most of his life, even though he was shaken by K. Gödel's proof of the impossibility of proving that mathematics is consistent.

He admired Gödel and praised him in strong terms: "Kurt Gödel's achievement in modern logic is singular and monumental—indeed it is more than a monument, it is a landmark which will remain visible far in space and time. . . . The subject of logic has certainly completely changed its nature and possibilities with Gödel's achievement." In a talk entitled "The Mathematician," speaking, among other things, of Gödel's work, he said: "This happened in our lifetime, and I know myself how humiliatingly easily my own values regarding the absolute mathematical truth changed during this episode, and how they changed three times in succession!"

He was Privatdozent at Berlin (1926–1929) and at Hamburg (1929–1930). During this time he worked mainly on two subjects, far from set theory but near to one another: quantum physics and operator theory. It is almost not fair to call them two subjects: due in great part to von Neumann's own work, they can be viewed as two aspects of the same subject. He started the process of making precise mathematics out of quantum theory, and (it comes to the same thing really) he was inspired by the new physical concepts to make broader and deeper the purely mathematical study of infinite-dimensional spaces and operators on them. The basic insight was that the geometry of the vectors in a Hilbert space has the same formal properties as the structure of the states of a quantum-mechanical system. Once that is accepted, the difference between a quantum physicist and a mathematical operator-theorist becomes one of language and emphasis only. Von Neumann's book on quantum mechanics appeared (in German) in 1932. It has been translated into French (1947), Spanish (1949), and English

(1955), and it is still one of the standard and one of the most inspiring treatments of the subject. Speaking of von Neumann's contributions to quantum mechanics, E. Wigner, a Nobel laureate, said that they alone "would have secured him a distinguished position in present day theoretical physics."

Princeton

In 1930 von Neumann went to Princeton University for one term as visiting lecturer, and the following year he became professor there. In 1933, when the Institute for Advanced Study was founded, he was one of the original six professors of its School of Mathematics, and he kept that position for the rest of his life. (It is easy to get confused about the Institute and its formal relation with Princeton University, even though there is none. They are completely distinct institutions. The Institute was founded for scholarship and research only, not teaching. The first six professors in the School of Mathematics were J. W. Alexander, A. Einstein, M. Morse, O. Veblen, J. von Neumann, and H. Weyl. When the Institute began it had no building, and it accepted the hospitality of Princeton University. Its members and visitors have, over the years, maintained close professional and personal relations with their colleagues at the University. These facts kept contributing to the confusion, which was partly clarified in 1940, when the Institute acquired a building of its own, about a mile from the Princeton campus.)

In 1930 von Neumann married Marietta Kövesi; in 1935 their daughter Marina was born. (In 1956 Marina von Neumann graduated from Radcliffe *summa cum laude,* with the highest scholastic record in her class. In 1972 Marina von Neumann Whitman was appointed by President Nixon to the Council of Economic Advisers.) In the 1930's the stature of von Neumann, the mathematician, grew at the rate that his meteoric early rise had promised, and the legends about Johnny, the human being, grew along with it. He enjoyed life in America and lived it in an informal manner, very differently from the style of the conventional German professor. He was not a refugee and he didn't feel like one. He was a cosmopolite in attitude and a U.S. citizen by choice.

The parties at the von Neumanns' house were frequent, and famous, and long. Johnny was not a heavy drinker, but he was far from a teetotaller. In a roadside restaurant he once ordered a brandy with a hamburger chaser. The outing was in honor of his birthday and he was feeling fine that evening. One of his gifts was a toy, a short prepared tape attached to a cardboard box that acted as sounding board; when the tape was pulled briskly past a thumbnail, it would squawk "Happy birthday!" Johnny squawked it often. Another time, at a party at his house, there was one of those thermodynamic birds that dips his beak in a glass of water, straightens up, teetertotters for a while, and then repeats the cycle. A temporary but firm house rule was quickly passed: everyone had to take a drink each time that the bird did.

He liked to drive, but he didn't do it well. There was a "von Neumann's corner" in Princeton, where, the story goes, his cars repeatedly had trouble. One often quoted explanation that he allegedly offered for one particular crack-up goes like this: "I was proceeding down the road. The trees on the right were passing me in orderly fashion at 60 miles an hour. Suddenly one of them stepped in my path. Boom!"

He once had a dog named "Inverse." He played poker, but only rarely, and he usually lost.

In 1937 the von Neumanns were divorced; in 1938 he married Klára Dán. She learned mathematics from him and became an expert programmer. Many years later, in an interview, she spoke about him. "He has a very weak idea of the geography of the house. . . . Once, in Princeton, I sent him to get me a glass of water; he came back after a while wanting to know where the glasses were. We had been in the house only seventeen years. . . . He has never touched a hammer or a screwdriver; he does nothing around the house. Except for fixing zippers. He can fix a broken zipper with a touch."

Von Neumann was definitely not the caricatured college professor. He was a round, pudgy man, always neatly, formally dressed. There are, to be sure, one or two stories of his absentmindedness. Klára told one about the time when he left their Princeton house one morning to drive to a New York appointment, and then phoned her when he reached New Brunswick to ask: "Why am I going to New York?" It may not be strictly relevant, but I am reminded of the time I drove him to his house one afternoon. Since there was to be a party there later that night, and since I didn't trust myself to remember exactly how I got there, I asked how I'd be able to know his house when I came again. "That's easy," he said; "it's the one with that pigeon sitting by the curb."

Normally he was alert, good at rapid repartee. He could be blunt, but never stuffy, never pompous. Once the telephone interrupted us when we were working in his office. His end of the conversation was very short; all he said between "Hello" and "Goodbye" was "Fekete pestis!" which means "Black plague!" Remembering, after he hung up, that I understood Hungarian, he turned to me, half apologetic and half exasperated, and explained that he wasn't speaking of one of the horsemen of the Apocalypse, but merely of some unexpected and unwanted dinner guests that his wife just told him about.

On a train once, hungry, he asked the conductor to send the man with the sandwich tray to his seat. The busy and impatient conductor said "I will if I see him." Johnny's reply: "This train is linear, isn't it?"

Speed

The speed with which von Neumann could think was awe-inspiring. G. Pólya admitted that "Johnny was the only student I was ever afraid of. If in the course of a lecture I stated an unsolved problem, the chances were he'd come to me as soon as the lecture was over, with the complete solution in a few scribbles on a slip of paper." Abstract proofs or numerical calculations—he was equally quick with both, but he was especially pleased with and proud of his facility with numbers. When his electronic computer was ready for its first preliminary test, someone suggested a relatively simple problem involving powers of 2. (It was something of this kind: what is the smallest power of 2 with the property that its decimal digit fourth from the right is 7? This is a completely trivial problem for a present-day computer: it takes only a fraction of a second of machine time.) The machine and Johnny started at the same time, and Johnny finished first.

One famous story concerns a complicated expression that a young scientist at the Aberdeen Proving Grounds needed to evaluate. He spent ten minutes on the first special case; the second computation took an hour of paper and pencil work; for the third he had to resort to a desk calculator, and even so took half a day. When Johnny came to town, the young man showed him the formula and asked him what to do. Johnny was glad to tackle it. "Let's see what happens for the first few cases. If we put $n = 1$, we get . . ."—and he looked into space and mumbled for a minute. Knowing the answer, the young questioner put in "2.31?" Johnny gave him a funny look and said "Now if $n = 2$, . . ." and once again voiced some of his thoughts as he worked. The young man, prepared, could of course follow what Johnny was doing, and, a few seconds before Johnny finished, he interrupted again, in a hesitant tone of voice: "7.49?" This time Johnny frowned, and hurried on: "If $n = 3$, then. . . ." The same thing happened as before—Johnny muttered for several minutes, the young man eavesdropped, and, just before Johnny finished, the young man exclaimed: "11.06!" That was too much for Johnny. It couldn't be! No unknown beginner could outdo him! He was upset and he sulked till the practical joker confessed.

Then there is the famous fly puzzle. Two bicyclists start twenty miles apart and head toward each other, each going at a steady rate of 10 m.p.h. At the same time a fly that travels at a steady 15 m.p.h. starts from the front wheel of the southbound bicycle and flies to the front wheel of the northbound one, then turns around and flies to the front wheel of the southbound one again, and continues in this manner till he is crushed between the two front wheels. Question: what total distance did the fly cover? The slow way to find the answer is to calculate what distance the fly covers on the first, northbound, leg of the trip, then on the second, southbound, leg, then on the third, etc., etc., and, finally, to sum the infinite series so obtained. The quick way is to observe that the bicycles meet exactly one hour after their start, so that the fly had just an hour for his travels; the answer must therefore be 15 miles. When the question was put to von Neumann, he solved it in an instant, and thereby disappointed the questioner: "Oh, you must have heard the trick before!" "What trick?" asked von Neumann; "all I did was sum the infinite series."

I remember one lecture in which von Neumann was talking about rings of operators. At an appropriate point he mentioned that they can be classified two ways: finite versus infinite, and discrete versus continuous. He went on to say: "This leads to a total of four possibilities, and, indeed, all four of them can occur. Or—let's see—can they?" Many of us in the audience had been learning this subject from him for some time, and it was no trouble to stop and mentally check off all four possibilities. No trouble—it took something like two seconds for each, and, allowing for some fumbling and shifting of gears, it took us perhaps 10 seconds in all. But after two seconds von Neumann had already said "Yes, they can," and he was two sentences into the next paragraph before, dazed, we could scramble aboard again.

Speech

Since Hungarian is not exactly a *lingua franca,* all educated Hungarians must acquire one or more languages with a popular appeal greater than that of their mother tongue. At home the von Neumanns spoke Hungarian, but he was perfectly at ease in German, and in French, and, of course, in English. His English was fast and grammatically defensible, but in both pronunciation and sentence construction it was reminiscent of German. His "Sprachgefühl" was not perfect, and his sentences tended to become involved. His choice of words was usually exactly right; the occasional oddities (like "a self-obvious theorem") disappeared in later years. His spelling was sometimes more consistent than commonplace: if "commit," then "ommit." S. Ulam tells about von Neumann's trip to Mexico, where "he tried to make himself understood by using 'neo-Castilian,' a creation of his own—English words with an 'el' prefix and appropriate Spanish endings."

He prepared for lectures, but rarely used notes. Once, five minutes before a non-mathematical lecture to a general audience, I saw him as he was preparing. He sat in the lounge of the Institute and scribbled on a small card a few phrases such as these: "Motivation, 5 min.; historical background, 15 min.; connection with economics, 10 min. . . ."

As a mathematical lecturer he was dazzling. He spoke rapidly but clearly; he spoke precisely, and he covered the ground completely. If, for instance, a subject has four possible axiomatic approaches, most teachers content themselves with developing one, or at most two, and merely mentioning the others. Von Neumann was fond of presenting the "complete graph" of the situation. He would, that is, describe the shortest path that leads from the first to the second, from the first to the third, and so on through all twelve possibilities.

His one irritating lecturing habit was the way he wielded an eraser. He would write on the board the crucial formula under discussion. When one of the symbols in it had been proved to be replaceable by something else, he made the replacement not by rewriting the whole formula, suitably modified, but by erasing the replaceable symbol and substituting the new one for it. This had the tendency of inducing symptoms of acute discouragement among note-takers, especially since, to maintain the flow of the argument, he would keep talking at the same time.

His style was so persuasive that one didn't have to be an expert to enjoy his lectures; everything seemed easy and natural. Afterward, however, the Chinese-dinner phenomenon was likely to occur. A couple of hours later the average memory could no longer support the delicate balance of mutually interlocking implications, and, puzzled, would feel hungry for more explanation.

Style

As a writer of mathematics von Neumann was clear, but not clean; he was powerful but not elegant. He seemed to love fussy detail, needless repetition, and notation so explicit as to be confusing. To maintain a logically valid but perfectly transparent and unimportant distinction, in one paper he introduced an extension of the usual functional notation: along with the standard $\phi(x)$ he dealt also with something denoted by $\phi((x))$. The hair that was split to get there had to be split again a little later, and there was $\phi(((x)))$, and, ultimately, $\phi((((x))))$. Equations such as

$$(\psi((((a)))))^2 = \phi((((a))))$$

have to be peeled before they can be digested; some irreverent students referred to this paper as von Neumann's onion.

Perhaps one reason for von Neumann's attention to detail was that he found it quicker to hack through the underbrush himself than to trace references and see what others had done. The result was that sometimes he appeared ignorant of the standard literature. If he needed facts, well-known facts, from Lebesgue integration theory, he waded in, defined the basic notions, and developed the theory to the point where he could use it. If, in a later paper, he needed integration theory again, he would go back to the beginning and do the same thing again.

He saw nothing wrong with long strings of suffixes, and subscripts on subscripts; his papers abound in avoidable algebraic computations. The reason, probably, is that he saw the large picture; the trees did not conceal the forest from him. He saw and he relished all parts of the mathematics he was thinking about. He never wrote "down" to an audience; he told it as he saw it. The practice caused no harm; the main result was that, quite a few times, it gave lesser men an opportunity to publish "improvements" of von Neumann.

Since he had no formal connections with educational institutions after he was 30, von Neumann does not have a long list of students; he supervised only one Ph.D. thesis. Through his lectures and informal conversations he acquired, however, quite a few disciples who followed in one or another of his footsteps. A few among them are J. W. Calkin, J. Charney, H. H. Goldstine, P. R. Halmos, I. Halperin, O. Morgenstern, F. J. Murray, R. Schatten, I. E. Segal, A. H. Taub, and S. Ulam.

Work Habits

Von Neumann was not satisfied with seeing things quickly and clearly; he also worked very hard. His wife said "he had always done his writing at home during the night or at dawn. His capacity for work was practically unlimited." In addition to his work at home, he worked hard at his office. He arrived early, he stayed late, and he never wasted any time. He was systematic in both large things and small; he was, for instance, a meticulous proofreader. He would correct a manuscript, record on the first page the page numbers where he found errors, and, by appropriate tallies, record the number of errors that he had marked on each of those pages. Another example: when requested to prepare an abstract of not more than 200 words, he would not be satisfied with a statistical check—there are roughly 20 lines with about 10 words each—but he would count every word.

When I was his assistant we wrote one paper jointly. After the thinking and the talking were finished, it became my job to do the writing. I did it, and I submitted to him a typescript of about 12 pages. He read it, criticized it mercilessly, crossed out half, and rewrote the rest; the result was about 18 pages. I removed some of the Germanisms, changed a few spellings, and compressed it into 16 pages. He was far from satisfied, and made basic changes again; the result was 20 pages. The almost divergent process continued (four innings on each side as I now recall it); the final outcome was about 30 typescript pages (which came to 19 in print).

Another notable and enviable trait of von Neumann's was his mathematical courage. If, in the middle of a search for a counterexample, an infinite series came up, with a lot of exponentials that had quadratic exponents, many mathematicians would start with a clean sheet of paper and look for another counterexample. Not Johnny! When that happened to him, he cheerfully said: "Oh, yes, a theta function . . . ," and plowed ahead with the mountainous computations. He wasn't afraid of anything.

He knew a lot of mathematics, but there were also gaps in his knowledge, most notably number theory and algebraic topology. Once when he saw some of us at a blackboard staring at a rectangle that had arrows marked on each of its sides, he wanted to know that what was. "Oh just the torus, you know—the usual identification convention." No, he didn't know. The subject is elementary, but some of it just never crossed his path, and even though most graduate students knew about it, he didn't.

Brains, speed, and hard work produced results. In von Neumann's *Collected Works* there is a list of over 150 papers. About 60 of them are on pure mathematics (set theory, logic, topological groups, measure theory, ergodic theory, operator theory, and continuous geometry), about 20 on physics, about 60 on applied mathematics (including statistics, game theory, and computer theory), and a small handful on some special mathematical subjects and general non-mathematical ones. A special number of the *Bulletin of the American Mathematical Society* was devoted to a discussion of his life and work (in May 1958).

Pure Mathematics

Von Neumann's reputation as a mathematician was firmly established by the 1930's, based mainly on his work on set theory, quantum theory, and operator theory, but enough more for about three ordinary careers, in pure mathematics alone, was still to come. The first of these was the proof of the ergodic theorem. Various more or less precise statements had been formulated earlier in statistical mechanics and called the ergodic hypothesis. In 1931 B. O. Koopman published a penetrating remark whose main substance was that one of the contexts in which a precise statement of the ergodic hypothesis could be formulated is the theory of operators on Hilbert space—the very subject that von Neumann used earlier to make quantum mechanics precise and on which he had written several epoch-making papers. It is tempting to speculate on von Neumann's reaction to Koopman's paper. It could have been something like this: "By Koopman's remark the ergodic hypothesis becomes a theorem about Hilbert spaces—and if that's what it is I ought to be able to prove it. Let's see now. . . ." Soon after the appearance of Koopman's paper, von Neumann formulated and proved the statement that is now known as the mean ergodic theorem for unitary operators. There was some temporary confusion, caused by publication dates, about who did what before whom, but by now it is universally recognized that von Neumann's theorem preceded and inspired G. D. Birkhoff's point ergodic theorem. In the course of the next few years von Neumann published several more first-rate papers on ergodic theory, and he made use of the techniques and results of that theory later, in his studies of rings of operators.

In 1900 D. Hilbert presented a famous list of 23 problems that summarized the state of mathematical knowledge at the time and showed where further work was needed. In 1933 A. Haar proved the existence of a suitable measure (which has come to be called Haar measure) in topological groups; his proof appears in the *Annals of Mathematics*. Von Neumann had access to Haar's result before it was published, and he quickly saw that that was exactly what was needed to solve an important special case (compact groups) of one of Hilbert's problems (the 5th). His solution appears in the same issue of the same journal, immediately after Haar's paper.

In the second half of the 1930's the main part of von Neumann's publications was a sequence of papers, partly in collaboration with F. J. Murray, on what he called rings of operators. (They are now called von Neumann algebras.) It is possible that this is the work for which von Neumann will be remembered the longest. It is a technically brilliant development of operator theory that makes contact with von Neumann's earlier work, generalizes many familiar facts about finite-dimensional algebra, and is currently one of the most powerful tools in the study of quantum physics.

A surprising outgrowth of the theory of rings of operators is what von Neumann called continuous geometry. Ordinary geometry deals with spaces of dimension 1, 2, 3, etc. In his work on rings of operators von Neumann saw that what really determines the dimension structure of a space is the group of rotations that it admits. The group of rotations associated with the ring of *all* operators yields the familiar dimensions. Other groups, associated with different rings, assign to spaces dimensions whose values can vary continuously; in that context it makes sense to speak of a space of dimension 3/4, say. Abstracting from the "concrete" case of rings of operators, von Neumann formulated the axioms that make these continuous-dimensional spaces possible. For several years he thought, wrote, and lectured about continuous geometries. In 1937 he was the Colloquium Lecturer of the American Mathematical Society and chose that subject for his topic.

Applied Mathematics

The year 1940 was just about the half-way point of von Neumann's scientific life, and his publications show a discontinuous break then. Till then he was a topflight pure mathematician who understood physics; after that he was an applied mathematician who remembered his pure work. He became interested in partial differential equations, the principal classical tool of the applications of mathematics to the physical world. Whether the war made him into an applied mathematician or his interest in applied mathematics made him invaluable to the war effort, in either case he was much in demand as a consultant and advisor to the armed forces and to the civilian agencies concerned with the problems of war. His papers from this point on are mainly on statistics, shock waves, flow problems, hydrodynamics, aerodynamics, ballistics, problems of detonation, meteorology, and, last but not least, two non-classical, new aspects of the applicability of mathematics to the real world: games and computers.

Von Neumann's contributions to war were manifold. Most often mentioned is his proposal of the implosion method for bringing nuclear fuel to explosion (during World War II) and his espousal of the development of the hydrogen bomb (after the war). The citation that accompanied his honorary D.Sc. from Princeton in 1947 mentions (in one word) that he was a mathematician, but praises him for being a physicist, an engineer, an armorer, and a patriot.

Politics

His political and administrative decisions were rarely on the side that is described nowadays by the catchall term "liberal." He appeared at times to advocate preventive war with Russia. As early as 1946 atomic bomb tests were already receiving adverse criticism, but von Neumann thought that they were necessary and (in, for instance, a letter to the *New York Times*) defended them vigorously. He disagreed with J. R. Oppenheimer on the H-bomb crash program, and urged that the U.S. proceed with it before Russia could. He was, however, a "pro-Oppenheimer" witness at the Oppenheimer security hearings. He said that Oppenheimer opposed the program "in good faith" and was "very constructive" once the decision to go ahead with the super bomb was made. He insisted that Oppenheimer was loyal and was not a security risk.

As a member of the Atomic Energy Commission (appointed by President Eisenhower, he was sworn in on March 15, 1955), having to "think about the unthinkable," he urged a United Nations study of world-wide radiation effects. "We willingly pay 30,000–40,000 fatalities per year (2% of the total death rate)," he wrote, "for the advantages of individual transportation by automobile." He mentioned a fall-out accident in an early Pacific bomb test that resulted in one fatality and danger to 200 people, and he compared it with a Japanese ferry accident that "killed about 1,000 people, including 20 Americans—yet the . . . fall-out was what attracted almost world-wide attention." He asked: "Is the price in international popularity worth paying?" And he answered: "Yes: we have to accept it as part payment for our more advanced industrial position."

Game Theory

At about the same time that he began to apply his analytic talents to the problems of war, von Neumann found time and energy to apply his combinatorial insight to what he called the theory of games, whose major application was to economics. The mathematical cornerstone of the theory is one statement, the so-called minimax theorem, that von Neumann proved early (1928) in a short article (25 pages); its elaboration and applications are in the book he wrote jointly with O. Morgenstern in 1944. The minimax theorem says about a large class of two-person games that there is no point in playing them. If either player considers, for each possible strategy of play, the *maximum* loss that he can expect to sustain with that strategy, and then chooses the "optimal" strategy that *minimizes* the maximum loss, then he can be statistically sure of not losing more than that minimax value. Since (and this is the whole point of the theorem) that value is the negative of the one, similarly defined, that his opponent can guarantee for himself, the long-run outcome is completely determined by the rules.

Mathematical economics before von Neumann tried to achieve success by imitating the technique of classical mathematical physics. The mathematical tools used were those of analysis (specifically the calculus of variations), and the procedure relied on a not completely reliable analogy between economics and mechanics. The secret of the success of the von Neumann approach was the abandonment of the mechanical analogy and its replacement by a fresh point of view (games of strategy) and new tools (the ideas of combinatorics and convexity).

The role that game theory will play in the future of mathematics and economics is not easy to predict. As far as mathematics is concerned, it is tenable that the only thing that makes the Morgenstern–von Neumann book 600 pages longer than the original von Neumann paper is the development needed to apply the abstruse deductions of one subject to the concrete details of another. On the other hand, enthusiastic proponents of game theory can be found who go so far as to say that it may be "one of the major scientific contributions of the first half of the 20th century."

Machines

The last subject that contributed to von Neumann's fame was the theory of electronic computers and automata. He was interested in them from every point of view: he wanted to understand them, design them, build them, and use them. What are the logical components of the processes that a computer will be asked to perform? What is the best way of obtaining practically reliable answers from a machine with unreliable components? What does a machine need to "remember," and what is the best way to equip it with a "memory"? Can a machine be built that can not only save us the labor of computing but save us also the trouble of building a new machine—is it possible, in other words, to produce a self-reproducing automaton? (Answer: in principle, yes. A sufficiently complicated machine, embedded in a thick chowder of randomly distributed spare parts, its "food," would pick up one part after another till it found a usable one, put it in place, and continue to search and construct till its descendant was complete and operational.) Can a machine successfully imitate "randomness," so that when no formulae are available to solve a concrete physical problem (such as that of finding an optimal bombing pattern), the machine can perform a large number of probability experiments and yield an answer that is statistically accurate? (The last question belongs to the concept that is sometimes described as the Monte Carlo method.) These are some of the problems that von Neumann studied and to whose solutions he made basic contributions.

He had close contact with several computers—among them the MANIAC (Mathematical Analyzer, Numerical Integrator, Automatic Calculator), and the affectionately named JONIAC. He advocated their use for everything from the accumulation of heuristic data for the clarification of our intuition about partial differential equations to the accurate long-range prediction and, ultimately, control of the weather. One of the most striking ideas whose study he suggested was to dye the polar icecaps so as to decrease the amount of energy they would reflect—the result could warm the earth enough to make the climate of Iceland approximate that of Hawaii.

The last academic assignment that von Neumann accepted was to deliver and prepare for publication the Silliman lectures at Yale. He worked on that job in the hospital where he died, but he couldn't finish it. His notes for it were published, and even they make illuminating reading. They contain tantalizing capsule statements of insights, and throughout them there shines an attitude of faith in and dedication to knowledge. While physicists, engineers, meteorologists, statisticians, logicians, and computers all proudly claim von Neumann as one of theirs, the Silliman lectures prove, indirectly by their approach and explicitly in the author's words, that von Neumann was first, foremost, and always a mathematician.

Death

Von Neumann was an outstanding man in tune with his times, and it is not surprising that he received many awards and honors. There is no point in listing them all here, but a few may be mentioned. He received several honorary doctorates, including ones from Princeton (1947), Harvard (1950), and Istanbul (1952). He served a term as president of the American Mathematical Society (1951–1953), and he was a member of several national scientific academies (including, of course, that of the U.S.). Somewhat to his embarrassment, he was elected to the East German Academy of Science, but the election didn't seem to take—in later years no mention is made of it in the standard biographical reference works. He received the Enrico Fermi award in 1956, when he already knew that he was incurably ill.

Von Neumann became ill in 1955. There was an operation, and the result was a diagnosis of cancer. He kept on working, and even travelling, as the disease progressed. Later he was confined to a wheelchair, but still thought, wrote, and attended meetings. In April 1956 he entered Walter Reed Hospital, and never left it. Of his last days his good friend Eugene Wigner wrote: "When von Neumann realized he was incurably ill, his logic forced him to realize that he would cease to exist, and hence cease to have thoughts. . . . It was heartbreaking to watch the frustration of his mind, when all hope was gone, in its struggle with the fate which appeared to him unavoidable but unacceptable."

Von Neumann was baptized a Roman Catholic (in the U. S.), but, after his divorce, he was not a practicing member of the church. In the hospital he asked to see a priest—"one that will be intellectually compatible," Arrangements were made, he was given special instruction, and, in due course, he again received the sacraments. He died February 8, 1957.

The heroes of humanity are of two kinds: the ones who are just like all of us, but very much more so, and the ones who, apparently, have an extra-human spark. We can all run, and some of us can run the mile in less than 4 minutes; but there is nothing that most of us can do that compares with the creation of the Great G-minor Fugue. Von Neumann's greatness was the human kind. We can all think clearly, more or less, some of the time, but von Neumann's clarity of thought was orders of magnitude greater than that of most of us, all the time. Both Norbert Wiener and John von Neumann were great men, and their names will live after them, but for different reasons. Wiener saw things deeply but intuitively; von Neumann saw things clearly and logically.

What made von Neumann great? Was it the extraordinary rapidity with which he could understand and think and the unusual memory that retained everything he had once thought through? No. These qualities, however impressive they might have been, are ephemeral; they will have no more effect on the mathematics and the mathematicians of the future than the prowess of an athlete of a hundred years ago has on the sport of today.

The "axiomatic method" is sometimes mentioned as the secret of von Neumann's success. In his hands it was not pedantry but perception; he got to the root of the matter by concentrating on the basic properties (axioms) from which all else follows. The method, at the same time, revealed to him the steps to follow to get from the foundations to the applications. He knew his own strengths and he admired, perhaps envied, people who had the complementary qualities, the flashes of irrational intuition that sometimes change the direction of scientific progress. For von Neumann it seemed to be impossible to be unclear in thought or in expression. His insights were illuminating and his statements were precise.

Paul Halmos claims that he took up mathematics because he flunked his master's orals in philosophy. He received his Univ. of Illinois Ph.D. under J. L. Doob. Then he was von Neumann's assistant, followed by positions at Illinois, Syracuse, M. I. T.'s Radiation Lab, Chicago, Michigan, Hawaii, and now is Distinguished Professor at Indiana Univ. . . . Professor Halmos' research is mainly measure theory, probability, ergodic theory, topological groups, Boolean algebra, algebraic logic, and operator theory in Hilbert space. . . .

Von Neumann: Only Human in Spite of Himself

Steve J. Heims

Steve J. Heims was born in Berlin, Germany, and received his Ph.D in physics from Stanford University. He has taught at Brandeis University and at Boston University and was a National Science Fellow in the history of science.

Hardy, in his book A Mathematician's Apology, *asserts that mathematicians can achieve a type of immortality through their theorems. If this is the case then John von Neumann is certain to be immortal. As Halmos relates at great detail in his article "The Legend of John von Neumann" (preceding this article), Von Neumann's reputation in mathematics was established by the 1930s, based mainly on his work in set theory, quantum theory, and operator theory. In the 1940s von Neumann switched to applied mathematics and wrote papers on statistics, shock waves, hydrodynamics, and meteorology. He almost single-handedly created the theory of games and is responsible for the von Neumann architecture that underlies all modern computers.*

Some of von Neumann's admirers in the sciences could play with his name and see in von Neumann the "new man" of the twentieth century—representative of the ideal scientific person, a model for what humanity should become in the future. Given that some could see in this remarkable genius a pattern for the future man and given that von Neumann could certainly view himself through the eyes of mathematical immortality, the reader may wonder how von Neumann reacted when he realized he was dying of cancer.

The following is a small excerpt of Heims's biography on Norbert Wiener and von Neumann. Heims's editorial orientation is clear from his title, "Von Neumann: Only Human in Spite of Himself." Nevertheless, the pathos of von Neumann's reaction to his imminent death is very apparent.

ASIDE FROM THE problem of the theoretic validity of science, von Neumann feared a future in which the basis for optimism would be undercut by a change in society's values: he once "expressed an apprehension that the values put on abstract scientific achievement in our present civilization might diminish: 'The interests of humanity may change, the present curiosities in science may cease, and entirely different things may occupy the human mind in the future.' "[1] In the 1950s we find von Neumann taking various opportunities to defend "science for science's sake," against the intrusion of philosophical considerations[2] and against the imposition mission-oriented objectives.[3] He was doing this even as his own work had become nearly entirely mission-oriented.

The possible future diminution of interest in science, which made von Neumann uneasy, was of course seen by others as promising liberation from scientific obscurantism and chauvinism, possibly through the intrusion of humanistic, social, and ethical dimensions into science. The two

Source: Excerpts from Steve J. Heims, in *John von Neumann and Norbert Weiner, from Mathematics to the Technologies of Life and Death* (Cambridge, Mass.: MIT Press, 1980), pp. 363–371. (Some references omitted.)

pillars of value neutrality and faith in the beneficence of all technological innovation could hardly hold up after World War II, and in fact Wiener had been one of those who emphasized that. If they crumbled, the true scientific virtues would also lose priority. These virtues had been promoted at the cost of others among the wide range of human possibilities. They had risen to prominence in the wake of the Newtonian and industrial revolutions. Both Wiener and von Neumann were specialized beings trained for mathematical or scientific work, and their limitations in other respects are also apparent. One may wonder whether the post-Hiroshima world does not require different virtues and an emphasis on development of other human possibilities. The rise of the "third world" indicates strong pressure in the direction of social justice; the growing likelihood of nuclear war exerts a pressure in the direction of nuclear disarmament; the limits of global resources create a pressure for conservation, awareness of ecological patterns, and restraints on the production of material luxuries; and political instabilities arising from rivalry among nation-states exert simultaneous pressures for global planning and decentralization. The assertion of the political and social dimension brings a focus on people instead of things, and subordinates science and technology to their human context. Precisely what the nature of the emerging values and virtues are or need to be is a topic deserving of a major inquiry of its own.

Some of von Neumann's scientific admirers have seen in him the "new man," the ideal type of future person, implied by his name. The burden of the vision as well as the historical evidence informing this book is that the direction of human development most valuable to the species in the late twentieth century is likely to be very different from that of the ideal scientific man. Von Neumann can with considerable justification be regarded as a superb and highly developed representative of a nineteenth-century vision of modern man, but if indeed a shift in the direction of civilization is occurring, it is inappropriate to view him as representative of what man should become in the future. The virtues and virtuosities of the ideal scientific man are particularly conducive to productive scientific work, but as a generalized ideal of conduct—and they have become that in a Western civilization—they are in conflict with a wider, more comprehensive humanness.

We are still in the midst of the scientific-technological ideals, values, and myths that have come to dominate Western civilization since the seventeenth century, and it is difficult to gain historical perspective. In the early feudal period in medieval France, a standard of masculine conduct emphasizing bravery, loyalty, fearlessness in battle, and gallantry toward conquered enemies was part of the code of conduct for warriors. In subsequent centuries this code of conduct evolved into the romantic ideal of chivalry, embodying these and still other virtues, and became a cultural myth. Whether or not many people were true to that ideal, it surely influenced their thinking and behavior. We hardly need Cervantes to show us, or the modern feminists to point out to us, that the chivalric ideal was in fact narrow, ludicrous at some points and cruel at others, and capable of being converted to the barbaric purposes of the Crusades. The "virtues" of knighthood do not constitute a comprehensive human ideal, but a very special, historically determined one. Similarly, the scientific virtues are limited, and their dominance may well be overcome by a more wholesome, comprehensive pattern of cultural values. Those who admired von Neumann as a "new man," more highly evolved than other men because of his logical-mental ability, were under the persistent spell of a scientific-technological system of values. Similarly, those who see the new postindustrial man in those experts such as von Neumann who wield influence in government through their professional advice are also thinking within the context of current technocratic, technological ideals. The idealization of these two kinds of "new man," derived by direct extrapolation from the presently dominant value structure, reflects a deep conservatism and a failure of imagination. Perhaps a modern Cervantes is needed to help our own age gain perspective.

The possibility that the scientific ideal will increase its dominance cannot be discounted. The cultural pressures in that direction, however, appear to contain self-destructive and dehumanizing elements, awareness of which tends to generate broader visions of what it means to be human.

Thus the scientific ideal is likely eventually to become incorporated into a broader vision of man and subordinated to it. It might go the way of the ideal of knighthood, surviving as a minor and increasingly decadent value within the culture even though a few individuals continue to be inspired by it.

Although the line separating science from technology is a blurred one, especially for someone like von Neumann who crossed back and forth frequently and easily, some people take distinctly different views of progress in "pure" science and progress in technology. I have suggested earlier that the coincidence of von Neumann's youth with Hungary's period of economic-industrial expansion helped to determine his optimism about and emotional commitment to technological progress, which his personal talents served to intensify. Moreover, we have seen that the events surrounding World War II—the Nazis' efficient extermination techniques and the use of the atomic bomb—did not suffice to undermine this optimistic outlook. What, then, was von Neumann's implicit or explicit philosophy of technology in the 1940s and 1950s? How about computer technology, which could be expected to alter society and with which von Neumann was so intimately involved?

It is clear from von Neumann's activities from late 1944 on, when he was first brought in as a consultant for the development of the ENIAC computer prototype, that he favored and helped bring about the most rapid possible rate of innovation and development in this field. In 1945, while the ENIAC was being built, von Neumann and his group were already designing a more advanced machine, the EDVAC, and von Neumann was proposing building a computer at Princeton that would leapfrog over the advances of the EDVAC. Moreover, he was eager to see his own machine surpassed. His complained to IBM executives and Navy brass that the rate of innovation is usually determined only by considerations as to "what the demand is, what the price is, whether it will be more profitable to do it in a bold way or in a cautious way, and so on" and recommends instead that it is sometimes desirable to write "specifications simply calling for the most advanced machine which is possible in the present state of the art."[4] Von Neumann meant what he said, according to Herman Goldstine:

> He and I were discussing ways to encourage industry to make bold new strides forward when suddenly he asked me whether it would not be a good idea to give contracts to both IBM and Sperry Rand to build the most advanced type computers that they would be willing to undertake. Out of this conversation grew a quite formal set of arrangements between the Atomic Energy Commissioner and each of these companies.[5]

Thus, some of von Neumann's desire for the maximization of progress in computer technology was propagated in the industrial world.

In any history of technology, certain major specific innovations in computer design must be attributed to von Neumann. But it would constitute a misunderstanding of the nature of technological change to regard his role as essential to computer development. For here, as is typical in technical change, numerous patent suits testify to the near-simultaneity of similar innovations by different research groups. In technological development, innovation must be supplemented by hardware implementation and by government or industrial interest in applying it; otherwise it lies fallow. Through his energetic advocacy, von Neumann did far more than most mathematicians would to bring about the hardware implementation and he also affected the field indirectly, in that "he was the man who, with his great prestige in the mathematical world, taking such an interest in these things, had the effect of stimulating many, many others to regard it as a very serious discipline."[6] Von Neumann's real impact on the computer was to speed the rate of advancement of the technology—more reliable and faster computers were available sooner because of him. This was something he wanted to and unquestionably did achieve. The same could be said concerning his role in weapons development and the nuclear armaments race. What is different in the case of weapons development is that the speed with which it took place precluded the possibility that gradually changing political conditions could obviate it.

In a 1955 article in *Fortune* magazine Atomic Energy Commission member von Neumann articulated a more general philosophy of technology as a basis for dealing with the crises created by technological progress.[7] It is possible that in such an article for the general reader his primary aim was to make official AEC policy (the nuclear arms race) palatable rather than to expose his personal views. Yet even with this qualification, the article is of some interest. He defines the crisis in terms of the finite size of the earth, which ultimately limits expansion into new geographical regions. For someone else this awareness might have led to a Wienerian emphasis on homeostasis or what today would be called an ecological consciousness, but for von Neumann it did not. He did concede that impact of some large-scale phenomena (he gives the melting of the polar ice caps as an example and alludes to nuclear weapons) might be felt worldwide and not limited to smaller political units. Nevertheless, he viewed the march of technological progress as inevitable and beyond human control, and dismissed as irrelevant efforts to impose any prohibitions on technical developments. Human beings simply had to make the necessary adaptations. Of course he knew firsthand how the U.S. government was deciding which technologies to develop. Even when they contain dangers, von Neumann insisted, new technologies are always ultimately constructive and beneficial. The scientist or technologist is consequently exonerated from responsibility. Although conceding in a vague, general way that major changes in political institutions would eventually be needed to accommodate the then-new weapons, he was adamantly opposed to any institutional changes that might slow down new technical developments. Von Neumann seemed to be religiously holding to the traditional view of the beneficence of technological progress, incorporating the new facts of nuclear weaponry without in any way significantly altering the classical views. As we have seen, in more private communications he had argued frankly that the sacrifice of human lives is an acceptable price to pay for technological progress.[8]

Von Neumann's apologetics for government policy seem to have coincided with his own views. Most of his erstwhile scientific friends and colleagues who were critical of his military involvement saw it only in personal terms or as a loss to mathematics or pure science. One of them, Warren McCulloch, wrote to von Neumann, "I regret, personally as well as scientifically, that honors and duties have ascended to your entertaining head. It would be good to see and hear you again—just for the fun of it."[9] From having read *Cybernetics*, if not from more direct conversation, von Neumann was aware of Wiener's views on technological progress, but he was obviously unconvinced. His *Fortune* article can be regarded as a public answer to Wiener.[10]

Von Neumann's all-embracing enthusiasm for scientific and technological progress seems relatively primitive for such a complex mind; the open vista may well have had a profound personal meaning to him. The idea of unending progress implies an optimistic view of time: it neglects limits, decay, demise. Von Neumann's whole commitment was to effective development of what the sociologist Jacques Ellul has labeled *la technique*, that which in English is suggested by the word "techno-logic." It specifically includes hardware technology as well as software techniques (such as decision theory and computer programs). It refers to a focus on instrumental rationality to achieve well-defined objectives and the creation or use of all kinds of mechanical and formal tools for this purpose. It has been persuasively argued by Ellul that when patterns of thinking and acting along the lines of techno-logic dominate an individual, he is deprived of essential elements of consciousness concerning what it means to be human. Does von Neumann the individual confirm Ellul's analysis?

The view of von Neumann as a paragon of science and as a technologist par excellence raises fundamental issues concerning the scientific community, technology, and our advancing but simultaneously deteriorating civilization. This is what makes von Neumann such a fascinating historical figure. On a personal psychological level he was deeply committed to unlimited technological progress, to be achieved with the greatest possible speed, a commitment related to the maintenance of an attitude of cheerful optimism, as if innovation could rejuvenate us, could save us from old age and death, that cyclical, unprogressive aspect of time. This was the irrational foundation that underlay von Neumann's sophisticated application of reason. On a sociological level his commitment was to providing high technology for that powerful group sometimes called the military-industrial complex, in effect strengthening the already great power of that group. Uninhibited by ethical considerations in his drive toward innovation, for better or for worse he chose to be "hard-boiled" concerning the exploitation of nuclear weaponry. By putting himself at the service of the conservative powers that be, he elaborated and helped to advance existing trends in weapons policy and could not help but become a symbol for these very trends.

Mircea Eliade has characterized the aspiration that lies at the heart of the West's technique- and technology-oriented civilization:

> We must not believe that the triumph of experimental science reduced to nought the dreams and ideals of the alchemist. On the contrary, the ideology of the new epoch, crystallized around the myth of infinite progress and boosted by the experimental sciences and the progress of industrialization which dominated and inspired the whole of the nineteenth century, takes up and carries forward—despite its radical secularization—the millenary dream of the alchemist. . . . On the plane of cultural history, it is therefore permissible to say that the alchemists, *in their desire to supersede Time,* anticipated what is in fact the essence of the ideology of the modern world.[11]

But the potential for fulfilling the dream of superseding time through science or technology is limited. Some mathematical theorems may be nearly timeless, but personal immortality is not the same as the long life of a theorem. Nevertheless, the ideal is dear in personal terms, and it is only human to maintain it as long as possible, however illusory it may be. If some circumstance, perhaps a painful confrontation with reality, destroys this illusion in someone whose attachment to it is especially strong, the victim of the experience of the harsh contradiction between illusion and reality may literally go crazy. If fortunate, the victim recovers from the experience.

In the summer of 1955 von Neumann slipped on a corridor in an office building and hurt his left shoulder. The injury led to the diagnosis of a bone cancer in August. It had already metastasized. His first response to the situation was optimistic, even stoic, and he worked harder than ever. To a remarkable extent he continued to meet the many demands made on him in his capacity as AEC commissioner and to pursue his other interests. There was a bitter irony in the situation in that von Neumann, who had waved away the cancer-producing effects of nuclear weapons tests, would himself contract this awful disease. He had increased his own chances of getting cancer by personally attending nuclear weapons tests and by staying at Los Alamos for long periods.

Before long his accustomed four or five hours of sleep did not suffice, and his enormous activity was forced into a slower pace. Yet death was unthinkable, inconceivable—so many plans and projects were still unfulfilled. The psychological stress was enormous, the more so as his special sense of invulnerability, which many of us share but to a lesser degree, was being challenged. To ease his spiritual troubles von Neumann sought not a Jewish Rabbi but a priest to instruct him in the Catholic faith. Morgenstern, among others, was shocked. "He was of course completely agnostic all his life, and then he suddenly turned Catholic—it doesn't agree with anything whatsoever in his attitude, outlook and thinking when he was healthy."[12] Whether his choice was prompted by his fondness for ritual, or his longing for immortality, or by some nostalgia for a world from which he had been excluded in childhood, or by a spiritual impulse to renounce the dynamo for the Virgin, or by entirely different reasons, in the spring of 1956 Father Strittmatter began coming to see him regularly.

But religion did not prevent him from suffering; even his mind, the amulet on which he always had been able to rely, was becoming less dependable. Then came complete psychological breakdown; panic; screams of uncontrollable terror every night. His friend Edward Teller said, "I think that von Neumann suffered more when his mind would no longer function, than I have ever seen any human being suffer."[13] And Eugene Wigner wrote,

> When von Neumann realized that he was incurably ill, his logic forced him to realize also that he would cease to exist, and hence cease to have thoughts. Yet this is a conclusion the full content of which is incomprehensible to the human intellect and which, therefore, horrified him. It was heart-breaking to watch the frustration of his mind, when all hope was gone, in its struggle with the fate which appeared to him unavoidable but unacceptable.[14]

Von Neumann's sense of invulnerability, or simply the desire to live, was struggling with unalterable facts. He seemed to have a great fear of death until the last, at least for as long as he could still communicate with Father Strittmatter.[15] No achievements and no amount of influence could save him from extinction now, as they always had in the past. Johnny von Neumann, who knew how to live so fully, did not know how to die.

Physiologically his brain was fully intact, not touched by the cancer, but he had been suffering steadily from physical pain.[16] And still the United States government depended on his thinking. As Admiral Lewis Strauss describes it, speaking of sometime after April 1956,

> Until the last, he continued to be a member of the [Atomic Energy] Commission and chairman of an important advisory committee to the Defense Department. On one dramatic occasion near the end, there was a meeting at Walter Reed Hospital where, gathered around his bedside and attentive to his last words of advice and wisdom, were the Secretary of Defense and his Deputies, the Secretaries of the Army, Navy and Air Force, and all the military Chiefs of Staff. The central figure was the young mathematician who but a few years before had come to the United States as an immigrant from Hungary. I have never witnessed a more dramatic scene or a more moving tribute to a great intelligence.[17]

Nevertheless, von Neumann was assigned an aide, Air Force Lt. Col. Vincent Ford, and air force hospital orderlies with top-secret security clearance, lest in his distraction he should babble "classified information." He died February 8, 1957, at the age of fifty-three.

Notes

1. *The American Mathematical Society,* 64, no. 3, pt. 2 (1958): p. 5.
2. "Method in the Physical Sciences," *Collected Works,* 6: 491.
3. Von Neumann, "The Role of Mathematics."
4. Von Neumann, "The NORC and Problems in High Speed Computing" (December 2, 1954), in *Collected Works,* 5: 238. Von Neumann was a consultant to IBM.
5. Goldstine, *Computer,* pp. 331–332.
6. Nicholas Metropolis, February 13, 1958, at hearings before the Subcommittee on Research and Development of the Joint Committee on Atomic Energy, 85th Congress second session, p. 634.
7. John von Neumann, "Can We Survive Technology?" *Fortune,* June 1955.
8. Memorandum to Lewis Strauss, dated Sept. 1955, in Strauss, *Men and Decisions* (Garden City, N.Y.: Doubleday, 1962), p. 441.
9. Warren McCulloch to von Neumann, December 15, 1955 (American Philosophical Society Archives).
10. He also dealt with some of Wiener's predictions, and made his own predictions concerning the impact of the new communications and control technologies on the economy, in a talk to the National Planning Association on December 12, 1955 (*Collected Works,* 6: 100). He begins, without mentioning Wiener directly, by saying that "there has been a great deal of talk . . . that something like a second industrial revolution is impending." In particular von Neumann speaks favorably of the experience of using automata in military decision making and anticipates their corresponding usefulness in economics.
11. Mircea Eliade, *The Forge and the Crucible* (New York: Harper & Row, 1971), pp. 172–173. (Originally published in French 1956.)
12. Morgenstern interview, May 11, 1970.
13. Teller speaking in the film *John von Neumann.*
14. Eugene Wigner, *Symmetries and Reflections* (Bloomington: Indiana University Press, 1967), p. 261 (written in 1957).
15. Reverend Anselm Strittmatter, interview with

the author, Washington, D.C., December 1, 1972.

16. Warren Shields, interview with the author, Boston, Mass., December 21, 1972. I am indebted to Dr. Shields for a medical description of von Neumann's last years.
17. Remarks of Lewis Strauss at von Neumann memorial dinner meeting, Catholic University, Washington, D.C., May 22, 1971. Cf. *Men and Decisions*, p. 236.

A Review of *John von Neumann and Norbert Wiener, from Mathematics to the Technologies of Life and Death*, by Steve J. Heims

Marshall H. Stone

Marshall Harvey Stone received his Ph.D. from Harvard in 1926 and has had academic appointments at Harvard, Yale, Columbia, and Chicago. He has been chairman of the mathematics department at Harvard and Columbia. As editor of the Mathematics Reviews from 1945 to 1950, Guggenheim fellow, member of the Institute for Advanced Study, President of the International Mathematics Union, or President of the American Mathematics Society, Stone has represented mathematics throughout the world.

The following book review must be read in connection with the previous article, an excerpt from Chapter 14 of Heims's book that describes the death of John von Neumann. The title of Chapter 14, "Only Human in Spite of Himself," clearly reveals the attitude of Heims toward his subject. We included the material because we felt that it said something rather important about mathematicians and their intense devotion to their subject. But on the same chapter Stone writes, ". . . in his (Heims's) final chapter on his subject . . . , he surpasses all bounds in an incredibly cruel and unfeeling description of von Neumann's tragic last days." Earlier in the same paragraph Stone writes, "The bare facts reported in this book are accurate enough so far as they go." It would seem that in the tragic death of John von Neumann, as in so many events in history, the bare facts are not enough.

Is Heims being cruel when he writes, "No achievements and no amount of influence could save him from extinction now, as they always had in the past. Johnny von Neumann, who knew how to live so fully, did not know how to die." His remarks certainly seem cruel but is there truth in this cruelty? There most certainly is. It is precisely because John von Neumann knew how to live that he did not know how to die. Von Neumann was, to all appearances, unsustained by any personal conviction of immortality. Given that belief would any rational being be pleased by the imminent prospect of permanent extinction? Would Steve J. Heims? Why is a stoic acceptance of the inevitability of death morally positive but von Neumann's refusal to ". . . go gentle into that good night" morally negative? Does this attitude say something about Heims and about the values that form the framework of his book? We think it does.

We also think that Heims's excerpt says a great deal about how many people currently feel about technology and its handmaiden, mathematics. Frankly, the editors of this anthology are more sympathetic to Stone than to Heims—which is exactly why we had better listen to Heims.

Source: Marshall H. Stone, a book review of *John von Neumann and Norbert Wiener, from Mathematics to the Technologies of Life and Death,* by Steve J. Heims, *Bulletin of the American Mathematical Society* 8:2 (March 1983): 395–399. (Some references omitted.)

BY ITS TITLE this book alerts the reader without any circumlocution that the author is not concerned primarily with writing biography but has set out to compose a contemporary morality play. His symbolic protagonists are Saint Norbert and Saint John Lucifer. The choice is apt. Saint Norbert shines forth as the valiant champion of a noble creed, of which the author, to be sure, is a passionate advocate. This creed is the "populist" philosophy of "socialized" science that has become so fashionable in some "liberal" and academic circles since World War II. Von Neumann's role is to exemplify the corruption and inhumanity hidden in the old belief in "science for science's sake," relieving the scientist of the responsibility for guaranteeing the consequences, proximate or even remote, of his discoveries. The overt acts used to identify von Neumann as an agent of evil were his work at Los Alamos on the atomic and hydrogen bombs and his service on the Atomic Energy Commission, acts treated as doubly suspect because they are alleged to reveal a lust for power that reflected his bourgeois origins. In the author's eagerness to sharpen the contrast he wants to draw between his two protagonists, he lets himself be betrayed into describing von Neumann's personality, career, and motivations in pejorative terms that can only be resented by those who knew him well. In this reviewer's opinion, the author has thereby done von Neumann a monstrous injustice in a single-minded resort to an argument ad hominem.

This unhappy outcome of the author's concentration upon the advancement of his creed is rooted in inadequate scholarly treatment of the biographical materials on which the book could and should have been based. In Wiener's case this does not have major importance because Wiener wrote so much autobiographical and semibiographical material that he is in little danger of being misunderstood or misinterpreted. Furthermore, he has found a competent biographer in Pesi Masani.[1] On the other hand, there is little personal, autobiographical, or philosophical material to be found in von Neumann's writings; and no serious biography of him has yet been undertaken so far as the reviewer knows. The bare facts reported in this book are accurate enough so far as they go. However, there are very important omissions and many over-facile interpretations offered in convenient support of the author's purposes. The author, who is a physicist rather than a mathematician, does not try to give full accounts of the mathematical achievements of his two protagonists. He is not to be faulted for this. However, the selections he makes, while appropriate and relevant to the theme of his book, will leave the reader with a woefully inadequate appreciation of the mathematical stature of these giants. The reader who wants to inform himself further can now consult the collected works of both men [von Neumann,[2] Wiener[3]]. The author is correct in judging that Wiener was the more intuitive, von Neumann the more analytical in his grasp of mathematics. However it seems to the reviewer that he greatly exaggerates this distinction and bases quite mistaken estimates of von Neumann's character and motivation upon it. Von Neumann's fruitful interest in logic, his liking for axiomatic presentations, and even his wonderful facility in doing mathematics in his head are used to brand him as "inhuman." Shades of Bertrand Russell! Indeed the author becomes so obsessed with this perception of von Neumann's "inhumanity" that in his final chapter on his subject, the fourteenth in the book, captioned "Only Human in Spite of Himself," he surpasses all bounds in an incredibly cruel and unfeeling description of von Neumann's tragic last days.

The reader will have little difficulty in identifying and appraising for himself those passages where the author resorts to tendentious interpretations in making his case against von Neumann. However, omissions are another matter altogether. If inadvertent, they may distort; if deliberate, they are inexcusable. There are important omissions in this book. The reviewer believes that the reader has a right to be warned of them and should have the privilege of compensating for them in his own way.

The author claims that von Neumann worked at Los Alamos and served on the Atomic Energy Commission because he had a bourgeois admiration of the powerful and a bourgeois ambition to acquire power. The appreciation for worldly success and pleasure in enjoying it are quite human attributes that could not have been lacking in von Neumann's character, despite his genuine modesty concerning his own achievements. Beyond this, the author's claims are assertions that need some justifying evidence to back them up. No such evidence is produced here. Nothing is said about the steps von Neumann took or may have taken to assure himself of a call to Los Alamos or an appointment to the A.E.C. Neither is anything said about the power von Neumann is supposed to have exercised in either place.

The path that led to Los Alamos was actually a rather round-about one for von Neumann. In the early days of America's preparation for World War II mathematicians were not eagerly sought by the authorities, military or civilian, as potential collaborators. Indeed, some of the leading mathematicians of the day banded together in order to convince the authorities of the usefulness of mathematics and mathematicians. A young mathematician, even one as brilliant as Wiener or von Neumann, had little chance of obtaining a war assignment by his own unaided efforts. He needed the sponsorship of some more prominent and influential mathematicians. Oswald Veblen of the Institute for Advanced Study in Princeton had an opportunity to recommend his young colleague von Neumann for a post, which as it happened, had nothing whatever to do with atomic physics or atomic weapons. The Navy Department desired to set up a mine warfare operations analysis group under the scientific direction of Francis Bitter, an M.I.T. physicist specializing in magnetism. Lt. Commander Bitter included several mathematicians in his group—von Neumann, J. L. Doob, the reviewer, and von Neumann's assistant J. W. Calkin, who had been a doctoral student of the reviewer. Von Neumann and Calkin were active in this group in 1942–43 until they were invited to Los Alamos. Doob remained there until the end of the war, while the reviewer left in 1943 for another assignment. In the Navy Department von Neumann worked mainly on shock waves and damage by explosives. In several conversations with this reviewer he outlined his view that this kind of problem was typical of a very broad mathematical problem—the solution of partial differential equations—that would require the use of computers to explore empirically their little-known behavior. In particular he was fond of pointing out that no theory of meteorology could be usable until the day when massive weather data from a large area could be processed by computers in an hour or so. In connection with his immediate studies, von Neumann and Calkin were sent to England to learn of the progress under way there. Presently it appeared that their special knowledge would be useful at Los Alamos and they moved there. The work they were first asked to do there seems to have been to participate in developing the implosion techniques for exploding an atom bomb, involving the principles of aero- and hydrodynamics. Beyond this point in history, the reviewer cannot go. It should be evident, however, that a serious biographer would have to make a determined effort to find out more about von Neumann's work at Los Alamos and about his role as a member of the A.E.C. Until it is possible to answer in some detail, even if incompletely, questions like those raised here, it will not be possible to comment intelligently upon his motives, his ambitions, or his alleged quest for power in relation to his war work. The reviewer's long acquaintance with von Neumann leads him to believe that von Neumann took too detached and at times too cynical a view of human affairs to harbor any deep illusions about power or its exercise. Von Neumann is said to have been a close student of Byzantine history. He can hardly have failed to learn some unforgettable lessons from the hours thus spent!

The truly astonishing omission from the author's biography of von Neumann is the almost complete lack of any reference to the dominating interest of von Neumann's scientific and intellectual life after World War II—from 1945 to 1957, the year of his death. No mathematician was closer to John von Neumann after 1944 than H. H. Goldstine. In his authoritative book on *The Computer from Pascal to von Neumann*[4] published in 1972 and reprinted as a soft-covered volume in 1977, Goldstine writes, "In thinking about von Neumann's contribution, I am of the opinion that he perhaps viewed his work on automata as his most important one, at least, the most important one in his later life. It not only linked up his early interest in logic with his later work on neurophysiology and on computers, but it had the potential of allowing him to make really profound contributions to all three fields through one apparatus.

It will always be a fundamental loss to science that he could not have completed his program in automata theory, or at least have pushed it far enough to make clear, for example, what his ideas were on a continuous model. He was never given to bragging or staking out a claim unless it deserved it; I am therefore confident that he had at least a heuristic insight into the model and at least some idea how it would interact on logics and neurophysiology. Finally, it is interesting to note that von Neumann worked on his theory of automata alone. This was in rather sharp distinction to most of his later work, where his practice almost always was to work with a colleague. Very possibly he wanted his automata work to stand as a monument to himself, as indeed it does."

This reviewer can corroborate the impression conveyed here by Goldstine. After a close friendship with von Neumann from 1927 to 1943, the reviewer had only infrequent meetings with him, such as a very lively and enjoyable luncheon at the Amsterdam Congress in 1954. Von Neumann had already received the first warnings of what was to be recognized eventually as a mortal illness. The next and last meeting took place when he was already confined to his bed in the hospital. His engineering associate Julian Bigelow was present during the conversation but took very little part in it. Von Neumann dwelt at some length on his deep desire to devote himself to a very ambitious program of work with computers, and to broadening the fields he had already started to open up. He expressed his wish to leave the Institute for Advanced Study and to move to the West Coast with facilities better suited to his plans than those available at the Institute. Had the opportunity been granted him, he would probably have overshadowed his earlier achievements. Unless this testimony about von Neumann's last and possibly brightest scientific goal is placed on record, no balanced view of him as scientist can be formed and no fair measure of his career or his motives established.

Notes

1. P. Masani, *Wiener, Norbert: His Life and Work,* Biographical article published in the "Encyclopedia of Computer Science and Technology, Vol. 14 (Jack Betzer, Albert Gitolzman, and Allen Kent, editors) Marcel Dekker, New York, 1980. (An expanded version is in preparation for publication by Birkhauser, Boston, as a book.)
2. J. von Neumann, *Collected Works* 6 vols (A. H. Taub, ed.) Pergamon Press.
3. N. Wiener, *Collected Works* in course of publication. Volumes I (1978), II (1971), and III (1982) have appeared. Volume IV is in preparation (Pesi Masani, editor) MIT Press, Cambridge, Mass.
4. H. H. Goldstine, *The Computer from Pascal to von Neumann,* Copyright (c) 1972 by Princeton University Press. The excerpt quoted is taken from the soft cover reprint of 1977, p. 285, and is included here by permission.

part three

The Development of Mathematics

THE HISTORY OF mathematics offers a number of clues as to how mathematics developed. In a curious way the entire history of the subject is a large example of the way in which one small area of mathematics grows. Indeed, the same pattern is repeated when a single idea takes shape in the mind of one individual. Consider the following. We know that mathematics began very nearly at the dawn of human consciousness. It started with the observation that there is pattern in human experience, that there are organized repetitions that can be used to order thoughts and make systematic an apparently random universe. In the same way, a single new mathematical idea begins with the realization that a seemingly diverse set of events is really connected by some pattern. When a single mind recognizes that pattern, gives it expression, and uses it to organize previously unrelated events, then mathematics has been created.

As a specific example, consider the idea of number. *It could spring from the observation that a bunch of spears and a bunch of animals are in some way related. If the bunches can be paired, individual by individual, then the two bunches are in some fundamental way similar. No one believes, of course, that spears and animals are the same. It is a fundamental property of the bunches that is the same. When this pattern is observed, the idea of number is not far away. Next, the "sameness" of these two collections is given a name. This name and others distinguish collections that can't be matched. When these names and their underlying collections are made systematic and when this system is extended to collections no one has ever observed, then mathematics is born.*

It should come as no surprise that there are human societies in such areas as New Guinea or the Amazon jungle where the idea of number does not exist. Yet even in such societies one cannot escape the illuminating presence of mathematical ideas. The very notion of matching, of pairing parents with children or individuals with possessions, is in itself a powerful mathematical idea. This primitive matching process is exactly how modern mathematics investigates infinite collections. The method of matching is used to create endless chains of infinite sets, each link in the chain a set incomparably more infinite, more endless than any that came before. In this we may find the greatest of all the paradoxes related to our understanding of mathematics: Even the most primitive of our attempts to impose order on the universe may be employed virtually without change to create a system of ideas that even the greatest modern mathematicians have not yet fully explored.

The following collection of essays looks at the development of mathematics from a variety of viewpoints. Included are the personal experiences of some of the best mathematicians of this century, relating how they created mathematics. There is a discussion of the currently unresolved dispute as to what is a proper foundation for modern mathematics. Also presented is an account of how mathematical ideas evolve in societies far less technological than our own. We hope that the reader can discern a pattern, possibly even a vaguely mathematical pattern, in all of this and that this pattern will provide some insight into the development of mathematics.

The Mathematician

John von Neumann

John von Neumann's biography is given in the article by P. R. Halmos in Part Two.

Since the life and times of John von Neumann have been discussed in some detail in Part Two of this volume, it would only try the reader's patience to repeat the details here. As was noted, his is one of the ten most noteworthy names in the history of mathematics in this century. In this essay von Neumann undertakes the task of exposing the essential nature of mathematics as a creative human activity. This article was written for general audiences and von Neumann's expository treatment of this complex subject makes it a model of clarity.

A portion of the article treats the controversies and paradoxes that have arisen in the study of the foundations of modern mathematics. While this treatment is not exhaustive it is an excellent introduction to these topics.

A DISCUSSION OF the nature of intellectual work is a difficult task in any field, even in fields which are not so far removed from the central area of our common human intellectual effort as mathematics still is. A discussion of the nature of any intellectual effort is difficult per se—at any rate, more difficult than the mere exercise of that particular intellectual effort. It is harder to understand the mechanism of an airplane, and the theories of the forces which lift and which propel it, than merely to ride in it, to be elevated and transported by it—or even to steer it. It is exceptional that one should be able to acquire the understanding of a process without having previously acquired a deep familiarity with running it, with using it, before one has assimilated it in an instinctive and empirical way.

Thus any discussion of the nature of intellectual effort in any field is difficult, unless it presupposes an easy, routine familiarity with that field. In mathematics this limitation becomes very severe, if the discussion is to be kept on a nonmathematical plane. The discussion will then necessarily show some very bad features; points which are made can never be properly documented, and a certain over-all superficiality of the discussion becomes unavoidable.

I am very much aware of these shortcomings in what I am going to say, and I apologize in advance. Besides, the views which I am going to express are probably not wholly shared by many other mathematicians—you will get one man's not-too-well systematized impressions and interpretations—and I can give you only very little help in deciding how much they are to the point.

In spite of all these hedges, however, I must admit that it is an interesting and challenging task to make the attempt and to talk to you about the nature of intellectual effort in mathematics. I only hope that I will not fail too badly.

The most vitally characteristic fact about mathematics is, in my opinion, its quite peculiar rela-

Source: John von Neumann, "The Mathematician," in Robert B. Heywood, *The Works of the Mind* (The University of Chicago Press, 1947), pp. 180–196.

tionship to the natural sciences, or, more generally, to any science which interprets experience on a higher than purely descriptive level.

Most people, mathematicians and others, will agree that mathematics is not an empirical science, or at least that it is practiced in a manner which differs in several decisive respects from the techniques of the empirical sciences. And, yet, its development is very closely linked with the natural sciences. One of its main branches, geometry, actually started as a natural, empirical science. Some of the best inspirations of modern mathematics (I believe, the best ones) clearly originated in the natural sciences. The methods of mathematics pervade and dominate the "theoretical" divisions of the natural sciences. In modern empirical sciences it has become more and more a major criterion of success whether they have become accessible to the mathematical method or to the near-mathematical methods of physics. Indeed, throughout the natural sciences an unbroken chain of successive pseudomorphoses, all of them pressing toward mathematics, and almost identified with the idea of scientific progress, has become more and more evident. Biology becomes increasingly pervaded by chemistry and physics, chemistry by experimental and theoretical physics, and physics by very mathematical forms of theoretical physics.

There is a quite peculiar duplicity in the nature of mathematics. One has to realize this duplicity, to accept it, and to assimilate it into one's thinking on the subject. This double face is the face of mathematics, and I do not believe that any simplified, unitarian view of the thing is possible without sacrificing the essence.

I will therefore not attempt to present you with a unitarian version. I will attempt to describe, as best I can, the multiple phenomenon which is mathematics.

It is undeniable that some of the best inspirations in mathematics—in those parts of it which are as pure mathematics as one can imagine—have come from the natural sciences. We will mention the two most monumental facts.

The first example is, as it should be, geometry. Geometry was the major part of ancient mathematics. It is, with several of its ramifications, still one of the main divisions of modern mathematics. There can be no doubt that its origin in antiquity was empirical and that it began as a discipline not unlike theoretical physics today. Apart from all other evidence, the very name "geometry" indicates this. Euclid's postulational treatment represents a great step away from empiricism, but it is not at all simple to defend the position that this was the decisive and final step, producing an absolute separation. That Euclid's axiomatization does at some minor points not meet the modern requirements of absolute axiomatic rigor is of lesser importance in this respect. What is more essential, is this: other disciplines, which are undoubtedly empirical, like mechanics and thermodynamics, are usually presented in a more or less postulational treatment, which in the presentation of some authors is hardly distinguishable from Euclid's procedure. The classic of theoretical physics in our time, Newton's *Principia,* was, in literary form as well as in the essence of some of its most critical parts, very much like Euclid. Of course in all these instances there is behind the postulational presentation the physical insight backing the postulates and the experimental verification supporting the theorems. But one might well argue that a similar interpretation of Euclid is possible, especially from the viewpoint of antiquity, before geometry had acquired its present bimillennial stability and authority—an authority which the modern edifice of theoretical physics is clearly lacking.

Furthermore, while the de-empirization of geometry has gradually progressed since Euclid, it never became quite complete, not even in modern times. The discussion of non-Euclidean geometry offers a good illustration of this. It also offers an illustration of the ambivalence of mathematical thought. Since most of the discussion took place on a highly abstract plane, it dealt with the purely logical problem whether the "fifth postulate" of Euclid was a consequence of the others or not; and the formal conflict was terminated by F. Klein's purely mathematical example, which showed how a piece of a Euclidean plane could be made non-Euclidean by formally redefining certain basic concepts. And yet the empirical stimulus was there from start to finish. The prime reason, why, of all Euclid's postulates, the fifth was questioned, was clearly the unempirical character of the concept of the entire infinite plane which intervenes there, and there only. The idea that in at least one significant sense—and in spite of all mathematico-logical analyses—the decision for or against Euclid may have to be empirical, was certainly present in the mind of the greatest mathematician, Gauss. And after Bolyai, Lobatschefski, Riemann, and Klein had obtained more abstracto, what we today consider the formal resolution of the original controversy, empirics—or rather physics—nevertheless, had the final say. The discovery of general relativity forced a revision of our views on the relationship of geometry in an entirely new setting and with a quite new distribution of the purely mathematical emphases, too. Finally, one more touch to complete the picture of contrast. This last development took place in the same generation which saw the complete de-empirization and abstraction of Euclid's axiomatic method in the hands of the modern axiomatic-logical mathematicians. And these two seemingly conflicting attitudes are perfectly compatible in one mathematical mind; thus Hilbert made important contributions to both axiomatic geometry and to general relativity.

The second example is calculus—or rather all of analysis, which sprang from it. The calculus was the first achievement of modern mathematics, and it is difficult to overestimate its importance. I think it defines more unequivocally than anything else the inception of modern mathematics, and the system of mathematical analysis, which is its logical development, still constitutes the greatest technical advance in exact thinking.

The origins of calculus are clearly empirical. Kepler's first attempts at integration were formulated as "dolichometry"—measurement of kegs—that is, volumetry for bodies with curved surfaces. This is geometry, but post-Euclidean, and, at the epoch in question, nonaxiomatic, empirical geometry. Of this, Kepler was fully aware. The main effort and the main discoveries, those of Newton and Leibnitz, were of an explicitly physical origin. Newton invented the calculus "of fluxions" essentially for the purposes of mechanics—in fact, the two disciplines, calculus and mechanics, were developed by him more or less together. The first formulations of the calculus were not even mathematically rigorous. An inexact, semiphysical formulation was the only one available for over a hundred and fifty years after Newton! And yet, some of the most important advances of analysis took place during this period, against this inexact, mathematically inadequate background! Some of the leading mathematical spirits of the period were clearly not rigorous, like Euler; but others, in the main, were, like Gauss or Jacobi. The development was as confused and ambiguous as can be, and its relation to empiricism was certainly not according to our present (or Euclid's) ideas of abstraction and rigor. Yet no mathematician would want to exclude it from the fold—that period produced mathematics as first class as ever existed! And even after the reign of rigor was essentially re-established with Cauchy, a very peculiar relapse into semiphysical methods took place with Riemann. Riemann's scientific personality itself is a most illuminating example of the double nature of mathematics, as is the controversy of Riemann and Weierstrass, but it would take me too far into technical matters if I went into specific details. Since Weierstrass, analysis seems to have become completely abstract, rigorous, and unempirical. But even this is not unqualifiedly true. The controversy about the "foundations" of mathematics and logics, which took place during the last two generations, dispelled many illusions on this score.

This brings me to the third example which is relevant for the diagnosis. This example, however, deals with the relationship of mathematics with philosophy or epistemology rather than with the natural sciences. It illustrates in a very striking fashion that the very concept of "absolute" mathematical rigor is not immutable. The variability of the concept of rigor shows that something else besides mathematical abstraction must enter into the makeup of mathematics. In analyzing the controversy about the "foundations," I have not been able to convince myself that the verdict must be in favor of the empirical nature of this extra component. The case in favor of such an interpretation is quite strong, at least in some phases of the discussion. But I do not consider it absolutely cogent. Two things, however, are clear. First, that something nonmathematical, somehow connected with the empirical sciences or with philosophy or both, does enter essentially—and its nonempirical character could only be maintained if one assumed that philosophy (or more specifically epistemology) can exist independently of experience. (And this assumption is only necessary but not in itself sufficient.) Second, that the empirical origin of mathematics is strongly supported by instances like our two earlier examples (geometry and calculus), irrespective of what the best interpretation of the controversy about the "foundations" may be.

In analyzing the variability of the concept of mathematical rigor, I wish to lay the main stress on the "foundations" controversy, as mentioned above. I would, however, like to consider first briefly a secondary aspect of the matter. This aspect also strengthens my argument, but I do consider it as secondary, because it is probably less conclusive than the analysis of the "foundations" controversy. I am referring to the changes of mathematical "style." It is well known that the style in which mathematical proofs are written has undergone considerable fluctuations. It is better to talk of fluctuations than of a trend because in some respects the difference between the present and certain authors of the eighteenth or of the nineteenth centuries is greater than between the present and Euclid. On the other hand, in other respects there has been remarkable constancy. In fields in which differences are present, they are mainly differences in presentation, which can be eliminated without bringing in any new ideas. However, in many cases these differences are so wide that one begins to doubt whether authors who "present their cases" in such divergent ways can have been separated by differences in style, taste, and education only—whether they can really have had the same ideas as to what constitutes mathematical rigor. Finally, in the extreme cases (e.g., in much of the work of the late-eighteenth-century analysis, referred to above), the differences are essential and can be remedied, if at all, only with the help of new and profound theories, which it took up to a hundred years to develop. Some of the mathematicians who worked in such, to us, unrigorous ways (or some of their contemporaries, who criticized them) were well aware of their lack of rigor. Or to be more objective: Their own desires as to what mathematical procedure should be were more in conformity with our present views than their actions. But others—the greatest virtuoso of the period, for example, Euler—seem to have acted in perfect good faith and to have been quite satisfied with their own standards.

However, I do not want to press this matter further. I will turn instead to a perfectly clear-cut case, the controversy about the "foundations of mathematics." In the late nineteenth and the early twentieth centuries a new branch of abstract mathematics, G. Cantor's theory of sets, led into difficulties. That is, certain reasonings led to contraditions; and, while these reasonings were not in the central and "useful" part of set theory, and always easy to spot by certain formal criteria, it was nevertheless not clear why they should be deemed less set-theoretical than the "successful" parts of the theory. Aside from the *ex post* insight that they actually led into disaster, it was not clear what *a priori* motivation, what consistent philosophy of the situation, would permit one to segregate them from those parts of set theory which one wanted to save. A closer study of the *merita* of the case, undertaken mainly by Russell and Weyl, and concluded by Brouwer, showed that the way in which not only set theory but also most of modern mathematics used the concepts of "general validity" and of "existence" was philosophically objectionable. A system of mathematics which was free of these undesirable traits, "intuitionism," was developed by Brouwer. In this system the difficulties and contradiction of set theory did not arise. However, a good fifty percent of modern mathematics, in its most vital—and up to then unquestioned—parts, especially in analysis, were also affected by this "purge": they either became invalid or had to be justified by very complicated subsidiary considerations. And in this latter process one usually lost appreciably in generality of validity and elegance of deduction. Nevertheless, Brouwer and Weyl considered it necessary that the concept of mathematical rigor be revised according to these ideas.

It is difficult to overestimate the significance of these events. In the third decade of the twentieth century two mathematicians—both of them of the first magnitude, and as deeply and fully conscious of what mathematics is, or is for, or is about, as anybody could be—actually proposed that the concept of mathematical rigor, of what constitutes an exact proof, should be changed! The developments which followed are equally worth noting.

1. Only very few mathematicians were willing to accept the new, exigent standards for their own daily use. Very many, however, admitted that Weyl and Brouwer were prima facie right, but they themselves continued to trespass, that is, to do their own mathematics in the old, "easy" fashion—probably in the hope that somebody else, at some other time, might find the answer to the intuitionistic critique and thereby justify them *a posteriori*.
2. Hilbert came forward with the following ingenious idea to justify "classical" (i.e., pre-intuitionistic) mathematics: Even in the intuitionistic system it is possible to give a rigorous account of how classical mathematics operate, that is, one can describe how the classical system works, although one cannot justify its workings. It might therefore be possible to demonstrate intuitionistically that classical procedures can never lead into contradictions—into conflicts with each other. It was clear that such a proof would be very difficult, but there were certain indications how it might be attempted. Had this scheme worked, it would have provided a most remarkable justification of classical mathematics on the basis of the opposing intuitionistic system itself! At least, this interpretation would have been legitimate in a system of the philosophy of mathematics which most mathematicians were willing to accept.
3. After about a decade of attempts to carry out this program, Gödel produced a most remarkable result. This result cannot be stated absolutely precisely without several clauses and caveats which are too technical to be formulated here. Its essential import, however, was this: If a system of mathematics does not lead into contradiction, then this fact cannot be demonstrated with the procedures of that system. Gödel's proof satisfied the strictest criterion of mathematical rigor—the intuitionistic one. Its influence on Hilbert's program is somewhat controversial, for reasons which again are too technical for this occasion. My personal opinion, which is shared by many others, is, that Gödel has shown that Hilbert's program is essentially hopeless.
4. The main hope of a justification of classical mathematics—in the sense of Hilbert or of Brouwer and Weyl—being gone, most mathematicians decided to use that system anyway. After all, classical mathematics was producing results which were both elegant and useful, and, even though one could never again be absolutely certain of its reliability, it stood on at least as sound a foundation as, for example, the existence of the electron. Hence, if one was willing to accept the sciences, one might as well accept the classical system of mathematics. Such views turned out to be acceptable even to some of the original protagonists of the intuitionistic system. At present the controversy about the "foundations" is certainly not closed, but it seems most unlikely that the classical system should be abandoned by any but a small minority.

I have told the story of this controversy in such detail, because I think that it constitutes the best caution against taking the immovable rigor of mathematics too much for granted. This happened in our own lifetime, and I know myself how humiliatingly easily my own views regarding the absolute mathematical truth changed during this episode, and how they changed three times in succession!

I hope that the above three examples illustrate one-half of my thesis sufficiently well—that much of the best mathematical inspiration comes from experience and that it is hardly possible to believe in the existence of an absolute, immutable concept of mathematical rigor, dissociated from all human experience. I am trying to take a very low-brow attitude on this matter. Whatever philosophical or epistemological preferences anyone may have in this respect, the mathematical fraternities' actual experiences with its subject give little support to the assumption of the existence of an *a priori* concept of mathematical rigor. However, my thesis also has a second half, and I am going to turn to this part now.

It is very hard for any mathematician to believe that mathematics is a purely empirical science or that all mathematical ideas originate in empirical subjects. Let me consider the second half of the statement first. There are various important parts of modern mathematics in which the empirical origin is untraceable, or, if traceable, so remote that it is clear that the subject has undergone a complete metamorphosis since it was cut off from its empirical roots. The symbolism of algebra was invented for domestic, mathematical use, but it may be reasonably asserted that it had strong empirical ties. However, modern, "abstract" algebra has more and more developed into directions which have even fewer empirical connections. The same may be said about topology. And in all these fields the mathematician's subjective criterion of success, of the worthwhileness of his effort, is very much self-contained and aesthetical and free (or nearly free) of empirical connections. (I will say more about this further on.) In set theory this is still clearer. The "power" and the "ordering" of an infinite set may be the generalizations of finite numerical concepts, but in their infinite form (especially "power") they have hardly any relation to this world. If I did not wish to avoid technicalities, I could document this with numerous set theoretical examples—the problem of the "axiom of choice," the "comparability" of infinite "powers," the "continuum problem," etc. The same remarks apply to much of real function theory and real point-set theory. Two strange examples are given by differential geometry and by group theory: they were certainly conceived as abstract, nonapplied disciplines and almost always cultivated in this spirit. After a decade in one case, and a century in the other, they turned out to be very useful in physics. And they are still mostly pursued in the indicated, abstract, nonapplied spirit.

The examples for all these conditions and their various combinations could be multiplied, but I prefer to turn instead to the first point I indicated above: Is mathematics an empirical science? Or, more precisely: Is mathematics actually practiced in the way in which an empirical science is practiced? Or, more generally: What is the mathematician's normal relationship to his subject? What are his criteria of success, of desirability? What influences, what considerations, control and direct his effort?

Let us see, then, in what respects the way in which the mathematician normally works differs from the mode of work in the natural sciences. The difference between these, on one hand, and mathematics, on the other, goes on, clearly increasing as one passes from the theoretical disciplines to the experimental ones and then from the experimental disciplines to the descriptive ones. Let us therefore compare mathematics with the category which lies closest to it—the theoretical disciplines. And let us pick there the one which lies closest to mathematics. I hope that you will not judge me too harshly if I fail to control the mathematical *hybris* and add: because it is most highly developed among all theoretical sciences—that is, theoretical physics. Mathematics and theoretical physics have actually a good deal in common. As I have pointed out before, Euclid's system of geometry was the prototype of the axiomatic presentation of classical mechanics, and similar treatments dominate phenomenological thermodynamics as well as certain phases of Maxwell's system of electrodynamics and also of special relativity. Furthermore, the attitude that theoretical physics does not explain phenomena, but only classifies and correlates, is today accepted by most theoretical physicists. This means that the criterion of success for such a theory is simply whether it can, by a simple and elegant classifying and correlating scheme, cover very many phenomena, which without this scheme would seem complicated and heterogeneous, and whether the scheme even covers phenomena which were not considered or even not known at the time when the scheme was evolved. (These two latter statements express, of course, the unifying and the predicting power of a theory.) Now this criterion, as set forth here, is clearly to a great extent of an aesthetical nature. For this reason it is very closely akin to the mathematical criteria of success, which, as you shall see, are almost entirely aesthetical. Thus we are now comparing mathematics with the empirical science that lies closest to it and with which it has, as I hope I have shown, much in common—with theoretical physics. The differences in the actual *modus procedendi* are nevertheless great and basic. The aims of theoretical physics are in the main given from the "outside," in most cases by the needs of experimental physics. They almost always originate in the need of resolving a difficulty; the predictive and unifying achievements usually come afterward. If we may be permitted a simile, the advances (predictions and unifications) come during the pursuit, which is necessarily preceded by a battle against some pre-existing difficulty (usually an apparent contradiction within the existing system). Part of the theoretical physicist's work is a search for such obstructions, which promise a possibility for a "break-through." As I mentioned, these difficulties originate usually in experimentation, but sometimes they are contradictions between various parts of the accepted body of theory itself. Examples are, of course, numerous.

Michelson's experiment leading to special relativity, the difficulties of certain ionization potentials and of certain spectroscopic structures leading to quantum mechanics exemplify the first case; the conflict between special relativity and Newtonian gravitational theory leading to general relativity exemplifies the second, rarer, case. At any rate, the problems of theoretical physics are objectively given; and, while the criteria which govern the exploitation of a success are, as I indicated earlier, mainly aesthetical, yet the portion of the problem, and that which I called above the original "break-through," are hard, objective facts. Accordingly, the subject of theoretical physics was at almost all times enormously concentrated; at almost all times most of the effort of all theoretical physicists was concentrated on no more than one or two very sharply circumscribed fields—quantum theory in the 1920's and early 1930's and elementary particles and structure of nuclei since the mid-1930's are examples.

The situation in mathematics is entirely different. Mathematics falls into a great number of subdivisions, differing from one another widely in character, style, aims, and influence. It shows the very opposite of the extreme concentration of theoretical physics. A good theoretical physicist may today still have a working knowledge of more than half of his subject. I doubt that any mathematician now living has much of a relationship to more than a quarter. "Objectively" given, "important" problems may arise after a subdivision of mathematics has evolved relatively far and if it has bogged down seriously before a difficulty. But even then the mathematician is essentially free to take it or leave it and turn to something else, while an "important" problem in theoretical physics is usually a conflict, a contradiction, which "must" be resolved. The mathematician has a wide variety of fields to which he may turn, and he enjoys a very considerable freedom in what he does with them. To come to the decisive point: I think that it is correct to say that his criteria of selection, and also those of success, are mainly aesthetical. I realize that this assertion is controversial and that it is impossible to "prove" it, or indeed to go very far in substantiating it, without analyzing numerous specific, technical instances. This would again require a highly technical type of discussion, for which this is not the proper occasion. Suffice it to say that the aesthetical character is even more prominent than in the instance I mentioned above in the case of theoretical physics. One expects a mathematical theorem or a mathematical theory not only to describe and to classify in a simple and elegant way numerous and *a priori* disparate special cases. One also expects "elegance" in its "architectural," structural makeup. Ease in stating the problem, great difficulty in getting hold of it and in all attempts at approaching it, then again some very surprising twist by which the approach, or some part of the approach, becomes easy, etc. Also, if the deductions are lengthy or complicated, there should be some simple general principle involved, which "explains" the complications and detours, reduces the apparent arbitrariness to a few simple guiding motivations, etc. These criteria are clearly those of any creative art, and the existence of some underlying empirical, worldly motif in the background—often in a very remote background—overgrown by aestheticizing developments and followed into a multitude of labyrinthine variants—all this is much more akin to the atmosphere of art pure and simple than to that of the empirical sciences.

You will note that I have not even mentioned a comparison of mathematics with the experimental or with the descriptive sciences. Here the differences of method and of the general atmosphere are too obvious.

I think that it is a relatively good approximation to truth—which is much too complicated to allow anything but approximations—that mathematical ideas originate in empirics, although the genealogy is sometimes long and obscure. But, once they are so conceived, the subject begins to live a peculiar life of its own and is better compared to a creative one, governed by almost entirely aesthetical motivations, than to anything else and, in particular, to an empirical science. There is, however, a further point which, I believe, needs stressing. As a mathematical discipline travels far from its empirical source, or still more, if it is a second and third generation only indirectly inspired by ideas coming from "reality," it is beset with very grave dangers. It becomes more and more purely aestheticizing, more and more purely *l'art pour l'art*. This need not be bad, if the field is surrounded by correlated subjects, which still have closer empirical connections, or if the discipline is under the influence of men with an exceptionally well-developed taste. But there is a grave danger that the subject will develop along the line of least resistance, that the stream, so far from its source, will separate into a multitude of insignificant branches, and that the discipline will become a disorganized mass of details and complexities. In other words, at a great distance from its empirical source, or after much "abstract" inbreeding, a mathematical subject is in danger of degeneration. At the inception the style is usually classical; when it shows signs of becoming baroque, then the danger signal is up. It would be easy to give examples, to trace specific evolutions into the baroque and the very high baroque, but this, again, would be too technical.

In any event, whenever this stage is reached, the only remedy seems to me to be the rejuvenating return to the source: the reinjection of more or less directly empirical ideas. I am convinced that this was a necessary condition to conserve the freshness and the vitality of the subject and that this will remain equally true in the future.

Some Thoughts on the History of Mathematics

Abraham Robinson

Abraham Robinson was born in Germany in 1918 and obtained a Ph.D. in mathematics from the University of London. He taught in England before emigrating to the United States where he taught at UCLA and Yale. The author of numerous research articles in mathematics and logic, he died in 1976.

Anyone who systematically studies the essays in this anthology must be impressed by the mathematicians' depth of feeling toward the philosophical foundation of their subject. In biography, in pedagogy, in any discussion that treats the nature of the subject, mathematicians are drawn almost mystically to questions of the meaning and validity of mathematics. In this virtually universal preoccupation with ultimate philosophical justification mathematicians are almost surely unique among modern technologists.

In this essay the question of meaning is expressed in a different and possibly more disturbing setting, that of the history of mathematics. The author is the distinguished American mathematician and logician Abraham Robinson. Robinson observes that the philosophical assumptions of mathematics are anything but constant over time. Yet if the assumptions of mathematics are not constant and if mathematics is nothing but the deductions obtained from its assumptions, then is mathematics anything at all? Notice that this line of thought is vastly different from the underlying assumptions of science. While our view of the physical universe may change radically, we assume that it was our understanding that was at fault and not the underlying physical reality that changed. Since mathematics is an exclusively human invention, however, there is no underlying object on whose reality one can depend. It is an interesting question and one with no easy solution.

1

The achievements of Mathematics over the centuries cannot fail to arouse the deepest admiration. There are but few mathematicians who feel impelled to reject any of the major results of Algebra, or of Analysis, or of Geometry and it seems likely that this will remain true also in the future. Yet, paradoxically, this iron-clad edifice is built on shifting sands. And if it is hard, and perhaps even impossible, to present a satisfactory viewpoint on the foundations of Mathematics today, it is equally hard to give an accurate description of the conceptual bases on which the mathematicians of the past constructed their theories. Some of the suggestions that we shall offer here on this topic are frankly speculative. Some may have been arrived at by comparing similar situations at different times in history, a procedure which is open to challenge and certainly should be used with great caution. Another preliminary remark which is appropriate here concerns the use of the word "real" with reference to mathematical objects. This term is ambiguous and has been stigmatized by some as meaningless in the present context. But the fundamental controversies on the significance of this word should not inhibit its use in a historical study, whose purpose it is to describe and analyze attitudes and not to justify them.

Source: Abraham Robinson, "Some Thoughts on the History of Mathematics," *Composito Math* 20 (1968): 188–193. Reprinted by permission of Sijthoff and Noordhoff.

2

It is commonly accepted that the beginnings of Mathematics as a deductive science go back to the Greek world in the fifth and fourth centuries B.C. It is even more certain that in the course of many hundreds of years before that time people in Egypt and Mesopotamia had accumulated an impressive body of mathematical knowledge, both in Geometry and in Arithmetic. Since this knowledge was recorded in the form of numerical problems and answers it is frequently asserted that pre-Greek Mathematics was purely "empirical." However, unless this expression is meant to indicate merely that pre-Greek Mathematics was not deductive and if it is to be taken literally, we are asked to believe, e.g., that the Mesopotamian mathematicians arrived at Pythagoras' theorem by measuring a large number of right triangles and by inspecting the numbers obtained as the squares of their side lengths. Is it not much more likely that these mathematicians, like their Greek successors, were already familiar with one of the arguments leading to a proof of Pythagoras' theorem by a decomposition of areas, but that no such proof was recorded by them since they regarded the reasoning as intuitively clear? To put it facetiously and anachronistically, if a Sumerian mathematician had been asked for his opinion of Euclid he might have replied that he was interested in *real* Mathematics and not in useless generalizations and abstractions. However, some major advances in Mathematics consisted not in the discovery of new results or in the invention of ingenious new methods but in *the codification of elements of accepted mathematical thought,* i.e. *in making explicit arguments, notions, assumptions, rules, which had been used intuitively for a long time previously.* It is in this light that we should look upon the contributions of the Greek mathematicians and philosophers to the foundations of Mathematics.

3

For our present discussion, the question whether the major contribution to the system of Geometry recorded in Euclid's *Elements* was due to Hippocrates or to Eudoxus or to Euclid himself is of no importance (except insofar as it may affect the following problem, for chronological reasons). However, it would be important to know to what extent the emergence of deductive Mathematics was due to the lead given by one of the Greek philosophers or philosophical schools of the fifth and fourth centuries. Is it true, as has been asserted by some, that the creation of the axiomatic method was due to the direct influence of Plato or of Aristotle or, as has been suggested recently by Á. Szabó, that it was a response to the teachings of the Eleatic school? In our time, the immediate influence of philosophers on the foundations of Mathematics is confined to those who are willing to handle technical-mathematical details. But even now, a general philosophical doctrine may, almost imperceptibly, affect the direction taken by foundational research in Mathematics in the long run. In classical Greece, the differentiation between Philosophy and Mathematics was less pronounced, but nevertheless, with the possible exception of Democritus, we do not know of any leading philosopher of that period who originated an important contribution to Mathematics as such. When Plato singled out Theaetetus in order to emphasize the generality of mathematical arguments he was, after all, referring to a real person who had died only a few years earlier, and he wished to take no credit for the achievement described by him. Nevertheless, by laying bare some important characteristics of mathematical thought, both he and Aristotle exerted considerable influence on later generations. Thus Aristotle, having studied the mathematics of the day, established standards of rigor and completeness for mathematical reasoning which went far beyond the level actually reached at that time. And although we may assume that Euclid and his successors were aware of the teachings of Plato and Aristotle, their own aims in the development of Geometry as a deductive science were less ambitious than Aristotle's program from a purely logical point of view. It is in fact well known that even in the domain of purely mathematical postulates Euclid left a number of glaring gaps. And as far as the laws of logic are concerned, Euclid confined himself to axioms of equality (and inequality) and did not include the rules of deduction which had already been made available by Aristotle. Thus Euclid, like Archimedes after him, was content to single out those axioms which could not be taken for granted or which deserved special mention for other reasons and then derived his theorems from those axioms *in conjunction with other assumptions whose truth seemed obvious, by means of rules of deduction whose legitimacy seemed equally obvious.* It would be out of place to ask whether Euclid would have been able to include in his list of postulates this or that assumption if he had wanted just as even today it would, in most cases, be futile to ask a working mathematician to specify the rules of deduction that he uses in his arguments. The chances are that the typical working mathematician would reply that he is willing to leave this task to the logicians and that, by contrast, his own intuition is sound enough to get along spontaneously. For example, when proving that any composite number has a prime divisor (*Elements,* Book VII, Proposition 31), Euclid appealed explicitly to the principle of infinite descent (which is a variant of the "axiom of induction") yet he did not include that principle among his axioms. By contrast, the axiom of parallels was included by Euclid (*Elements,* Book I, Postulate 5) because though apparently true, it was not intuitively obvious. Similarly "Archimedes axiom" was included by Archimedes (*On the sphere and cylinder,* Book I, Postulate 5) because although required for developing the method of exhaustion, it was not intuitively obvious either. In fact, Euclid did not accept this axiom at all explicitly but instead introduced a definition (*Elements,* Book V, Definition 5) which implies that he did not wish to exclude the possibility that magnitudes which are non-Archimedean relative to one another actually exist, but that he deliberately confined himself to Archimedean systems of magnitudes in order to be able to develop the theory of proportions and, to some extent, the method of exhaustion.

4

From the beginnings of the axiomatic method until the nineteenth century A.D. axioms were regarded as statements of fact from which other statements of fact could be deduced (by means of legitimate procedures and relying on other obvious facts, see above). However, there is in Euclid an element of "constructivism" which, on one hand, seems to hark back to pre-Greek Mathematics and, on the other hand, should strike a chord in the hearts of those who believe that Mathematics has been pushed too far in a formal-deductive direction and who advocate a more constructive approach to the foundations of Mathematics. And although the first three postulates of the *Elements,* Book I can be interpreted as purely existential statements, the "constructivist" tenor of their actual style is unmistakable. Moreover, the cautious formulations of the second and fifth postulates seems to show a trace of the distaste for infinity that we find already in Aristotle. In addition, there are, of course, scattered through the *Elements* many "propositions" which are actually constructions.

5

Euclid's geometry was supposed to deal with real objects, whether in the physical world or in some ideal world. The definitions which preface several books in the *Elements* are supposed to communicate what object the author is talking about even though, like the famous definition of the point and the line, they may not be required in the sequel. The fundamental importance of the advent of non-Euclidean geometry is that by contradicting the axiom of parallels it denied the uniqueness of geometrical concepts and hence, their reality. By the end of the nineteenth century, the interpretation of the basic concepts of Geometry had become irrelevant. This was the more important since Geometry had been regarded for a long time as the ultimate foundation of all Mathematics. However, it is likely that the independent development of the foundations of the number system which was sparked by the intricacies of Analysis would have deprived Geometry of its predominant position anyhow.

An ironic fate decreed that only after Geometry had lost its standing as the basis of all Mathematics its axiomatic foundations finally reached the degree of perfection which in the public estimation they had possessed ever since Euclid. Soon after, the codification of the laws of deductive thinking advanced to a point which, for the first time, permitted the satisfactory formalization of axiomatic theories.

6

In the twentieth century, Set Theory achieved the position, once occupied by Geometry, of being regarded as the basic discipline of Mathematics in which all other branches of Mathematics can be embedded. And, within quite a short time, the foundations of Set Theory went through an evolution which is remarkably similar to the earlier evolution of the foundations of Geometry. First the initial assumptions of Set Theory were held to be intuitively clear being based on natural laws of thought for whose codification Cantor, at least, saw no need. Then Set Theory was put on a postulational basis, beginning with the explicit formulation of the least intuitive among them, the axiom of choice. However, at that point the axioms were still supposed to describe "reality," albeit the reality of an ideal, or Platonic, world. And finally, the realization that it is equally consistent either to affirm or to deny some major assertions of Set Theory such as the continuum hypothesis led, in the mid-sixties, to a situation in which the belief that Set Theory describes an objective reality was dropped by many mathematicians.

The evolution of the foundations of Set Theory is closely linked to the development of Mathematical Logic. And here also we can see how, in our own time, advances have been made through the codification of notions (such as the truth concept) which were used intuitively for a long time previously. And again it may be left open whether the postulates of a system deal with real objects or with idealizations (e.g. the rules of formation and deduction of a formal language). And there is every reason to believe that the codification of intuitive concepts and the reinterpretation of accepted principles will continue also in future and will bring new advances, into territory still uncharted.

Added March 20, 1968:

In an article published since the above lines were written ("Non-Cantorian Set Theory," *Scientific American,* vol. 217, December 1967, pp. 104–116) Paul J. Cohen and Reuben Hersh compare the development of geometry and set theory and anticipate some of the points made here.

The Origin and Growth of Mathematical Concepts

R. L. Wilder

Raymond Louis Wilder was born in 1896 in Massachusetts. He received his doctorate in 1923 from the University of Texas. He was a member of the National Academy of Sciences and also president of the American Mathematical Society from 1955 to 1956 and president of the Mathematical Association of America from 1965 to 1966. He received the Lester R. Ford Award from the Mathematical Association of America in 1973.

Where do mathematical ideas originate? What is their source and what makes them grow? Are they necessary facts embedded like raisins in porridge that we must inevitably encounter and deal with, or are they individual creations as peculiar as the paint splatters of Jackson Pollock or the painted bison of neolithic man? While such questions elude definitive resolution, R. L. Wilder provides insights into some aspects of them in this article.

Wilder originally delivered this essay as an address at a 1952 meeting of the American Association for the Advancement of Science. He discusses both the historical facts concerning the development of the major modern mathematical ideas and the philosophical basis of their origin and necessity.

1. Introduction

According to A. N. Whitehead[1] (p. 20), "The science of Pure Mathematics, in its modern developments, may claim to be the most original creation of the human spirit"; a statement with which probably few mathematicians would quarrel. A layman, however, might and probably would take this to mean that modern mathematics is something which has already been created; a creation, that is, which has already been accomplished and is now safely embalmed with remains on view in any good library.

Of course, as mathematicians we know that mathematics is in no such static shape; that it is, on the contrary, a dynamic affair, changing even from day to day. Our late colleagues who were in the forefront of mathematical creation about the turn of the century, 1900, would be amazed to see what mathematics is like today; indeed, many of them, I'd wager, would not like some of the modern developments, probably taking the same attitude toward them that some mathematicians of the late 19th century took toward Cantor's innovations regarding the infinite. Moreover, those of us who are active today would probably feel the same way toward the mathematics of the year 2000, if we were in some way able to view it. We ourselves probably have not sufficient perception of the changes going on in mathematics at the present time; it is well known that the participants in great social changes are usually unaware of them. And there is some evidence that our awareness of the process of mathematical change, although not so at variance with the facts as that of the layman, is still so defective in some ways as to lead to unfortunate but avoidable situations. This is partly due, I suppose, to our being so

Source: R. L. Wilder, "The Origin and Growth of Mathematical Concepts," *Bulletin of the American Mathematical Society* 59 (1953): 423–448. (Some references omitted.)

busy creating new mathematics that we have little time or patience to view our behavior from the outside and study its characteristics. And even when we do so, we seem prone to take such a specialized angle from which to make our observations that we get only a partial perspective.

2. Motives

Poincaré remarked, in one of his numerous essays[2] (p. 376), that "mathematical science must reflect on itself," and supported this dictum with several studies of the psychological aspects of mathematical creation. These studies inspired others of a similar nature, among which might be noted especially the little book[3] by Hadamard on the *Psychology of Invention in the Mathematical Field,* published 7 years ago.

However, the angle from which these studies were made, that of psychology, can furnish only part of the general picture. Studies of a psychological nature are significant in that they analyze the mental processes by which the individual mathematician uses the materials and tools present in his culture to make new constructs. But they fail to make proper connection with the cultural stream in which they are imbedded. The mathematician is not an isolated entity grinding out new ideas, with everything coming out and nothing going in. Rather he is making new syntheses out of the concepts that are going in. Young men who have just received the doctorate usually seem to have at least an intuitive recognition of this, for how often do we hear them express their fears of having to accept positions where there are few or no mathematicians who have interests similar to their own; or, if they find it necessary to accept such positions, they worry over whether the libraries, by which they mean the means of keeping contact with their fellow workers through the medium of journals, are good or not. The popularity of such centers as Göttingen in the twenties, and now the Institute for Advanced Study, furnishes further evidence of the gregariousness of the mathematician. And of course there are our societies and associations, through whose meetings and journals we may exchange ideas.

I believe, therefore, that even if it only leads to a rational understanding and appreciation of these matters, a study of these *cooperative* features of mathematical creation, or what we might call a study of the "group dynamics" of mathematical creation, is warranted. Because of its high level of abstraction, we are inclined to look upon mathematics as strictly an *individual* activity. In his address opening the recent International Congress of Mathematicians in Cambridge[4] (p. 125), Oswald Veblen comments that "Mathematics is terribly individual. Any mathematical act, whether of creation or apprehension, takes place in the deepest recesses of the individual mind." However, he goes on to say, "Mathematical thoughts must nevertheless be communicated to other individuals and assimilated into the body of general knowledge. Otherwise they can hardly be said to exist." Note that last sentence, by the way—"Otherwise they can hardly be said to exist." Obviously Veblen is aware that while, in one of its aspects, mathematical creation is, as he says, "terribly individual," in its other aspects it is not an individual affair at all. Later, in the same address, we find an affirmation of the necessity for mathematicians forming such associations as we have today: "The resultant organizations of various kinds have accomplished many important things known to us all. Of these accomplishments I am sure that the most important is the maintenance of a set of standards and traditions which enable us to preserve *that coherent and growing something which we call Mathematics.*" [Italics mine.]

To summarize my motives, then, I wish to inquire, above the individual level, into the manner in which mathematical concepts originate, and to study those factors that encourage their formation and influence their growth. I think that much benefit might be derived from such an inquiry. For example, if the individual working mathematician understands that when a concept is about to make its appearance, it is most likely to do so through the medium of more than one creative mathematician; and if, furthermore, he knows the reasons for this phenomenon, then we can expect less indulgence in bad feelings and suspicion of plagiarism to ensue than we find in notable past instances. Mathematical history contains numerous cases of arguments over priority, with nothing settled after the smoke of battle has cleared away except that when you come right down to it practically the same thing was thought of by someone else several years previously, only he didn't quite realize the full significance of what he had, or did not have the good luck to possess the tools wherewith to exploit it. Coolidge, in his *A History of Geometrical Methods,* remarks[5] (p. 122), "It is a curious fact in the history of mathematics that discoveries of the greatest importance were made simultaneously by different men of genius." This is quite true, except that there is nothing "curious" about it, nor is it confined to mathematics. And it is exactly what one should expect if he is acquainted with the manner in which concepts evolve.

I shall have to be brief and incomplete in my presentation here, giving only a general description of my study. In particular, I shall pay little attention to questions such as "What is a concept?" or to the different type levels of concepts, important though these are. I shall take it for granted that we know what we mean by such terms as "the group concept," "the concept of limit," and the like—these are well defined today. Other things which we label "concepts" are not at all well defined. And of course some concepts embrace other concepts, such as that of the calculus embracing the concept of limit of a set of real numbers.

Furthermore, in the case histories which I shall presently mention, you will probably be impressed with the fact that in tracing the roots of a concept under discussion, we find invariably that some of them go back to Greece. It should not be inferred from this that they *began* in Greece. The fact is, that these roots have many smaller tendrils that reach out into pre-Greek or contemporary cultures, and I am not enough of a historian to have their identities, if such exist, at my disposal. This will not, however, affect my main conclusions in the least. I do want to warn, however, against the impression that these roots necessarily start in Greece.

In reviewing such case histories, I am of course going over much material which is well known. What we get out of such material, however, is strongly influenced by our orientation, and in repeating these historical details it is with the purpose of bringing out aspects that are not ordinarily emphasized and frequently not observed.

3. Case Studies

Now how do mathematical concepts originate; where do they come from? I suppose we might, without thinking very seriously about our answer, reply, "By taking thought;" or "by shutting one's self in his study, getting out a pencil or piece of chalk, and going to work." This kind of reply is reminiscent of the answer which someone made to the question, "How does one write novels?", viz., "By pressing the seat of one's pants to a chair for three hours at a time and writing."

Like most pat answers, this is not, on further reflection, quite acceptable. To be sure, concepts don't get born without someone taking thought. We must not forget, however, that gregarious aspect of mathematical activity which I mentioned above. Not only does this have its influence on the formation of concepts, but it must be conceded an equal partner, at least, in the process.

A concept doesn't just pop up full grown "like Venus from the waves," although it may seem to, to the individual mathematician who does the conceiving. Usually its elements are lying in what, if the term were not already in technical mathematical use, one might call the *mathematical continuum,* but which we might better call the *mathematical culture stream.* As the eminent historian of science, George Sarton, comments[6] (p. 36): "—creations absolutely *de novo* are very rare, if they occur at all; most novelties are only novel combinations of old elements, and the degree of novelty is thus a matter of interpretation, which may vary considerably according to the historian's experience, standpoint, or prejudices. The determination of an event as the 'first' is not a special affirmation relative to that event, but a general negative proposition relative to an undetermined number of unknown events."

3a. Number and Geometry Concepts. The absurdity of always striving for a point of origin is nowhere better exemplified than in the origins of mathematics itself. We are unable here to find any "names" by which to label innovations even if we would. And this, I am certain, is not due simply to absence of adequate written record. The accompanying figure is supposed to represent the origin and growth of the earliest mathematical concepts, those of *number* and *geometry*.

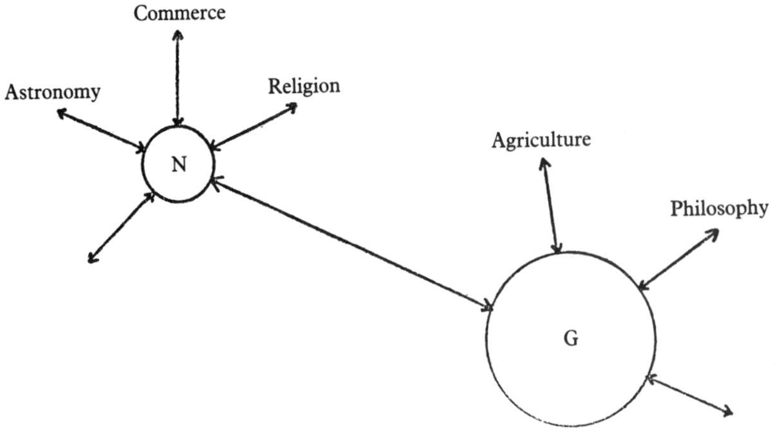

Just when these began to merit being called *concepts* we don't know, but enough evidence exists to warrant the conclusion that this occurred at no particular date, since each gradually emerged from the cultures of the Middle East and Greece; and of course they may have emerged in other cultures, as the number concept did in the Mayan culture. The short arrows represent such influences as those of a Babylonian priesthood facing the necessity of assessing taxes and keeping records of a host of tax-gatherers, the desire to compute the quantity of seeds necessary to sow a given piece of land, and other social needs. From such matters to the concept of a number—"four" for example, as contrasted with "four men," "four dogs"—must have been a long evolution. Certainly no one has suggested a prehistoric Archimedes leaping out of his bath or primitive swimming pool and running through the forest primeval shouting, "I've got it—the number four, the number four!" Similarly, geometry did not begin with Euclid.

3b. Analytic Geometry. Or to take a very elementary, but historically important example, in histories of mathematics one finds serious discussion of whether Fermat or Descartes should be credited with the "invention" of analytic geometry, sometimes accompanied by a choice between the two and an extended argument supporting the choice.

In such cases, I am inclined to feel that the historian is unconsciously influenced by that principle of popular folklore: "everything has a beginning." And conceding this, no historian worth his salt can fail to produce one. In some cases, particularly in that of a very specialized concept, such as that of the "theta-Fuchsian series" whose genesis in the mind of Poincaré is so interestingly described in his article on *Mathematical Creation*[2] (p. 387), to speak of a "beginning" is perhaps justified—although even in such instances one could without doubt, given sufficient evidence, trace the inception of the notion from prior concepts. But in the case of broader concepts, such as "analytic geometry," we seem prone to error due to the vagueness of the concept.

It is interesting to note the way in which Coolidge handles the question in his *A history of Geometrical Methods*. To quote[5] (p. 117): "The opinion is currently held among mathematicians that analytic geometry sprang full-armed from the head of Descartes as did Athene from that of Zeus. . . . There is much to be said in favor of this thesis, but . . . another opinion is certainly possible. The fact is that in inquiring into the origin of analytic geometry we run into a difficulty that lies at the bottom of a good proportion of our disputes in this Vale of Tears. What do we mean by the words 'analytic geometry'? Till that is settled, it is futile to inquire as to who discovered it." That the essence of the subject, however we define it, was known to the Greeks, such as Apollonius, seems beyond dispute; but as a coordinate geometry with a workable algebra like that which we associate with the notion, it evolved through the works of Fermat and Descartes as well as various predecessors. Had the Greeks possessed a good algebra, the story might have been different.

3c. Calculus. Everyone is familiar with the way the basic postulate of popular folklore that everything has a beginning has operated in the case of the Calculus. The dispute over its so-called "authorship" between Newton and Leibniz and their followers is, I suppose, the classical example of useless disputation attributable to the impulse to conform to the postulate. In a recent article,[7] A. Rosenthal makes clear the indebtedness of both Newton and Leibniz to their forerunners as well as their contemporaries; and that no matter how you define it, the calculus is a product of a slow evolution that has been recorded as far back as the Greeks. And although I suspect that the author privately feels that it makes little sense to speak of the invention or the discovery of the calculus as an actual temporal event, he tries to justify the thesis that Newton and Leibniz originated the concept, by affording a definition.

After remarking that "there certainly was an extensive development of the theory of integration and differentiation in the period immediately before Newton and Leibniz," he asks[7] (p. 83), "What then was missing at that time,"—i.e., at the time when Newton and Leibniz entered the picture. He points out that one "very important point still missing was the general fact that differentiation and integration are inverse processes, that is, the so called fundamental theorem of integral calculus." And after some discussion as to whether various mathematicians, who evidently came very close to it, really recognized it or not, he discloses that it was known in full generality by Isaac Barrow, the teacher of Newton at Cambridge. So he repeats, "What more remained to be done?" replying, "What had to be created was just the *Calculus*, a general symbolic and systematic method of analytic operations, to be performed by strictly formal rules, independent of the geometric meaning. . . . it is just this Calculus which was established by Newton and Leibniz, independent of each other and using different types of symbolism." Although this is not really a *definition* in the true sense, at least it is an attempt to fix the general nature of just what it was that Newton and Leibniz did do. Whether it is a good thing to try to assign an originator to as broad a concept as the calculus, even when it is impossible to make the originator *unique*, is another question.

It should be noted that, as in the case of analytic geometry, if the Greeks had possessed enough symbolic material, they might have progressed further with the calculus. Also that despite the lack of a rigorous theory of limits, that which most agree to call "calculus" appeared nevertheless with Leibniz and Newton, but essentially as an operational apparatus. The theory of limits, and calculus as we know it, had still to evolve.

It is interesting to compare Coolidge's comments[5] (p. 38). He speaks of the Greek geometers' "procedure in attacking problems in computation which led naturally to the integral calculus. If we seek the root of the difference between their method of attack and ours, it lies partly in the greatly increased flexibility of our notation, partly in their lack of a clear concept of a limit." That is, a suitable notation was lacking—a suitable *tool*—and a necessary *concept*, that of limit, had yet to evolve. Since the concept of limit was not developed until after the work of Newton and Leibniz, this seems to place the emphasis again upon the Greek lack of an algebraic tool, as the root of the difference between the Greek calculus and that of Leibniz and Newton.

3d. The Curve Concept. Finally, I would like to view the evolution of a narrower, more specialized concept, that had its origin in antiquity, and which not only continued its growth in modern times but contributed in a noteworthy manner to a large portion of modern mathematics. It is not, I am sure, generally realized how great an influence this concept has had on the development of concepts which today contain hardly a trace of the notion. I refer to the concept of *curve*.

The accompanying chart, which may be referred to during the discussion, is supposed to set forth the principal lines of influence, and it is to these that I shall confine myself. It would be impossible to sketch here the history of the concept in full detail. It has been a common mathematical notion, as a rule purely intuitive, since the time of the Greeks, at least. And it is one of those mathematical notions that have been adopted by the layman. Usually it has been just a conventional categorical term, and seems to have continued so throughout the time when the study of curves was resumed analytically during the Renaissance. The first successful attempts at formulating precise definitions of the concept were not made by geometers, but by analysts! So far as I have been able to find, Georg Cantor was the first to be credited with a definition, but apparently what he really did[8] (p. 194) was to determine those set-theoretic properties that constitute the *continuous*, as opposed to the discrete, aspect of a curve—or as a topologist would express it, he defined the topological notion of *continuum*. But a square together with its interior is a continuum, and we would hardly call this a curve! However, if the continuum lies in the plane, all we need to do is require that it contain no such square and we get what came to be called a "cantorian line." And although satisfactory only as a definition of the concept of plane curves, the cantorian line coincides exactly with what most topologists feel is the natural definition.

The next most notable attempt at definition was also not made by a geometer, but by C. Jordan in 1887. In the first (1882) volume of his *Cours d'Analyse,* he speaks of a plane curve as given by any two functions $x = f(t)$, $y = g(t)$, where the range of t is the real number interval $[0,1]$. Although no mention of continuity is made, if he was familiar with Cantor's proof[8] (p. 122 ff.) published in 1878, 4 years earlier, of the (1–1)-correspondence between the points of such an interval and the points of the square plus its interior, he must have realized that without it, this definition could not be satisfactory. However, in an appendix ("Note") to the 3rd volume of this work, published 5 years later, in repeating this definition, he added[9] (vol. 3, p. 587), "If the functions are continuous, the curve will be called *continuous*." ([The] reference to the first volume of the second edition, 1893, is the one customarily cited as the "origin" of the concept "continuous curve.") Cantor had noted that the (1–1)-correspondence just referred to was impossible in case the functions were continuous, and Jordan's definition of *continuous* curve lay between the two extremes—it does not require (1–1)-ness, although the functions are of course single-valued, and it does require continuity.

It is not difficult to surmise where Jordan got this notion; it was common practice to define curves parametrically in the books on calculus. For instance, the parametric representation of the cycloid is given in Cauchy's famous calculus, published in 1826. Evidently what Jordan did was to formulate a general concept which was already current. I don't know who first attached Jordan's name to the notion, as a matter of fact. The Polish topologists were already doing so during the 'teens of the present century, and most writers of the twenties assume he was the first to formulate it explicitly. However, I have found no evidence that either Jordan's contemporaries or his immediate successors associated his name with the concept, which seems to be further evidence that the notion was generally current among analysts.

But to get on with the story: Geometers, possibly becoming suddenly conscious of their inadequacy, or perhaps jealous of their prerogatives, pounced upon the notion. I am confident that Jordan had no idea of usurping their natural rights; that all he had in mind was to make precise certain analytic notions that were already current, for his own special purposes. For instance, he showed that for a curve to be rectifiable, it is necessary and sufficient that the functions defining it be of bounded variation. And for the proof of his famous theorem, the Jordan Curve Theorem, he stipulated that the continuous curve was to have no double points except at the end points of the interval of definition of the variable t. For these purposes, his definition was a good one, and still is.

As is well known, it was only three years after Jordan's formulation until the Italian logician and geometer Peano published, in 1890, his famous

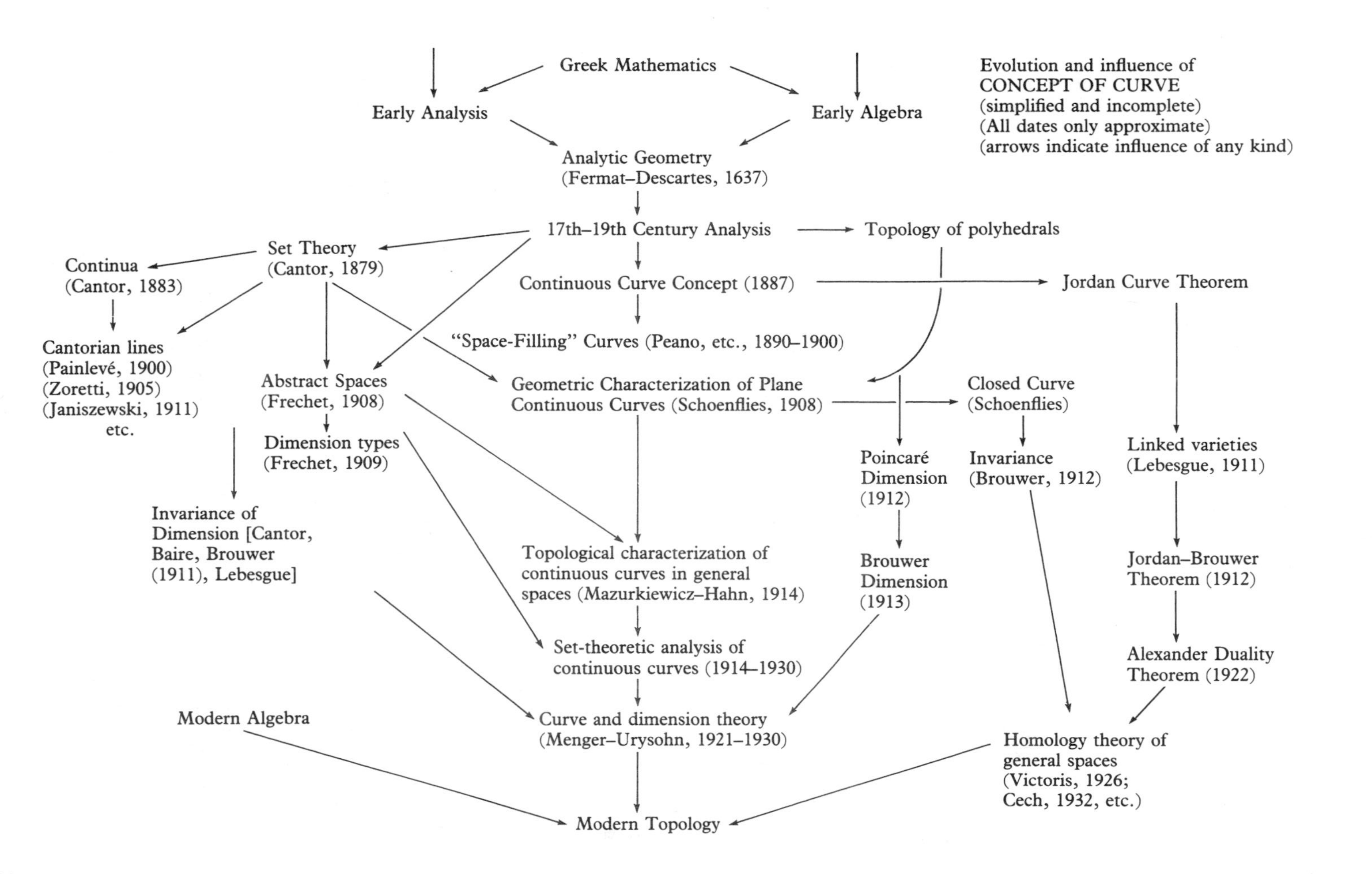
Greek Mathematics
Early Analysis
Early Algebra
Evolution and influence of
CONCEPT OF CURVE
(simplified and incomplete)
(All dates only approximate)
(arrows indicate influence of any kind)
Analytic Geometry
(Fermat–Descartes, 1637)
17th–19th Century Analysis
Topology of polyhedrals
Set Theory
(Cantor, 1879)
Continua
(Cantor, 1883)
Continuous Curve Concept (1887)
Jordan Curve Theorem
"Space-Filling" Curves (Peano, etc., 1890–1900)
Cantorian lines
(Painlevé, 1900)
(Zoretti, 1905)
(Janiszewski, 1911)
etc.
Abstract Spaces
(Frechet, 1908)
Geometric Characterization of Plane
Continuous Curves (Schoenflies, 1908)
Closed Curve
(Schoenflies)
Dimension types
(Frechet, 1909)
Poincaré
Dimension
(1912)
Invariance
(Brouwer, 1912)
Linked varieties
(Lebesgue, 1911)
Invariance of
Dimension [Cantor,
Baire, Brouwer
(1911), Lebesgue]
Topological characterization of
continuous curves in general
spaces (Mazurkiewicz–Hahn, 1914)
Brouwer
Dimension
(1913)
Jordan–Brouwer
Theorem (1912)
Set-theoretic analysis of
continuous curves (1914–1930)
Alexander Duality
Theorem (1922)
Modern Algebra
Curve and dimension theory
(Menger–Urysohn, 1921–1930)
Homology theory of
general spaces
(Victoris, 1926;
Cech, 1932, etc.)
Modern Topology

space-filling curve. Hilbert and E. H. Moore followed suit soon after. None of them mentions Jordan's name, incidentally, perhaps further evidence that the concept was probably not considered as his personal property! Their examples show, of course, that Jordan's definition was not a satisfactory formulation of the precise notion of curve as it was intuitively conceived.

However, fortunately for the development of modern mathematics, the fact that the geometer found Jordan's analytic definition wanting did not mean that it was doomed to oblivion so far as geometry was concerned; even though it came to be recognized that in addition to including configurations we don't want to call "curves," it excluded some we do want to call curves (such as the well known continuum formed from sin $1/x$ and its limit points on the y-axis). That it included among its special cases the general Euclidean space element—for of course the cube and its interior in three and higher dimensions are continuous curves in Jordan's sense—was of intense geometric interest. Thus it developed that interest in the continuous curve was focused on the nature of its geometric counterpart; if this counterpart was not what we call "curve" in geometry, just what configurations do come within its scope?

The first systematic investigation of this question was made by Schoenflies, who gave a complete answer for the planar continuous curves. His work, incidentally, shows a remarkable perception of the nature of the problem as well as ingenuity in selecting the right methods for solving it. Realizing that it could not be solved by any of the methods common to the older forms of geometry, he called upon the new set-theoretic ideas introduced by Cantor and the (apparently unrelated) topology of polyhedrals of Riemann and Poincaré.

Among the first to be influenced by Schoenflies' work was Brouwer, probably best known for his Intuitionism, but among topologists and analysts famous for the remarkable work which he did on the theory of manifolds, especially as regards mapping and fixed point theorems, during the second decade of the present century. Brouwer was impressed not only by Schoenflies' positive results, but by certain fundamental errors which, due to faulty intuition, subtly crept into Schoenflies' work. For in his study of the Jordan Curve Theorem, Schoenflies conceived the idea of separating, from its tie-up to the topological images of the circle, the property of a curve being "common boundary of two domains in the plane" and called any configuration having this property a "closed curve." That a closed curve so defined could not be the common boundary of *more than* two domains seemed perfectly natural to assume. Only it happens not to be true, as Brouwer was able to prove. Also, Schoenflies was unable to give any proof of the invariance of the closed curve under topological transformations. Brouwer also supplied this, using a method of proof which contained the central idea of the extension, to general metric spaces, of the homology theory of polyhedrals which had been initiated by Riemann, Poincaré, and others, and extended in this country by Veblen and his colleagues.

But I must not get too far ahead. The threads of influence leading from Jordan's definition begin now to entangle with one another and with threads from other sources. About the time when Brouwer was writing his dissertation on the law of the excluded middle, with perhaps little thought of topology, Fréchet was working on his dissertation on "abstract spaces." This tendency to abstract from the Euclidean situation, which of course goes back, in its roots, to Fréchet's predecessors but was given clear-cut realization by him, was to influence the treatment of the continuous curve problem in the following manner: It had proved impossible, with the tools available at the time, to extend Schoenflies' methods to 3-space or higher dimensions. Here the point of view is relative to the imbedding space, the Euclidean space of 3 or higher dimensions, rather than intrinsic. The introduction of the abstract space idea led to the problem of *intrinsic characterization*, by geometric methods, of the continuous curve. In order to eliminate the separate functions defining the coordinates in Jordan's definition, one recognized that an equivalent idea is to consider the curve as defined by a single function, $f(t)$, single-valued and continuous over the interval $[0, 1]$, but the values of the function to be in any space allowing of the definition of limit. From this point of view the problem became manageable and received a solution, independently, at the hands of Hahn and Mazurkiewicz. Using the topological notion of local connectedness, which had actually been introduced (presumably unknown to them) three years earlier by Pia Nalli in another connection, they characterized continuous curves as compact metric continua that are locally connected—a thoroughly intrinsic definition, as contrasted with the results of Schoenflies.

This was about the time of the start of World War I, and the simultaneity of these results did not become known until after the War. There followed, not only in Poland and Austria, but in this country where the students of E. H. Moore had become interested in the new ideas in the foundations of geometry, a fairly complete analysis, especially during the period up to 1930, of the concept of continuous curve from the set-theoretic point of view.

Now it is characteristic of mathematical concepts that they often tend to be swallowed, as for instance in the process of generalization, by more embracing concepts. As a rule, this happens subsequent to the formulation and investigation of the concept. The number concept is a simple example of this. But in the case of the curve concept, the assimilation was to be coincident with the formulation of the larger concept. And so, for a few minutes, I shall turn to the concept of *dimension*. I cannot take time to review its ancient history—remarks analogous to those which I made concerning the concept of curve hold here, such as its use by the Greeks, its adoption into popular parlance, etc. As we approach modern times, attempts to formulate a dimension concept begin to crop out in the literature, often in the works of analysts, whose formulation was usually based on the number of parameters involved in the description. Cantor gave the body blow to this when he showed that a single real parameter suffices, if one neglects continuity. The resulting investigations—and these included the Jordan formulation of curve as well as the Peano results already mentioned—especially the problem of the invariance of dimension under bicontinuous (1–1)-transformations, are too detailed for me to go into here. They played their part, however, in the evolution of the dimension concept. On the left-hand side of the chart will be found some indications of the miscellaneous work done in this direction—note the names of Baire, Brouwer, and Lebesgue accompanied by the date 1911. In addition certain dimension definitions or attempts at definitions are also indicated, and there is a notation regarding "cantorian lines," which were studied by both Zoretti and Janiszewski among others.

Now I do not maintain that any particular mathematician read any of the work of these men, and was directly influenced thereby. I call attention to their work chiefly as a proof that the notions of curve (or "line" as it was frequently called) and dimension were evolving in the mathematical culture stream.

To the right of the center of the chart you will notice the notation "Poincaré dimension, 1912." It seemed appropriate to place this over to the other side of the chart for several reasons; the most important being that it leads quite naturally to the concept which was finally generally accepted as embodying the precise formulation of the intuitive notion of curve and dimension. Another reason for setting this apart from the material on the left side of the chart is that although Poincaré did the most important kind of work in laying the foundations of the topology of polyhedrals, as is well known he had little acquaintance with or use for the theory of sets. Consequently, when he tried to indicate just what the dimension of a space is, in such essays as that on *The Notion of Space,* for instance, he thought globally instead of in the small. And so his definition is an "in the large" definition—which may suffice for a homogeneous space, but most topological spaces are not homogeneous. However, the terminology in which it is formulated, such as "continua" and "separating" or "cutting," were actually set-theoretic. As a result, the definition was not given in precise mathematical terms. Nevertheless, anyone who happened to notice it and who had the suitable tools in hand could make it precise and a better approximation to a satisfactory definition. And this is exactly what Brouwer did the very next year, in 1913.

Brouwer's paper attracted little attention, and when, in the early twenties, a satisfactory theory of curve and dimension was announced by Menger and Urysohn, evidently neither of these gentlemen knew of the work of Brouwer, nor of each other's work. It is now known, of course, that all three theories are equivalent for a wide class of spaces—the locally connnected, separable metric spaces.

The arrows on the chart, which, as the key in the upper right-hand corner states, are supposed to represent influence of any kind, need perhaps some further explanation. That the curve and dimension theory of Menger and Urysohn was influenced by the various set-theoretic activities of their predecessors is, I think, too obvious to need further comment. As I have already pointed out, it is probable that lack of contact with these could have prevented Poincaré from setting up some kind of a precise definition of dimension. The arrows coming in from the sides, however, are not so obvious. Just how much, if any, of the works on dimension problems which preceded them were actually read by Menger or Urysohn, I don't know. In the introduction to his formal presentation of his work published in 1925, Urysohn refers to the work of Zoretti on cantorian lines, for instance, and not only adopts the same term but entitles his monographs *Sur les Multiplicités Cantoriennes;* he also remarks that his dimension definition is quite in conformity with the desires that motivated Poincaré. But even if no such direct and obvious influences existed, there is still the indirect influence that may have been exercised through the medium of others. Menger frequently made gracious acknowledgment to his teacher, Hans Hahn, not, so far as I know, for specific suggestions regarding definitions but for general guidance and inspiration; Urysohn's association with Alexandroff is well known.

But my time is running out, and I cannot give further details. I have already mentioned how the Brouwer proof of invariance of the Schoenflies closed curve contained the germ of the application of the homology theory of polyhedrals to general spaces—a fact which led Vietoris, who acknowledges his debt to Brouwer, to extend this theory to general metric spaces in order to establish an intrinsic form of the analogous higher dimensional invariance. Let me also call attention to the arrows on the extreme right of the chart leading from the Jordan Curve Theorem to work of Lebesgue and Brouwer in higher dimensions, and thence to the Alexander Duality Theorem (which was motivated by the desire to obtain a general higher-dimensional extension of the Jordan-Brouwer separation theorem), and hence into algebraic aspects of modern topology. Incidentally, I have said very little about the role played by algebra in these matters; but its influence is mainly quite modern, and my intent has been not to sketch a history of curve theory, but to indicate as well as I can, in the time at my disposal, the manner in which the curve concept has influenced various other concepts right down to the inception of modern topology. Of course other concepts played a part too, but as I said before the threads of influence become quite tangled, and it is better to follow one at a time.

Before leaving the chart, let me point out, to prevent misunderstanding, the "simplified and incomplete" notation in the upper right-hand corner. Many things have been omitted, not because they are unknown (although there are many such), but to avoid complicating the diagram. There are, for example, geometric contacts that are omitted. If only in the way of providing certain perspective, geometric concepts exercised an influence on set theory, for example, and on curve and dimension theory.

4. Some Inferences from Case Histories

Now what can be inferred from such observations as these? We can hardly *prove* anything, in the sense that we prove a mathematical theorem, any more than we can prove that certain methods of teaching mathematics are the best we might use. In such matters we are dealing with a kind of "metamathematics," although not in the technical sense in which the term is being used, to be sure. We can, however, make certain inferences of value.

4a. Factors Influential in Concept Formation.

For the sake of completeness we first note the existence of the evolution of mathematical concepts from the general cultural environment—the nonmathematical environment, that is. This was not only operative in the origin of the primitive concepts concerning number and geometry, but has continued throughout mathematical history. It is going on today, as anyone who has worked on war or government projects can bear witness. And although it deserves closer study, I shall not give it any more attention here, but shall pass on to factors operative within mathematics. So from now on, in my conclusions, I shall treat mathematics as an organism which, although influenced by an outside environment, is going to be of interest only from the standpoint of its internal structure.

I have just used the term "organism." This was to emphasize the *cooperative* character of mathematical creation, as contrasted with the individual aspect. As Veblen observed, mathematics is "terribly individual." But we don't work in a vacuum. Neither, it is true, do we work in teams like the experimental scientists do, although we do collaborate in small groups of two or three—rarely in greater number. I have in mind, however, also those types of cooperation which are not so obvious—especially the type of cooperation I imply when I make the statement that a Newton can carry on only from the level which the mathematics of his time has reached. This is not peculiar to mathematics, of course—Beethoven was as much indebted to the musical developments of his predecessors as to the more obvious inventions of the musical scale in which he wrote and the instruments for which he composed. But I think that we mathematicians have not been properly conscious of the fact. When, for example, a genius of the stature of Leibniz expends so much energy in collecting evidence of the priority of his ideas over those of Newton, is it not clear that he is not properly aware of the debt owed to those on whose shoulders he stood? Or, when a concept has germinated and gradually emerged over a period of anywhere from 50 to 2000 years, and then is suddenly brought into full light through the inventive powers of two, or possibly three or four mathematicians—and when, moreover, historical research shows that others were nearly ready to bring out the idea—does it make sense to call the event a *coincidence?*

This is not to deprive brilliant intellects that make final syntheses of concepts of any credit. Without them, concepts would never get born. But it is equally true that without the cooperation which they have received from innumerable other workers, many of whom may be gone and forgotten, the material from which they made their syntheses, and the tools they used, would not have been available. And by properly giving credit where credit is due, perhaps more modesty in the effort to establish priority will make everyone concerned a little happier.

Of course many mathematicians have sensed the organic character of mathematics. One of the most recent examples I have noticed of this was in an article by H. Weyl entitled *A Half-Century of Mathematics*. When half-way through this article he pauses and exclaims, "The constructions of the mathematical mind are at the same time free and necessary. The individual mathematician feels free to define his notions and to set up his axioms as he pleases. But the question is, will he get his fellow-mathematicians interested in the constructs of his imagination? We can not help feeling that certain mathematical structures which have evolved through the combined efforts of the mathematical community bear the stamp of a necessity not affected by the accidents of their historical birth. Everybody who looks at the spectacle of modern algebra will be struck by this complementarity of freedom and necessity."

The individual mathematician cannot do otherwise than preserve his contact with the mathematical culture stream; he is not only limited by the state of its development and the tools which it has brought forth, but he must accommodate his desires to those concepts that have reached a state where they are ready for synthesis.

Analytic geometry could not be synthesized in the manner which Fermat and Descartes are usually credited with doing, until both the necessary geometry and algebra were at hand; nor until their predecessors had carried the synthesis nearly to fruition. The same remark holds for calculus. If we interpret the achievement of Newton and Leibniz as the creation of a symbolic method which synthesized those ideas of the calculus which were already in existence, then their dependence on the cooperation of others is clear. And it seems doubtful that Menger and Urysohn would have, independently, arrived at the precise notions of dimension and curve which they formulated, without the preliminary development of not only the topological tools necessary for their formulation, but of such concepts as that of *localization* which were not, apparently, possessed by Poincaré. And that a concept of dimension was trying to break through to the mathematical consciousness is evidenced by the related problems that had been treated during the preceding decades, as well as by its ultimate formulation by two independent workers (as in the case of analytic geometry and calculus).

As I said before, let me emphasize that I do not make these observations with any intent of taking away credit where credit is due. Furthermore, we must beware of looking backward from the heights which we have now scaled and concluding that the tasks of our predecessors were easy. Not long ago I overheard a group of young topologists discussing the state of topology about 1920, particularly the fertile field for research which was opening up at that time in the investigation of continuous curves. They seemed quite in agreement that they had been born 30 years too late—that it is a much harder task to find a subject for a dissertation now than it was then. As I was one of those who had been born 30 years sooner, I felt compelled to point out what the situation really was like at that time. I could tell them, for instance, of one young mathematician, who had just received the Ph.D. degree, and who decided to forsake topology since it was obviously all worked out! None of us really knew of the possibilities at that time. And much energy had to be expended on the invention of concepts which now seem almost trivial, and of new methods of attack.

The greatest factor in the evolution of concepts today is probably what I would call *conceptual contacts;* on the individual level, it is what we might call a "meeting of minds"; on the group level, it is the diffusion of concepts. Examples of the former, the meeting of individual minds, are to be found in the many cases of mathematicians, who, due to the political conditions prevailing in Europe the past 40 years, moved to new mathematical centers and established contacts with men whose ideas and methods often found fruitful syntheses with their own. Similar contacts are, of course, being made today through the medium of grants to foreign mathematicians enabling them to visit mathematical centers in this country—as well as in the reverse process wherein members of our group visit abroad. Diffusion of concepts on the group level is of that type which was exemplified in the past by the fusion of algebra and geometry to form analytic geometry; or by the diffusion into geometry of set-theoretic methods in order to provide satisfactory formulations of the curve and dimension concepts; or in the recent past by such fusions as those of algebra and topology to form algebraic topology or topological algebra, depending on which aspect of the fusion you place emphasis. Concepts such as these are not the result of the chance meeting of two mathematicians at a mathematical gathering or because of an invited lectureship, as was the case in some instances which I can recall; but resulted from the gradual building up, by many independent workers, of a host of component concepts. The manner in which such syntheses as these are brought about needs, I imagine, no further amplification. For the individual mathematical discipline they form the analogue, within the field of mathematics, of the interplay between mathematics and its environment in the general culture. They point up, of course, the desirability and the necessity, if we would encourage healthy mathematical growth, of inter-university cooperation on a more enduring basis than is provided by the meetings of societies; even though the latter do foster individual contacts and undoubtedly play an important part in developments on the higher group level.

Not far removed from the types of concept formation already mentioned is that which consists in the observation of similar patterns in several different branches of mathematics, or, at a lower level, in several special cases. The classic example is probably the concept of abstract group. The Jordan definition of continuous curve illustrates the formulation of a concept from observation of its special cases commonly used in analysis. And again this is a kind of concept-formation which, within mathematics, is the analogue of the type of concept-formation which has for its basis the observation of mathematical patterns in the external environment, particularly in physics.

Also of a related character is the introduction of new tools from other branches of mathematics. The concept of the general continuous curve space, formulated independently by Hahn and

Mazurkiewicz, making use of the tools provided by abstract space concepts, is one example. Another example is the introduction of the axiomatic method into algebra with results which have been remarked upon by a number of writers, particularly Weyl. A current example is the use of metamathematical tools to establish theorems about the theorems in a given branch or branches of mathematics.

This reminds one of certain methods, which, because of their wide applicability, have earned the right to special notice as concept-builders. The *axiomatic method* is one of the most notable of these. I cannot resist recalling here what Poincaré, in his article entitled *The Future of Mathematics,* conceived as the chief use to which the method would be put[2] (p. 382). He devoted less than half a page to it; remember that when he wrote, the method was still quite new in its modern sense. After commenting that Hilbert had obtained "the most brilliant results" with it, he observed that the problem of providing axiomatic foundations for various parts of mathematics would be very "restricted," and "there would be nothing more to do when the inventory should be ended, which could not take long. But when," he continued, "we shall have enumerated all, there will be many ways of classifying all; a good librarian always finds something to do, and each new classification will be instructive for the philosopher." It would seem that Poincaré believed that the method was of little importance for mathematical creation. I wonder what his opinion of it would be today, if he could see what a tool for research it has become?

Related to the axiomatic method, in that it provides a special field for its use, is the process of altering of concepts already formulated. This is the kind of thing we do when we replace an axiom, such as the parallel axiom of Euclidean geometry for instance, by a contradictory—in the case mentioned replacing the parallel axiom by one of the types of non-parallel axioms. Here one has to be especially careful, of course, to preserve contact with established theories—or as Weyl put it, mindful of whether one will "get his fellow-mathematicians interested in the constructs of his imagination."

That this process operates on the group level as well as on the individual level is shown by the case already cited, that of non-Euclidean geometry. When this began to break through to the individual mathematical consciousness, it had already smoldered for years in the fires of Saccheri's and others' determination to prove its logical impossibility. When it did find expression, it found it in several places. As Coolidge remarks[5] (p. 73), "The outstanding effect of a comparison of the work of Lobachevski and Bolyai is surprise at their likeness. Both start with the parallel angle, both note the Euclidean nature of the geometry on the limiting surface, both develop the formula for corresponding lengths on parallel limiting surfaces, both note the relation to spherical trigonometry and the independence of this from the parallel axiom." And then when Gauss revealed that he had obtained almost identical results earlier, is it to be wondered that Bolyai suspected Lobachevski of plagiarism and exclaimed, "It can hardly be possible that two, even perhaps three, persons, knowing nothing of one another, have achieved almost complete the same results at about the same time, even if by different paths." It is a pity that he could not know that not only was it possible, but quite probable, since the evolution can be clearly traced in the mathematical and philosophical thought of the preceding centuries.

Then there is the well known and much abused generalization! That this operates on the individual level needs not to be elaborated on, since we all use it in our individual research. However, it does operate on the group level. For instance, that kind of generalization which results from the splitting of concepts may take place on the group level; I suppose the splitting of the concept of the real number system into its structural or topological aspects, and its operational or algebraic aspects, furnishes an example of this. Splitting of concepts, that is, may be due to the gradual evolution of the mathematical organism rather than to the conscious act of an individual mathematician—even though the final syntheses may be the acts of individuals.

4b. Life Span of a Concept. Having discussed some of the modes by which concepts originate, let us turn for a moment to the life of the concept, following its exact formulation. The examples of the precise formulation of the topological character of continuous curve by Hahn and Mazurkiewicz, and that of the general curve and dimension theory by Menger and Urysohn, are instructive here. It seems that the formulation of a new concept may result in a distinct branch, or sub-branch of mathematics as the case may be—a kind of "specialty," we might call it. This may become so popular that one is not in the "swing" in his particular field unless he joins in the investigation of the new concept. Mathematics is as subject to fashions as any other aspect of man's behavior! Sooner or later, however, the so-called specialty is developed to such an extent that either it loses its popularity or achieves immortality through contact with other concepts. Thus we find in the cases mentioned that as soon as the related widespread investigations had resulted in solutions of the main problems, the fundamental parts of the theories were either absorbed in new concepts or made part of the general framework, both as to tools and materials, of general topology. The theory of continuous curves contributed not only such notions as that of local connectedness and its properties to the study of other theories, such as that of Lie groups, but was absorbed in a new concept, that of lc^n spaces, whose definition became possible because of the extension of homology tools to general spaces in the decade succeeding the investigations on continuous curves. Like comments can be made concerning dimension theory, whose fundamental notions are now part of the general equipment of a topologist; and which was absorbed in a more general theory of dimensions utilizing homological notions, and is constantly used in applications of topology. I presume that a more elementary example may be furnished by elementary geometry as the Greeks knew it. I imagine that it had already reached its zenith and was on the wane when Euclid wrote his encyclopaedic work—the historians can correct me if I am wrong about this. Here, if ever, was a case of a mathematical theory being embalmed. When mathematical archeologists, so to speak, resurrected it, it seems to have stimulated only futile attempts to prove the parallel postulate or to trisect the angle and square the circle. Only new perspective that comes with conceptual contacts and new tools could infuse life into it. And, of course, this came in the realization of the validity of the non-Euclidean geometries, and the introduction of methods of algebra and analysis which resulted in the modern form of analytic geometry, and a surge upward in analysis following the rise of the calculus.

To summarize, then, a concept, having only a limited range of possibilities, seems to have a certain life span like that of a star, going through a period of tremendous activity and then waning. Without the infusion of, or fusion with, new conceptual material, it is probably destined to die. But there is insufficient evidence to judge of this, since fortunately the new material is usually ready. New tools may be introduced from other parts of mathematics or even from the same branch, or even suggested by cultural elements outside of mathematics; or the concept may join with other concepts to form new ones more fruitful than the old; I have mentioned examples of all these possibilities. I cannot, as a matter of fact, cite offhand any case of a mathematical concept that has died in the sense that it no longer has any use whatsoever. The accumulated body of knowledge to which it has led is sooner or later seized by workers in other branches of mathematics, or even outside of mathematics, and put to work—often in a way of which its devotees never dreamed.

An interesting fact, which is probably a result of this, is that whereas there may seem to emerge from all this the spawning of a host of narrow specialties, a process of unification is continually going on. It is as though mathematics strikes out in all visible directions, but conscious that it will never do to become too diverse, pauses periodically to consolidate and unify the gains made before making new advances.

5. Import to the Individual Mathematician

Since I have spent so much time in this talk on the group level, I should like to make a comment or two about the implications to the individual.

Some of these I mentioned in my introduction and will not repeat here. I should like to make one amplification of my previous remarks concerning the benefits that may accrue to the individual research worker, especially those who are starting their careers. I have been continually impressed, since beginning this study, with the manner in which it brings out not only the advisability and necessity of keeping in touch with the mathematical culture stream through the media of journals and personal contacts—something we already habitually do—but with the suggestive ideas that may be found by exploring the works of the past. I am not advocating here a mathematical "great books program," but just a little browsing now and then in the original attempts at formulations of commonly used concepts, as well as of concepts not yet satisfactorily defined. They often contain suggestions which were impossible of fulfillment at the time when written, because the tools for exploiting them were not available. It is not difficult to imagine that the dimension theory of Poincaré, which was published in a philosophical journal, might, if mathematics had developed in different fashion, not yet be susceptible to precise formulation; and would, in that case, be lying ready for the moment to arrive when the suitable tools become available.

For a similar reason, it seems that every new generation of mathematicians should give attention to the famous unsolved problems—for one never knows when the tools adequate for a solution may have evolved.

6. Future Developments

It is time I brought this talk to a close. In doing so, I hope you will be so kind as to let me indulge in a little speculation. Up to this point I believe I have stuck to facts as well as one can in a domain where, as I said before, proofs are not possible in the mathematical sense. What I want to say now I shall freely label "speculation," and you can take it for what it seems worth to you.

Whenever I reflect on the changes that have occurred in mathematics during the past 40 years, I invariably recall that statement of Spengler whose *Decline of the West,* published in 1918, caused such a commotion in the intellectual world, namely: That in the concept of an abstract group mathematics achieved its "last and conclusive creation." And his prediction that "the time of the *great* mathematicians is past. Our tasks today are those of preserving, rounding off, refining, selection—in place of big dynamic creation, the same clever detailwork which characterized the Alexandrian mathematics of late Hellenism." Could any prophecy have been demonstrated fallacious more quickly and conclusively than this one has been? Also I recall a statement in Struik's little history that "toward the end of the 18th century some of the leading mathematicians expressed the feeling that the field of mathematics was somehow exhausted. The laborious efforts of Euler, Lagrange, D'Alembert, and others had already led to the most important theorems; the great standard texts had placed them, or would soon place them, in their proper setting; the few mathematicians of the next generation would only find minor problems to solve." And this was before 1800!

If I were given to the vice of prophecy, I would not hestitate to risk my reputation as a prophet with the prediction that mathematics, in this year of our Lord 1952, is only reaching out toward maturity, and that given encouragement it will, during the next 50 years, yield new concepts and methods which will revolutionize the subject. I envy those young men who are only on the threshold of their mathematical careers, for they will be possessed of powers that will put their elders to shame. I have already noticed, and no doubt some of you have, evidence of greater powers among recent recruits to the ranks of mathematical research. The future will no doubt see not only an increase in this, but a corresponding improvement will also be noticed in the ranks of undergraduates. I am one of those who believe that genius is not a rare occurrence—genius seems rare chiefly because of lack of opportunity. And in mathematics opportunity is measured to considerable extent by the quantity and magnitude of its concepts and the power of its methods. There is every indication that these are now approaching an all time high. To take a leaf from my own experiences: For some 30 years I have witnessed the growth of topology, sometimes halting, sometimes feverish, from a mere infant to a size where we don't even attempt to give it explicit definition any more. Yet today, I am amazed at the power and ingenuity of the new methods and concepts that are being introduced. And topology is not unique in this.

But it is not only for this reason that I feel as I do about the future of mathematics. I am also a firm believer in the evolutionary character of mathematical development. Concepts are not stable; they continually grow even though their outcroppings are discrete events. This we have noted in the evolution of the concepts of number and curve. And when a concept has evolved so far as to achieve a precise formulation in the mathematical framework then available, its further growth or evolution seems virtually determined by the existence of those logical laws and methods which every mathematician observes and uses. It is much as in the case of a particular axiomatic system, which, once set up, will lead almost inevitably to a certain set of theorems as its logical consequences. Or, to draw once more upon our case history of the concepts of curve and dimension, it is quite clear that the concepts of curve and dimension were already evolving in the mathematical culture stream before being precisely formulated. It is a nice experiment, by the way, to take a group of graduate students who have been trained in basic *modern* methods, especially in topology, but who have little or no acquaintance with dimension theory, and to give them Poincaré's notions about dimension. If they are any good, they will unfailingly grind out the concept of dimension which corresponds to these notions, just as Poincaré might have if he had had the necessary tools. This would not, of course, be the form which Menger and Urysohn gave. Similarly we can find, without reading too much into the prior facts, the evolving forms of concepts such as the calculus and analytic geometry. Their eventual emergence seems to be virtually a certainty. The ways of thought in which we are all trained, and which we all share more or less, seem as much of a guarantee of certain eventual conceptions being synthesized, from given basic material, as do the logical laws for the theorems of an axiomatic system.

As in any evolutionary process, however, the environment cannot be ignored. Just as the individual mathematician does not work in a vacuum, and is influenced by the work of his predecessors and coworkers, so does mathematics itself not evolve in a vacuum. If world crises occur, as we have ample reason to fear they may, they may negate entirely what I said above about the future of mathematical creation. That interplay between evolving concepts and the minds of those individual mathematicians who achieve their syntheses can be interrupted by forces from without. The process of mathematical creation can cease either from stagnation, as from lack of diffusion and the resultant mixing of concepts, or from stifling incidental to major disruptions of society. The former we have no longer any cause to fear; with modern means of communication, and other aids to cultural contacts, our field is more alive than ever before. With good fortune, we may escape the latter, and then I think the prophecy that I said I would make, if I were a prophet, will come true!

Notes

1. A. N. Whitehead, *Science and the Modern World*, Pelican Mentor Book, N.Y., 1948.
2. H. Poincaré, *The Foundations of Science*, Lancaster, Science Press, 1946.
3. J. Hadamard, *The Psychology of Invention in the Mathematical Field*, Princeton University Press, 1945.
4. O. Veblen, *Opening Address*, Proceedings of the International Congress of Mathematicians, Providence, American Mathematical Society, vol. I, 1952, pp. 124–125.
5. J. L. Coolidge, *A History of Geometrical Methods*, Oxford, 1940.
6. G. Sarton, *The Study of the History of Science*, Cambridge, Harvard University Press, 1936.
7. A. Rosenthal, *The History of Calculus*, Amer. Math. Monthly, vol. 58 (1951) pp. 75–86.
8. G. Cantor, *Gesammelte Abhandlungen*, ed. by E. Zermelo, Berlin, Springer, 1932.
9. C. Jordan, *Cours d'Analyse*, Paris, Gauthier-Villars, vol. 1, 1882; vol. 3, 1887.

A Dialogue on the Applications of Mathematics

Alfréd Rényi

Alfréd Rényi began publishing his works in 1948 and continued to do so until his sudden death in 1970. He began his studies at the University of Debrecen and then went on to the University of Budapest. He was a visiting professor at Stanford, Michigan State, and the University of North Carolina. For fifteen years he was the director of the Mathematical Institute of the Hungarian Academy of Sciences. He was editor of *Studia Scienticrum Mathematicerum Hungarian* and on the editoral board of *Acta Mathematica, Annales Science Math., Publicationes Math., Math Lapok, Zeitschrift für Wahrscheinlichheitstheorie, Journal of Applied Probability,* and *Journal of Combinatorial Analysis.*

Mathematicians love mathematics for its beauty and order. Society tolerates mathematics because it is so very useful. The second fact is far more important to the survival of the subject than is its capacity to stimulate the aesthetic feelings of its devotees. The most famous mathematician of antiquity was Archimedes. He understood very well the importance of applying mathematics and was by no means above using his skills to aid any of the princes who were his sponsors. The following article is an imaginary dialogue between Archimedes and his patron Hieron on the subject of applying mathematics.

ARCHIMEDES Your Majesty! What a surprise at this late hour! To what do I owe the honor of a visit from King Hieron to my modest home?

HIERON My dear friend Archimedes, this evening we had a dinner in my palace to celebrate the great triumph of our small city, Syracuse, over the mighty Romans. I invited you, but your place remained empty. Why didn't you come, you to whom above all we owe today's victory? Your huge, concave brazen mirrors set afire ten of the twenty big ships of the Romans; they sped like fiery torches out of the harbor in the southwest gale; all went down before reaching the open sea. I could not go to sleep without thanking you for delivering our city from the enemy.

ARCHIMEDES They may come back, and we are still surrounded on the mainland.

HIERON We shall speak about that later. First let me hand you a present, the best I can give.

ARCHIMEDES A wonderful masterpiece indeed!

HIERON This tray is of pure gold; you may test it with your method, you will find no trace of silver in it.

ARCHIMEDES The reliefs show the adventures of Odysseus, I assume. In the middle I see the unsuspecting Trojans pulling the giant wooden horse into their city—I always wondered whether the Trojans used some sort of compound pulley to accomplish that. Of course, the horse stood on wheels, but the road to the city must have been rather steep.

Source: Alfréd Rényi, "A Dialogue on the Applications of Mathematics," in *Dialogues on Mathematics* (Holden-Day, Inc., 1967), pp. 28–47.

HIERON My dear Archimedes, by Zeus, forget your pulleys for a moment. You know how astonished I was when by yourself you launched the heavy ship I wanted to send to King Ptolemy, simply by turning the handle of your triple pulley. But have a look at the other scenes on the tray.

ARCHIMEDES I recognize the Cyclops, and Circe as she changes the companions of Odysseus into pigs, and here the concert of Sirens to which Odysseus listens while he is chained to the mast of his ship (if you look at his face you can almost hear the enticing song). And there is Odysseus in the netherworld, meeting the shadow of Achilles, and here he is frightening the charming Nausicaä and her girls, and finally, of course, the scene where Odysseus, disguised as an old beggar, spans his bow and squares his account with the suitors—a marvelous piece of art. I thank you, my gracious king; this is truly a king's gift.

HIERON It was the best piece in my treasure-house, but you deserve it. I chose it not only for its beauty and value, but for a third reason. What you did today for Syracuse can be compared only with the trick of Odysseus. Both are triumphs of a sharp mind over brute force.

ARCHIMEDES You make an old man blush. But let me remind you again that the war is not over yet. Would you like to hear the advice of an old man?

HIERON I even order you, as your king, to tell me your opinion frankly.

ARCHIMEDES This is the moment you should make peace with the Romans; since the war began, we have never been in such a favorable bargaining position. If Marcellus does not send his envoy to you before midnight, you should send yours to him before dawn, and make peace before the sun sets again. Marcellus is eager to withdraw the troops which besiege the city because he needs to use them against Hannibal. Moreover, if he can reach an agreement tomorrow, he can report to Rome a victory, if only a diplomatic one, and not just the sad news that half his fleet is lost. When the report about today's battle reaches Rome, the Romans will be so furious that they will not be satisfied with anything less than total victory.

HIERON Your analysis is correct. As a matter of fact, I received a message from Marcellus this evening in which he offered peace and withdrawal of his troops under certain conditions. If you knew these conditions, you would be less keen on making a deal with the Romans.

ARCHIMEDES What does Marcellus want?

HIERON Well, of course he wants a lot of gold and silver. He also wants ten new ships for the ten we sank today, and further that all our forts be demolished except one in which a garrison of Roman soldiers would be stationed. He wants us to declare war on Carthage, and finally he demands my son Gelon, my daughter Helena, and you as hostages. He promises, however, that no harm will be done to the city and its inhabitants as long as we adhere to the treaty.

ARCHIMEDES Perhaps he will not insist on everything, although he will insist that you hand me over to him.

HIERON You speak coolly about this. By all the gods of Olympus, as long as I am alive I will not place my children in the hands of the enemy, nor will I give you to them. I do not mind the gold and the ships, he can have them. But what I dislike most about his conditions is that if we fulfill them, we shall be completely at his mercy. What guarantees are there that he will keep the treaty? He does not give me any hostages.

ARCHIMEDES Take care not to question whether he will stick to his word; Romans are sensitive about their honor, at least during negotiations. But perhaps you can avoid giving him your children.

HIERON And what about you? Would you be ready to make this sacrifice for your city?

ARCHIMEDES Is that a question or a request?

HIERON A question only, of course. Do you want to know what I replied to Marcellus?

ARCHIMEDES You have answered already?

HIERON Yes. I accepted all his conditions except that of giving you as a hostage; but I agreed to give my children as hostages only under the condition that he send me two of his children as hostages. As for you, I told him that your age does not permit your living in camp. However, knowing that he does not really want you as a hostage but only wants your wisdom, I promised that you would write him a full description of all your inventions which are of military importance.

ARCHIMEDES I will never write anything about my inventions concerning warfare.

HIERON Why not? If there is peace, we will not need them any more. Explain why you refuse to write about your inventions.

ARCHIMEDES If you have the patience to listen to my reasons, I shall do so.

HIERON I am ready to listen. I want to remain awake and wait for Marcellus' answer.

ARCHIMEDES Then we have a lot of time because it will take Marcellus awhile to formulate his answer. It will sound like a whip.

HIERON Do you think he will discontinue the negotiations?

ARCHIMEDES Of course. You contested his honor. He will never forgive that, and there will be no agreement.

HIERON You may be right.

ARCHIMEDES I have always admired your artful diplomacy and your psychological insight into your opponents' hearts. But this time you neglected this art.

HIERON I have to admit it. Perhaps I was too drunk with wine and victory. But what is done is done. Still, I want to hear your reasons.

ARCHIMEDES Though the question becomes an academic one, nevertheless, I shall explain my point of view. You compared my machines with the wooden horse of Troy. Well, your comparison is really very close, but in quite a different sense. Odysseus used the wooden horse to smuggle himself and some Greek soldiers into Troy. I used my machines to smuggle an idea into the public mind of the Greek world, the idea being that mathematics—not only its elements, but also its most subtle parts—can be applied successfully to practical purposes. I must confess that I hesitated quite a lot before doing this because I hate war and murder. But the war was here anyway, and this was the only way to make myself understood. I have tried other ways, but in vain. May I remind you that some years ago when I invented a pump to take the water out of your mines so that the people who work there should not wade to their hips in it, you were not interested. The supervisor of your mines told me he did not care about the legs of slaves getting wet—they were not made of salt; these were his words. And do you remember when I proposed to make a machine to irrigate your fields? I was told that slave work was cheaper. And when I proposed to use the force of steam to drive the mills of King Ptolemy, what was his answer? He said that the mills which served his ancestors would serve him as well. Shall I remind you of other examples? There were at least a dozen others. All my endeavors to show the world what mathematics can do for them in peace were in vain. But as war approached, suddenly you remembered my pulleys, cogwheels and levers. In peacetime everybody regarded my inventions as toys, unworthy of a serious grown-up citizen, still less of a philosopher. Even you, who always supported me and helped me to realize my ideas, did not take them quite seriously; you showed them to your guests to entertain them, but that was all. Then the war came and the Roman ships closed the harbor; I ventured a casual remark, that by throwing stones on them with a catapult we could drive them away; you jumped at the idea. I could not take back what I had said, and had to go ahead. Once I started on this road, I had to continue. But my feelings about it were mixed from the start. I was, of course, happy that my inventions were not ridiculed anymore, and that at last I had a chance to show the world that mathematics in action really was. But this was not the sort of action by which I wanted to prove the practical value of mathematical ideas. I saw men killed by my machines and this made me feel guilty. I made a solemn oath to Athena that I would never tell the secret of my war machines to anybody either by word or in writing. I tried to soothe my conscience by telling myself that the news about Archimedes defeating the Romans by mathematics would reach all corners of the Greek-speaking world, and this would be remembered even when the war was over and the secrets of my war-machines were buried with me.

HIERON It is true, my dear Archimedes; I am getting letters from kings who are my friends asking about your inventions.

ARCHIMEDES And what do you answer them?

HIERON I tell them that these questions cannot be answered as long as the war continues.

ARCHIMEDES I hope you understand now what my reasons are for not publicizing my secrets. I succeeded in keeping them even from those who carry out my plans. Each man knows about some detail only. I am glad you never asked me questions because I would have refused to answer.

HIERON But now I shall ask you some questions. Don't be afraid, I shall not ask for your secrets, only about the underlying general principles.

ARCHIMEDES I think I can answer such questions without breaking my oath.

HIERON Before I begin, I want to ask you something else. Why was it so important to you that your ideas about the usefulness of mathematics be accepted?

ARCHIMEDES Perhaps I was a fool, but I thought that I could change the course of history. I was worried about the future of our Greek world. I thought that if we applied mathematics on a large scale—after all, mathematics is a Greek invention, and I think the best achievement of the Greek spirit—we might save our Greek way of life. Now I realize that it is too late. The Romans will conquer not only Syracuse but all other Greek cities too; our time is over.

HIERON Even were that the case, our Greek culture would not be lost: the Romans would take it over. Look how they try to imitate us already. They copy our statues, translate our literature—and you see that Marcellus is already interested in your mathematics.

ARCHIMEDES The Romans will never really understand it. They are too practical-minded, and they are not interested in abstract ideas.

HIERON They are certainly interested in its practical uses.

ARCHIMEDES But these things cannot be separated. One has to be a dreamer of dreams to apply mathematics with real success.

HIERON That sounds rather paradoxical. I thought that to apply mathematics one should first of all have a good practical sense. This leads me to my first question. What really is the secret of the new science which you invented—let us call it applied mathematics? And what is the main difference between your applied mathematics and that sort of mathematics—let us call it pure mathematics—which is taught in schools?

ARCHIMEDES I am sorry to disappoint you. There exists no other kind of mathematics besides that which your teachers taught you, and not without success, as I recall. Applied mathematics, as an art which is different and separated from mathematics as such, does not exist! My secret is so well hidden because it is no secret at all; its very obviousness is its best disguise. It is hidden like a golden coin thrown into the dust of the street.

HIERON Do you mean that your marvelous machines are based on the sort of mathematics every educated man knows?

ARCHIMEDES You are getting nearer the truth.

HIERON Could you give me an example?

ARCHIMEDES Well, let us take as an example the mirrors which did such a good job today. What I did was simply to remember a well-known property of the parabola: take any point P of a parabola, connect it with the focus, and draw through P a line which is parallel to the axis. These two lines form equal angles with the tangent to the parabola in the point P. You can find this theorem in the books of my distinguished colleagues of Alexandria.

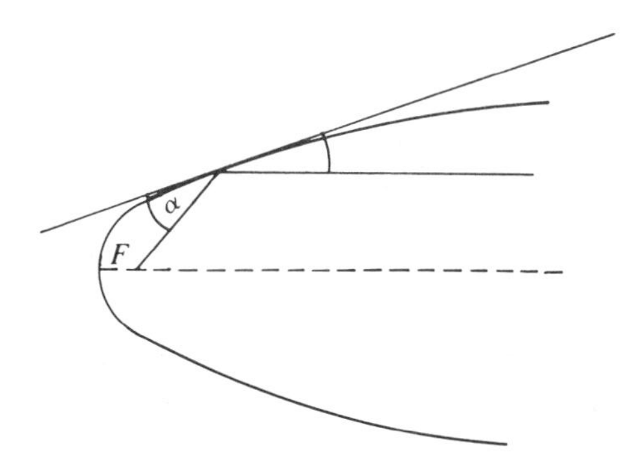

HIERON It is hard to believe you have destroyed half the fleet of Marcellus by this theorem, which is one of hundreds of similar geometrical propositions. I remember it vaguely, though I forget its proof.

ARCHIMEDES Probably when you heard one of its ingenious proofs you understood it, and perhaps even admired its beauty and elegance, but that was all. Some mathematicians went one step further; they explored some of its purely geometrical consequences, or invented new proofs, but there they stopped. I simply went just one step further; I looked at its non-mathematical consequences also.

HIERON I thought you invented some new law of optics.

ARCHIMEDES Optics is, after all, nothing but another branch of geometry. What I used from optics, the law of reflection of a ray, has also been known a long time.

HIERON Do you mean that in order to apply mathematics one does not need to get new mathematical results, only to fit together a practical situation and its mathematical counterpart, some well-known mathematical proposition?

ARCHIMEDES No, it is not quite so simple. It often happens that the theorem which one needs does not exist, and one has to find and prove it oneself. But even if this is not necessary, to find the mathematical counterpart—as you call it (I prefer to call it a mathematical model)—of a practical situation is not the same as matching gloves. First of all, one can construct many mathematical models for the same practical situation, and one has to choose the most appropriate, that which fits the situation as closely as practical aims require (it can never fit completely). At the same time, it must not be too complicated, but still must be mathematically feasible. These are, of course, conflicting requirements and a delicate balancing of the two is usually necessary. You have to approximate closely the real situation in every respect important for your purposes, but lay aside everything which is of no importance for your actual aims. A model need not to be similar to the modeled reality in every respect, only in those which really count. On the other hand, the same mathematical model can be used to fit quite different practical situations. For instance, I also used the properties of parabolas to construct catapults, because the track of a stone thrown by a catapult can be approximated to some extent by a parabola. I also used parabolas to compute how deep a ship will dip into the sea under the weight of its load. Of course, the cross section of a ship has not exactly the same shape as a parabola, but a more realistic model would not have been mathematically manageable. The results were, nevertheless, in fairly good agreement with the facts. Especially I was able to find out under what conditions a ship would be able to stand upright when buffeted by the wind and the waves, because its center of gravity tends to be in the deepest possible position. In trying to describe such a complicated situation, even a very rough model may be useful because it gives at least qualitatively correct results, and these may be of even greater practical importance than quantitative results. My experience has taught me that even a crude mathematical model can help us to understand a practical situation better, because in trying to set up a mathematical model we are forced to think over all logical possibilities, to define all notions unambiguously, and to distinguish between important and secondary factors. Even if a mathematical model leads to results which are not in accordance with the facts, it may be useful because the failure of one model can help us find a better one.

HIERON It seems to me that applied mathematics is similar to warfare: sometimes a defeat is more valuable than a victory because it helps us to realize the inadequacy of our arms or of our strategy.

ARCHIMEDES Now you have really grasped the essential point.

HIERON Tell me something more about your mirrors.

ARCHIMEDES I have told you the basic idea already. After I hit on the idea of using the mentioned property of parabolas, I had to solve the question of how to grind and polish a metal mirror into the form of a concave paraboloid of revolution, but I would prefer not to speak about this. Of course, I also had to select an appropriate alloy.

HIERON Without intruding into your secrets—it is clear that besides the properties of the parabola you also had to know a lot about metals and the art of dealing with them. This shows, it seems to me, that the knowledge of mathematics is not sufficient if somebody wants to apply it. Isn't a man who wants to apply mathematics in a position similar to that of a man who wants to ride two horses at the same time?

ARCHIMEDES I would change your simile slightly: he who wants to apply mathematics is like a man who wants to harness two horses to his carriage. This is not so difficult to do. Some knowledge of the horses as well as of the carriage is, of course, needed, but any of your coachmen has such knowledge.

HIERON Now I am quite confused: every time I think that applied mathematics is mysterious, you show me that it is really quite simple; but when I become convinced that the whole thing is really simple, you point out that it is much more complicated than I imagined.

ARCHIMEDES Its principles are obvious, but the details are sometimes quite involved.

HIERON I do not understand yet what you mean by a mathematical model. Tell me more about this.

ARCHIMEDES Do you remember the sphere I constructed some years ago to imitate the motions of the sun, the moon and the five planets, the one by which it was possible to show how the eclipses of the sun and the moon happen?

HIERON Of course, it is one of the things in the palace that I show to all my visitors; everybody thinks that it is marvelous. Is this a mathematical model of the universe?

ARCHIMEDES No, I would call it a physical model. Mathematical models are invisible, they exist only in our mind, and can be expressed only by formulae. A mathematical model of the universe is that which is common to the real universe and to my physical model. In the physical model, for instance, each planet is a tiny ball about the size of an orange. In my mathematical model of the universe the planets are represented simply by points.

HIERON I think I am beginning to understand what you mean by a mathematical model. But let us return to the simile about horses. The art of harnessing horses to a carriage and driving them is quite different from that of breeding and raising them. Isn't the art of applying mathematics quite distinct from that of finding and proving theorems?

ARCHIMEDES You are, of course, right about horses, though the man who has raised a horse usually knows the most about it and can drive it better than anybody else. As regards mathematics, I pointed out earlier that in order to be able to apply it successfully one has to have a deep understanding of it, and if somebody wants to apply mathematics in an original way, he has to be a creative mathematician. Conversely, a concern with applications can aid in pure mathematical research.

HIERON How is that possible? Could you give an example?

ARCHIMEDES Perhaps you remember that some time ago I was very interested in a question of mechanics, namely in finding the center of gravity of a body. The results which I obtained about centers of gravity helped me to build machines, and also they helped me to prove new geometrical propositions. *I have developed a peculiar method which consists of investigating geometrical problems by means of mechanical considerations concerning centers of gravity. This method is of a heuristic character; this means it does not furnish exact proofs. Many theorems first became clear to me by using this method of reasoning. Of course, as the method does not furnish actual demonstrations, I afterwards proved the theorems I conjectured by means of my mechanical method by the traditional methods of geometry. It is much easier to supply the proof if one has previously acquired some knowledge of the question through mechanical analogies, and thus knows what must be proved.*

HIERON Tell me one theorem which you have found in this strange way.

ARCHIMEDES *The area of any segment of a parabola is four-thirds the area of the triangle which has the same base and height.* After finding the result with my method, I found a proof along traditional lines too.

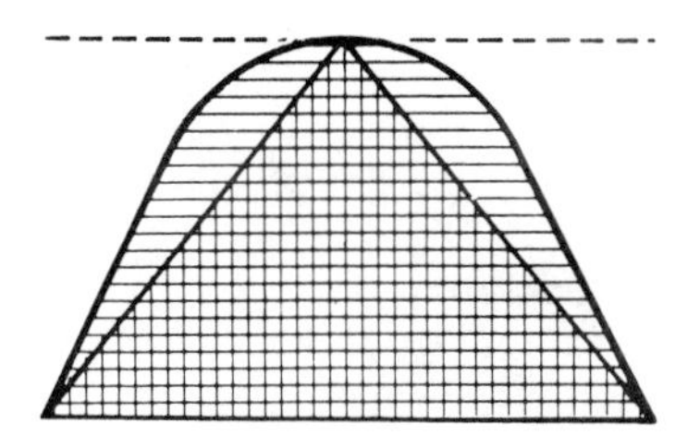

HIERON If you found this theorem by mechanics, why did you bother about the geometrical proof?

ARCHIMEDES When I first discovered my method, the results which I got with its help were not all correct; later, by analyzing the cases in which the method misled me, I developed it so far that now it never misleads me. But still I cannot prove that every result I get this way is really true; maybe somebody will prove this some day, but until that time I do not have complete confidence in the method.

HIERON But are strict proofs in applied mathematics really necessary? After all, as you said, the mathematical model is only an approximation of reality. If you use approximately correct formulae, your results will be still approximately close to reality and they can never be absolutely correct anyway.

ARCHIMEDES You are mistaken, my king. Just because the mathematical model is only an approximation and there is always a certain discrepancy with the facts, one has to take care not to increase this discrepancy further by a careless use of mathematics. One has to be as accurate as possible. By the way, in regard to approximations, there is a common misunderstanding that using approximations means departing from mathematical precision. Approximations have a precise theory, and results about approximations—for instance, inequalities—have to be proved as rigorously as identities. Perhaps you remember the approximations which I gave for the area of a circle with given diameter; I proved them with a rigor usual in geometry.

HIERON What other results did you find by your mechanical method?

ARCHIMEDES This method also led me to discover that *the volume of a sphere is two-thirds the volume of a circumscribed cylinder.*

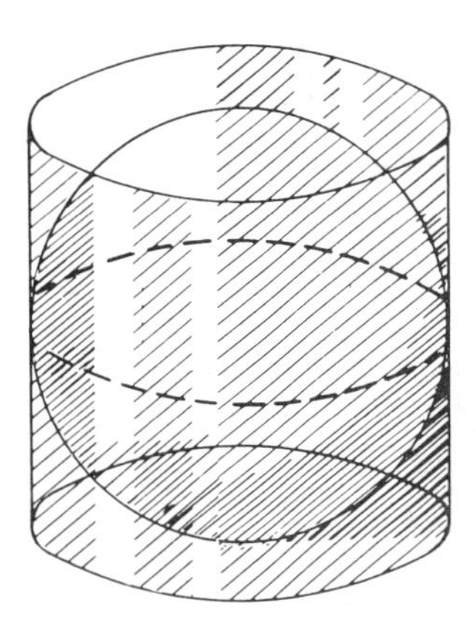

HIERON I heard you say that when you die, you want this theorem to be inscribed on your gravestone. Does it mean that you consider this your most outstanding conclusion?

ARCHIMEDES I think the method itself is more important than any of the particular results I got with its help. Do you remember when I told you once, speaking about levers: *"Give me a place to stand and I can move the earth"?* Of course, there exists no such fixed point in the world. However, in mathematics, one does have such a fixed point to lean on, namely, axioms and logic. To apply mathematics to problems of the real world means to move the earth from the fixed point of mathematics.

HIERON You always speak about applying mathematics, but all the examples you give are applications of geometry. In regard to geometry, I now see how it can be applied. For instance, the functioning of a machine depends on the form and size of its components; the track of a stone thrown by your catapult is a curve, approximately a parabola, as you said. But what about other branches of mathematics, say number theory? I can hardly imagine how it can have any practical importance. Of course, I am not speaking about the elements of arithmetic which are clearly used in every sort of computation; I mean things like divisibility, prime numbers, least common multiple and other similar topics.

ARCHIMEDES Well, if you connect two cogwheels each with a different number of teeth, the least common multiple comes in inevitably. Does this simple example convince you? Recently I got a letter from my friend Eratosthenes of Cyrene in which he wrote about a simple but ingenious method—he calls it the sieve method—which he invented to find prime numbers. Thinking about his method, I made a sketch of a machine which realizes his idea. This machine works with a set of cogwheels: when you turn it a number of times, say n, and look into a hole, if the view is clear, the number n is prime; but if the view is closed, then n is composite.

HIERON That is really amazing. When the war is over, you must build this machine. My guests will love it.

ARCHIMEDES If I am alive, I shall certainly do so. It will show that machines can solve mathematical problems. Perhaps it will help mathematicians to realize that, even from their own point of view, they may gain something from studying the relation of mathematics and machines.

HIERON Speaking about gains, I recall a story about Euclid. *One of his students studying geometry asked Euclid, "What shall I gain by learning these things?" Whereupon Euclid called his slave and said, "Give him a coin, since he wants to gain from what he learns."* It seems to me that this story shows Euclid thought it unnecessary for a mathematician to bother about the practical use of his results.

ARCHIMEDES I have, of course, heard the anecdote, but you will be surprised when I tell you that I sympathize completely with Euclid. In his place I would have said something similar.

HIERON Now I am confused again. Up to now you spoke enthusiastically about the applications of mathematics, and now you agree with the purists who think that the only reward which a scientist should expect is the pleasure of knowledge.

ARCHIMEDES I think you and most people misunderstand the story about Euclid. It does not mean that he was not interested in the practical consequences of mathematical results and that he considered them unworthy of a philosopher. This is pure nonsense; he has written, as you certainly know, a book called *Phaenomena,* about astronomy, and a book on optics, and he is probably the author of the book *Catoptrica* too, which I used in constructing my mirrors; he was also interested in mechanics. As I understand the story, Euclid wanted to emphasize the remarkable fact that mathematics rewards only those who are interested in it not only because of its rewards but also because of itself. Mathematics is like your daughter, Helena, who suspects every time a suitor appears that he is not really in love with her, but is interested in her only because he wants to be the son-in-law of the king. She wants a husband who loves her for her own beauty, her wit and charm, and not for the wealth and power which he can get by marrying her. Similarly, mathematics reveals its secrets only to those who approach it with pure love, for its own beauty. Those who do this are, of course, also rewarded with results of practical importance. But if somebody asks at each step, "What can I profit by this?" he will not get far. You remember I told you that the Romans would never be really successful in applying mathematics. Well, now you see why; they are too practical-minded.

HIERON I think we should learn from the Romans, then we would have better chances in fighting them.

ARCHIMEDES I do not agree. If we try to win by giving up the ideas we stand for and by imitating our opponents, we are lost before the battle begins. Even if we could win the war in this way, it would not be worthwhile; such a victory is worse than a defeat.

HIERON Let us not speak about the war but return to mathematics. Tell me, how do you construct your mathematical models?

ARCHIMEDES It is difficult to explain this in general terms. Perhaps a simile will help. A mathematical model of a real situation is something like its shadow on the screen of the mind.

HIERON It seems to me that your philosophy is the exact opposite of Plato's. He says that real things are the shadows of ideas, while if I correctly understood the meaning of your words, you are saying that ideas are the shadows of reality.

ARCHIMEDES The two points of view are not so far from each other as it seems. Plato was puzzled by the correspondence between mathematical ideas and reality, and he thought that the main task of philosophy was to explain this correspondence. Up to that point, I agree with him completely. I do not agree with his explanation, but at least he saw the problem clearly and tried to work out one of the logically possible answers. But I think we have to leave philosophy and return to the facts of life—I hear somebody knocking at the door. I shall open it.

HIERON Let me do it. It must be my envoy with Marcellus' answer. Here is the message.

ARCHIMEDES What is his answer?

HIERON Read it yourself.

ARCHIMEDES Let me see. "Marcellus sends his greetings to King Hieron and announces that he will conquer Syracuse before the new moon; then King Hieron will realize that a Roman keeps his word."

HIERON Now what do you think about that?

ARCHIMEDES His Greek is really not bad. As for the contents, they are what I expected.

HIERON Truly, your prediction was as correct as if you had found it by your method.

ARCHIMEDES Well, at least we know now what to expect.

HIERON I must go, I need some sleep. Tomorrow we have to prepare ourselves for a new attack. Thank you for this interesting talk.

The Crisis in Contemporary Mathematics

Errett Bishop

Errett Bishop was born and educated in the United States and obtained a Ph.D. in mathematics from the University of Chicago. He has taught at the University of California at Berkeley and at the University of California at San Diego. He has been associated with the Institute for Advanced Study and was awarded a Sloan Fellowship. Errett Bishop died in April 1983.

To the nonmathematician the idea that mathematics is capable of undergoing a crisis seems absurd. The notion that mathematicians could be engaged in long, emotional, and currently unresolved disputes as to the very nature of the subject may seem as nonsensical as would the air conditioning of hell to a seventeenth-century Calvinist. The "fourness" of two plus two is to many a paradigm for all that is fixed, certain, and unassailable in human experience. But such a supposition is false, and that fact is discouraging and yet exciting. Mathematicians now vigorously debate the merits and validity of many assertions that are part of the very foundations of modern mathematics.

Near the end of the nineteenth century the German mathematician Georg Cantor formulated the theory of sets. This powerful methodology both simplified and unified a remarkable number of diverse mathematical ideas. By the beginning of this century set theory was almost universally accepted as the single unifying concept upon which all mathematics could be based. The simplicity, power, and apparent universality of set theory allowed mathematicians to create generalizations and extensions of traditional ideas that were both breathtaking in scope and unbelievably complex.

Much of the new mathematics thus created was so arcane and so abstract, however, that a few mathematicians began to question whether ideas so remote from individual human experience had any real meaning. Aside from this philosophical objection a number of ugly formal paradoxes emerged, sprouting like weeds in the soil of set theory.

A variety of solutions were put forward to meet these objections; one of the most popular was the suggestion that they be ignored. Not everyone doing mathematics at that time, however, felt that this suggestion was philosophically honest, and among the most successful of the counterproposals was that devised by the Dutch mathematician L. E. J. Brouwer, beginning in 1907. Brouwer had been a successful and prolific contributor to the mathematics of set theory and formal deduction. His increasing concern over the validity of his own formal creations led him to narrowly redefine what was acceptable in making deductions from set theory. This reformulation of what is mathematically valid is now called intuitionism. At present relatively few mathematicians would classify themselves as intuitionists; the vast majority carry on as though the objections of Brouwer and his followers had never been made. The majority are more or less followers of the formalist program of the German mathematician David Hilbert. Hilbert recognized the problems of set theory and attempted to rescue it by refining the reasoning process. Hilbert's program preserved virtually all of the mathematics that had been done up to the early twentieth century, which explains its popularity.

Source: Errett Bishop, "The Crisis in Contemporary Mathematics," *Historia Mathematica* 2 (1975): 507–517. Reprinted by permission of Academic Press. (References omitted.)

The author of this essay, Errett Bishop, was an effective and well-known proponent of the minority intuitionist or constructivist position. In his view a majority of modern mathematicians practice their craft in a slipshod and philosophically dangerous fashion. He believed that a number of the most basic assumptions currently used in mathematical reasoning are at best vague to the point of having no meaning or at worst false. What necessarily follows if Bishop is correct is that a large part of modern mathematics is either false, nonsensical, or at the very least needs to be redemonstrated.

THERE IS A crisis in contemporary mathematics, and anybody who has not noticed it is being willfully blind. *The crisis is due to our neglect of philosophical issues*. The courses in the foundations of mathematics as taught in our universities emphasize the mathematical analysis of formal systems, at the expense of philosophical substance. Thus it is that the mathematical profession tends to equate philosophy with the study of formal systems, which require knowledge of technical theorems for comprehension. They do not want to learn yet another branch of mathematics and therefore leave the philosophy to the experts. As a consequence, we prove these theorems and we do not know what they mean. The job of proving theorems is not impeded by inconvenient inquiries into their meaning or purpose. In order to resolve one aspect of this crisis, emphasis will have to be transferred from the mechanics of the assembly line which keeps grinding out the theorems, to an examination of what is being produced.

The product (i.e., the concepts, theorems and techniques) of this assembly line can be evaluated from at least three distinct standpoints: pure, applied (physical sciences), and applied (data processing). Today, I wish to concentrate mainly on pure mathematics, although the crisis certainly extends further.

As pure mathematicians, we must decide whether we are playing a game, or whether our theorems describe an external reality. Assuming that it is no game, we must be as clear as possible about what objects we are describing, and what it is that we are saying about those objects. The basic point here is already made in full force by considering the question "What do we mean by an integer?" It is clear that the integer 3 differs in quality from the integer

$$99^{99^{99}}$$

which in turn differs in quality from the integer that is defined to be 1 if the four color theorem is true and is 0 otherwise. So then, there are at least three possibilities: (1) an integer may mean one that we can actually compute, (2) one that we can compute in principle only, or (3) one that is not computable by known techniques, even in principle.

To my mind, it is a major defect of our profession that we refuse to distinguish, in a systematic way, between integers that are computable in principle and those that are not. We even refuse to do mathematics in such a way so as to permit one to make the distinction. Many mathematicians do not even find the distinction interesting. Of course, the distinction between computable and non-computable, or constructive and non-constructive is the source of the most famous disputes in the philosophy of mathematics, and will continue to be the central issue for many years to come.

Now, the little that I have read in the history of the philosophy of mathematics has left me with an overwhelming impression: that the history of the philosophy of mathematics is very dangerous. I am surprised that this point has only been made in passing at this meeting. I think that it should be a fundamental concern to the historians that what they are doing is potentially dangerous. The superficial danger is that it will be and in fact has been systematically distorted in order to support the status quo. And there is a deeper danger: it is so easy to accept the problems that have historically been regarded as significant as actually *being* significant.

For example, there is a problem of the *truth* of the statement that every bounded monotone increasing sequence of real numbers converges. People sometimes ask me whether I believe that this or some similar statement is untrue. My answer is that it is not possible to answer the question until they tell me what interpretation they wish to attach to the statement. The statement is true when interpreted *classically* and false when interpreted *constructively*. Thus what are historically regarded as problems about truth are actually problems about meaning. I believe that if we agree on the meaning of such statements, then we can settle the question of their truth relatively easily.

There is only one basic criterion to justify the philosophy of mathematics, and that is, does it contribute to making mathematics more meaningful. It is not true that this criterion is commonly accepted. In fact, the philosophical criterion that most mathematicians prefer is that it enables them to prove more theorems and to be more secure about the theorems that they have already proved.

A very brief review of the central historical controversy about the nature of *mathematics* will be sufficient for me to discuss with you what I believe to be the important philosophical issues in the philosophy of mathematics today.

The controversy to which I refer is the grand dispute between Kronecker, Brouwer, H. Weyl and perhaps a few others on the one hand, who gave us techniques for deepening the meaning of mathematics, and Hilbert and others, who to a great extent rejected their discoveries. Hilbert feared that his cherished theorems, his paradise, would be taken away. In fact, the threat was never real. The only real threat was to label Hilbert's theorems for what they were, and to try to make some of them more meaningful; but perhaps Hilbert did not realize this. This controversy raged during the late nineteenth and early twentieth centuries, finally having been resolved, in the opinion of many, in favor of Bourbaki. I shall contend that had the disputants been less dogmatic and more thoughtful, then we might all be more thoughtful mathematicians today, and in fact perhaps be doing mathematics which is in some respects quite different from the mathematics that we are doing.

Perhaps, the most logical place to begin looking at this dispute is with Cantor, because the paradoxes that arose in Cantorian set theory are what I think really provoked the crisis and the re-examination of the foundations of mathematics that took place. They gave the problem an urgency.

People reacted to these contradictions in basically two ways: (1) Cantor himself, Hilbert, Russell and a host of others, seemed to believe that the Cantorian ideas were essentially correct, and that the task of philosophy was to secure them for posterity by analyzing the source of the contradictions and in some way insuring that the same thing would not happen again; (2) Brouwer, Weyl, Borel and possibly Poincaré, took the contradictions as indicating that something was fundamentally wrong with Cantor's ideas and possibly even with pre-Cantorian mathematics. Of course, Kronecker is a special case, because even before this particular set of contradictions, Kronecker had said that something was wrong with the classical theory of the real numbers, as it had been developed during his lifetime.

It seems to me that the disputants in this controversy missed the point. The point is not whether a particular statement is true, but what do we *mean* by the statement. They should have been asking the question "What is a set?" and "What do we mean by the set of all sets?" instead of asking whether or not the set of all sets really existed.

So, the wrong question was asked. It is fascinating to speculate what would have happened if they had asked the right question, that is, what do these things mean, not whether these things are true or false. I am going to reconstruct history, and tell you what might have happened and what I wish had happened, if the disputants had been more concerned with communicating to one another rather than justifying themselves and putting each other down. It is important to remember that Brouwer and Hilbert understood the propositions of Cantorian set theory in different ways. They attached different meanings to Cantorian set theory, so it was necessary that one of them should reject it and the other accept it.

A similar situation undoubtedly held in the dispute between Kronecker and Weierstrass, about the validity of the real number system as it existed in those days. So there was a violation of the general philosophical principle not to discuss questions of truth until one settles questions of meaning. I think that Brouwer made a valiant attempt to say explicitly what meaning he attached to every mathematical theorem. For example, an integer to Brouwer (in my interpretation of Brouwer's philosophy) is either an integer in decimal notation or a method that in principle will lead after a *finite* number of steps to an integer in decimal notation. Again there is this notion of computability: if the integer is not given directly to be sure that it is finitely computable. This is as far as Brouwer could possibly have gone in expressing himself on this subject. What was an integer to Hilbert? As far as I know he never discussed the point.

Perhaps Brouwer should not have denounced the mathematics that Hilbert wished to do as meaningless, even though Hilbert did not go to the pains that Brouwer did in saying what he meant by his mathematics. Perhaps he should have said to Hilbert, "I have told you what I mean by these things to the best of my ability; now you tell me what these things mean to you!" This would have been the first step in my reconstruction: for Brouwer to have taken this approach in his dealings with Hilbert on the philosophical question.

Then it is fascinating to try to anticipate what Hilbert's response would have been if Brouwer had approached him in this way. I can think of three possibilities:

1. He could have said, "I cannot discuss that. The most I can do is to tell you the rules for doing mathematics and the meaning is then to be found in the rules plus whatever additional personal meaning you wish to read into it." If this had been the answer, then the designation of formalist that has been attached to Hilbert is indeed justified.
2. Possibly Hilbert would have responded by a description of the inductive construction of the Cantorian universe, in as much detail and with as much care as it was possible for him to give. I doubt if he would have done so.
3. Possibly Hilbert would have responded to Brouwer as follows: "I understand your explanation and the meaning that you attach to the objects and statements of mathematics, with the exception perhaps of your theory of choice sequences, which however can be omitted without significantly affecting the mathematics that you would be doing. In my opinion, you have a valid and consistent point of view, but there are other points of view, and I do not think that you should reject them as being meaningless. In your system of mathematics, everything ultimately reduces to finite computation within the set of integers. Let us extend your mathematics by allowing infinite computations. For most purposes, one particular kind of infinite computation will suffice: the examination of a sequence of integers to determine whether or not they all vanish. With this extended mathematics, I shall rest content." It would not have been possible for Hilbert to have preserved the Cantorian paradise within this extended mathematics, which allows in addition this one infinite computation consisting of the examination of a sequence of integers. But I suspect that what Hilbert really wanted to preserve was his own mathematics and other mathematics of the same sort. This certainly would have been possible under the system that I proposed that Hilbert propose. If he really wished to preserve the full Cantorian paradise, he would have been compelled to introduce other infinite computations.

Let us go back to Brouwer. Assuming that Hilbert had responded as I told you that he should have done, Brouwer might then have replied as follows: "I appreciate your motivation, but I will not permit you to introduce any new computations into mathematics. The object constructed by an infinite computation is inherently different from an object constructed by a finite computation, as I have already told you many times. Your proposal would make it impossible to distinguish between them in a systematic way. However, there is a course which I believe will satisfy both of us. Let LPO (limited principle of omniscience) denote the statement that *it is possible to make an infinite computation of the type we described, that is, searching a sequence of integers to see whether they all vanish.* Then if we need LPO to prove a theorem, simply develop your mathematics in my system as the implication LPO implies whatever the theorem happens to be. You will be able to do your mathematics in my system without any loss of meaning and without any essential change in the method you have already been using. For me to do my mathematics in your system would entail a significant loss of meaning. Since we could both work in my system and pursue what we want to do and I cannot work in your system, please defer to me and accept my system." Hilbert would then have accepted Brouwer's proposal and mathematics would not be where it is today.

Where would mathematics be today if all this had come about? We would accept the meanings of "or," "there exists" and all the other connectives and quantifiers, as defined by Brouwer, not as defined classically. In particular, negation, disjunction, and existence would have their meanings changed. We would improve on Brouwer's definition of set in a way that I do not want to go into here. Classical mathematics would go on entirely as before except that every theorem would be written as an implication, either LPO → A or some extended version of an infinite computation implying A. So Hilbert's Cantorian paradise would remain intact within Brouwer's system. Those mathematicians who still believe in the Cantorian paradise as representing ultimate truth, would not be forced to taste forbidden fruit. On the other hand, when they saw that other mathematicians were tasting the fruit and thriving on the diet, they might decide that there was no reason to hold out. Of course, new vistas would be opened up, and it might transpire that Hilbert's paradise was not so perfect after all.

This is all very abstract. I want to take a concrete instance and illustrate what we might be doing. Unfortunately, I am about the only one who is doing it now and so I must apologize for choosing a concrete instance from my own mathematics. This theorem is not crucial, but it is an efficient demonstration of the sort of thing I am talking about, namely, the classical theorem that a function of bounded variation defined on the unit interval has a derivative almost everywhere. Now you might say that it does have a derivative almost everywhere and I would not disagree with you. But, if I wanted to talk to you in your own language, I would say compute the derivative. I suspect that many of you would answer, "I do not know how to compute the derivative." Some of you might say "I do not care" and others might really care and simply not know. Or, you might reply: "If you care whether you can compute the derivative, go ahead and consider the question, but do not try to change the whole system of mathematics just because you want to consider questions of whether you can compute things."

My point is that you *cannot* consider questions of whether you can compute things systematically and do a good job of it, unless you *do* change the whole system of mathematics. Now this does not seem to be true in things like number theory. One can do *ad hoc* constructivism in number theory and I do not think that it has posed any problems to do it that way. One simply cannot do *ad hoc* constructivism in analysis and develop good general theorems which correspond to the theorems of classical mathematics. In fact, there have been fewer analysts interested in constructive questions than there have been number theorists, and I suggest that it is because the classical system has tied their hands.

So let us see what we are going to do with this theorem, call it "A," that a function of bounded variation has a derivative almost everywhere. The classical "proof" actually proves A′, which is the theorem LPO → A.

The harder problem is to prove A without using LPO. You simply cannot. Brouwer could have easily shown that there is no hope of actually computing the derivative of a function of bounded variation, essentially because the derivative does not necessarily get approximated when the function gets approximated.

Since there is no hope of proving Theorem A, you might think that the constructivist mathematician should then rest content. He knows that he cannot get Theorem A in his system, and the classical mathematicians have already given him LPO → A. However, constructivism is not that trivial. This is what I mean by saying that accepting Brouwer's system would deepen the meaning of mathematics. Even though we cannot prove A, we still think that the implication LPO → A is ugly. So what can we do if we do not like the implication and we cannot prove the theorem? We can get an implication which is natural and reflects the nature of the problem. LPO is a general hypothesis *not* related at all to the structure of this particular theorem in any special way. Let us replace the left hand side of this implication by some statement which is naturally and, after we give it to you, obviously involved with the conclusion on the right hand side. Let f be the function of bounded variation whose derivative we wish to compute. Let $B(f)$ be the statement that we can *compute the total variation of* f. (I put it in this way for the benefit of the non-constructivists in the audience.) The theorem that I want to state is that *if we can compute the total variation then we can compute the derivative:* $B(f) \rightarrow A(f)$. This is a much stronger, much more natural and much more useful result than LPO → A. It is about the best that we can hope to do if you think about it; you cannot hope to get anything better than that.

[At this point Garrett Birkhoff gave another variant, the Jordan decomposition of a function of bounded variation. You can break such a function down into decreasing and increasing functions, so that your theorem would say that you can constructively prove that an increasing function is everywhere differentiable, because you then know the variation. Bishop agreed, pointing out that just because you can compute the derivative almost everywhere, does not guarantee that you can decompose it.]

In addition, we have a very nice corollary, generalizing it in an essential way, not trivially. We get the fact that an indefinite integral of an integrable function has a derivative equal to that function almost everywhere. This is because you can compute the total variation of the indefinite integral which is, of course, equal to the integral of the absolute value of the function.

This is the kind of mathematics that we might be doing. We might be taking many classical theorems and doing exactly this sort of thing to them if history had taken the course that I have discussed.

Actually, the development of this particular example should not stop here, because whenever you have a theorem: B → A, then you suspect that you have a theorem: B is approximately true → A is approximately true. So there should be an even further development of this theory, namely to say what we mean for B to be approximately true, and then we have conjectured an implication, which we should try to prove. I have not done this, but it occurred to me while preparing this talk that the conjecture is clear enough. I shall not take the time to present it here.

In a way, the imaginary dialogue that I presented here might be regarded as a historical investigation if you believe as I do that it shows how two titanic figures such as these might have reached an accommodation that would have changed the course of mathematics in a profound way, had they spoken to each other with less emotion and more concern for understanding each other.

Instead, Hilbert tried to show that it was all right to neglect computational meaning, because it could ultimately be recovered by an elaborate formal analysis of the techniques of proof. This artificial program failed.

A more recent attempt at mathematics by formal finesse is non-standard analysis. I gather that it has met with some degree of success, whether at the expense of giving significantly less meaningful proofs I do not know. My interest in non-standard analysis is that attempts are being made to introduce it into calculus courses. It is difficult to believe that debasement of meaning could be carried so far.

Many mathematicians regard the theory of computation as a branch of recursive function theory. It is true that many constructivists, for instance the school of Markov in Russia, are recursivists. Brouwer, of course was not. The recursive constructivists seem to be motivated by the desire to avoid such vague terms as "rule" and "set." Their mathematics is forbiddingly involved and laborious, a great price to pay for the precision they hope to attain. My personal opinion is that they have not attained any additional precision. Perhaps any attempt to make the notion of "rule" more precise is futile. It is clear that the concept of a set, in its full generality, can be avoided to a very great extent, again however, at the price of awkward complications. More research is needed on this point.

In my opinion, the positive contributions of recursive function theory to both constructive mathematics and the more concrete aspects of the theory of computation are the construction of counterexamples, but here again impressions are somewhat misleading. The methods of Brouwer, now largely neglected, are more suitable for providing counterexamples in most cases of interest than are the methods of recursive function theory.

That is all I want to say about pure mathematics. I would like to consider next another very interesting question that has occupied many people: what does the constructivist point of view entail for the applications of mathematics to physics? My own feeling is that the only reason mathematics is applicable is because of its inherent constructive content. By making that constructive content explicit, you can only make mathematics more applicable. Hermann Weyl seems to have had an opposite opinion. For him, the utility of mathematics extended even to that part of mathematics that was not inherently computational. I hesitate to disagree with Weyl, but I do. It is a very serious subject for investigation; it would be interesting and worthwhile to settle this point.

I have one final concern to express today. Perhaps the most critical problem in applied mathematics is what to do about the over-mathematization of our society. The scientists who developed the atom bomb would like to feel that they were not responsible for its use. Those of us who teach calculus etc. would like to feel that we are not responsible for the inappropriate uses to which our instruction is put and I am not talking about the construction of bombs. In these days, mathematics is being applied to psychology, to economics, etc. in a very thoughtless way. We need a philosophy, if that is the right word, of when mathematics is applicable and when it is not. In the meantime, I tell my students that I doubt the validity of many of the applications of mathematics to the non-physical sciences that are presented in the text books, and that more important than being able to do mathematics is to be sure the applications are meaningful.

I want to discuss today one fundamental reason why mathematics is so often applied so thoughtlessly: the arrogance of mathematicians. I have experienced this arrogance ever since I began work in the philosophy of mathematics and I am sure that you historians have experienced it too. People tell me in so many words that when I was proving theorems, I was doing something original and worthwhile; but when I started to think about philosophical questions, I could not possibly be doing anything deep. This prejudice, that all good work must be technical in the mathematical sense, has made economists, sociologists, etc. feel inferior, as if they should mathematicize, very often to the detriment of the real *meaning* of their work.

Discussion

Aspray and Moore asked Bishop to comment on the work of Fitting, Troelstra and Kreisel, who have also worked on Brouwer's ideas. The following discussion ensued.

BISHOP Intuitionism was transmuted by Heyting from something which was anti-formal to something which is formal. When one speaks today of intuitionism, one is talking of all sorts of formal systems (studied by the logicians). That's not what Brouwer had in mind.

MOORE So you see yourself more the follower of Brouwer than Heyting or Kreisel are? (Bishop concurred.)

KLINE You did not indicate where one should stand on LPO (the limited principle of omniscience) described in your paper. Should it be accepted? Should one opt for a more limited assumption? Or should one not accept it and follow the intuitionists? Or is this a personal question? After all, one can prove more with LPO than without it.

BISHOP It is personal, because it is not going to affect our mathematics. Write the theorems that need LPO in their proofs as implications, and be careful not to use LPO for results that can be achieved without the use of it.

MACKEY Would you please justify your use of the word "crisis"? What terrible things are going to happen if we ignore what you're telling us? To put the question differently, let us compare this with the relationship of mathematics to physics. Consider the foundational question in physics: what is the real mathematics that the physicists are doing? The physicists don't care; they go ahead and say that they can get the kind of results they want—and do. But we don't tell the physicists that they are having a big crisis. How do you compare these situations?

BISHOP Meaning in physics is different from meaning in mathematics. I am not a physicist; but physicists have told me that the sort of meaning that is appropriate to physics is *not* to ask whether the mathematics in question is rigorous. Rather, it involves the relations of the results to the real world.

MACKEY Brouwer had a point, but my reaction is that I don't want to think about these questions. I have faith that what I am doing will have some kind of meaning—no matter what the status of these questions is.

BISHOP You can keep your attitude; but why can't you give me the kind of cooperation that Brouwer was willing to give Hilbert in my imaginary dialogue? Such cooperation will not harm your attitude. Mathematicians have cut themselves off from a large portion of mathematics which many, including myself, have thought to be meaningful because of their refusal to adopt a system that would cost them nothing.

BIRKHOFF I think I have an answer to both Kline's and Mackey's queries. If mathematicians would admit that they don't know the answer to these fundamental questions, e.g. whether LPO or the Axiom of Choice are true under all circumstances, and would keep an open mind about them, the situation would be better. I think this is what Bishop is urging. We should keep track of our assumptions, and keep an open mind.

FREUDENTHAL Bishop's thesis, that there is a crisis in mathematics, is not new. There has always been a crisis in mathematics. The present is not any different from other times in mathematics. For example, before Cauchy and Gauss complex numbers were considered a crisis in mathematics.

DIEUDONNÉ There is no crisis in mathematics. Mathematics has never been as prosperous as it has been in the last ten years. Never before had we proved so many new and powerful theorems. I just want to work in the way Gauss, Riemann, and Poincaré worked; I want nothing else.

ABHYANKAR My paper is in complete sympathy with Bishop's position.

KAHANE I agree partly with Bishop, partly with Dieudonné. I have to respect Bishop's work; but I find it boring. Perhaps it is boring to me because the constructivists do not have a unified consistent language.

BISHOP Most mathematicians feel that mathematics has meaning, but it bores them to try to find out what it is. You are typical of most mathematicians.

KAHANE I feel that Bishop's appreciation has more significance than my lack of appreciation.

DREBEN It has often been said that the main reason for the development of mathematical logic has been the paradoxes of Cantorian set theory. That is historically false. Frege, the greatest logician since Aristotle and the creator of the foundations of mathematical logic, that is, quantification theory and a totally formalized language for mathematics, had nothing to do with Cantorian paradoxes. The main reason for the development of pure mathematical logic, first by Frege and then by Russell (and Whitehead), was philosophical. Both Frege and Russell were motivated in their early work primarily by a desire to refute Kant. What is historically and philosophically interesting is that each of them took essentially the same technical path in order to refute Kant's thesis about the nature of pure arithmetic and its relation to logic; yet they came up with different conclusions. This might be taken as evidence for Wittgenstein's position that no technical result will ever really resolve any technical philosophical problem. Kant held that logic is analytic but arithmetic is synthetic *a priori*. Frege thought that in his *Grundlagen der Arithmetik*, and later in his *Grundgesätze der Arithmetik*, he had shown (by "reducing" it to logic) that pure arithmetic is analytic *a priori*, contrariwise. Russell in his classical period up to 1912 believed that the "logical reduction" had shown Kant to be right about arithmetic but wrong about logic; that is, since arithmetic was "derivable" from logic, logic had to be synthetic *a priori*. Of course, both Frege and Russell held that Kant had too narrow a conception of logic and was wrong in thinking that arithmetic and hence mathematics rested on extralogical modes of reasoning. The *epistemic* nature of logic and pure mathematics were what seemed important to them.

Mathematical Problems

David Hilbert

No individual has had a greater impact on the mathematics of this century than David Hilbert. Born in Konigsberg of East Prussia (now Kaliningrad of the Soviet Union) in 1862 to a middle-class German family, Hilbert had a rather uneventful childhood. He enrolled in the local university in 1882 and completed his studies in 1885 for which he was awarded the doctorate in mathematics. In 1895 he was appointed professor at Göttingen, a position he held until his death in 1943. At Göttingen he was instrumental in establishing a mathematical institute that was in its time the most important in the world.

Hilbert's impact on modern mathematics was far greater than just the total of his own personal research. He was instrumental in the establishment of an agenda for what is significant in mathematics. The method Hilbert used for establishing an agenda was to propose lists of problems whose solutions would in his view make significant progress in mathematics. The priorities set by Hilbert in his problem lists still exert substantial influence on how mathematicians judge the importance of each other's work. Hilbert was also concerned about the foundations of mathematics and formulated in 1915 what has become known as the formalist position. To the extent that they can be characterized at all, most current mathematicians could be described as formalists and do mathematics exactly as Hilbert thought it should be done.

The most famous of Hilbert's lists is the list of twenty-three problems presented at the International Congress of Mathematics in Paris (see the Appendix) in the year 1900. This essay is the text of Hilbert's address to that Congress, omitting the list of those twenty-three notable puzzles.

WHO OF US would not be glad to lift the veil behind which the future lies hidden; to cast a glance at the next advances of our science and at the secrets of its development during future centuries? What particular goals will there be toward which the leading mathematical spirits of coming generations will strive? What new methods and new facts in the wide and rich field of mathematical thought will the new centuries disclose?

History teaches the continuity of the development of science. We know that every age has its own problems, which the following age either solves or casts aside as profitless and replaces by new ones. If we would obtain an idea of the probable development of mathematical knowledge in the immediate future, we must let the unsettled questions pass before our minds and look over the problems which the science of today sets and whose solution we expect from the future. To such a review of problems the present day, lying

Source: Excerpt from David Hilbert, "Mathematical Problems," *Bulletin of the American Mathematical Society* 8 (1901–02): 437–479. Reprinted by permission of the American Mathematical Society. (References omitted.) Translated, with the author's permission, by Dr. Mary Winston Newson. The original appeared in the *Göttingen Nachrichten,* 1900, pp. 253–297, and in the *Archiv der Mathematik und Physik,* 3d ser., vol. 1 (1901), pp. 44–63 and 213–237.

at the meeting of the centuries, seems to me well adapted. For the close of a great epoch not only invites us to look back into the past but also directs our thoughts to the unknown future.

The deep significance of certain problems for the advance of mathematical science in general and the important role which they play in the work of the individual investigator are not to be denied. As long as a branch of science offers an abundance of problems, so long is it alive; a lack of problems foreshadows extinction or the cessation of independent development. Just as every human undertaking pursues certain objects, so also mathematical research requires its problems. It is by the solution of problems that the investigator tests the temper of his steel; he finds new methods and new outlooks, and gains a wider and freer horizon.

It is difficult and often impossible to judge the value of a problem correctly in advance; for the final award depends upon the gain which science obtains from the problem. Nevertheless we can ask whether there are general criteria which mark a good mathematical problem. An old French mathematician said: "A mathematical theory is not to be considered complete until you have made it so clear that you can explain it to the first man whom you meet on the street." This clearness and ease of comprehension, here insisted on for a mathematical theory, I should still more demand for a mathematical problem if it is to be perfect; for what is clear and easily comprehended attracts, the complicated repels us.

Moreover a mathematical problem should be difficult in order to entice us, yet not completely inaccessible, lest it mock at our efforts. It should be to us a guide post on the mazy paths to hidden truths, and ultimately a reminder of our pleasure in the successful solution.

The mathematicians of past centuries were accustomed to devote themselves to the solution of difficult particular problems with passionate zeal. They knew the value of difficult problems. I remind you only of the "problem of the line of quickest descent," proposed by John Bernoulli. Experience teaches, explains Bernoulli in the public announcement of this problem, that lofty minds are led to strive for the advance of science by nothing more than by laying before them difficult and at the same time useful problems, and he therefore hopes to earn the thanks of the mathematical world by following the example of men like Mersenne, Pascal, Fermat, Viviani and others and laying before the distinguished analysts of his time a problem by which, as a touchstone, they may test the value of their methods and measure their strength. The calculus of variations owes its origin to this problem of Bernoulli and to similar problems.

Fermat had asserted, as is well known, that the diophantine equation

$$x^n + y^n = z^n$$

(x, y and z integers) is unsolvable—except in certain self-evident cases. The attempt to prove this impossibility offers a striking example of the inspiring effect which such a very special and apparently unimportant problem may have upon science. For Kummer, incited by Fermat's problem, was led to the introduction of ideal numbers and to the discovery of the law of the unique decomposition of the numbers of a circular field into ideal prime factors—a law which today, in its generalization to any algebraic field by Dedekind and Kronecker, stands at the center of the modern theory of numbers and whose significance extends far beyond the boundaries of number theory into the realm of algebra and the theory of functions.

To speak of a very different region of research, I remind you of the problem of three bodies. The fruitful methods and the far-reaching principles which Poincaré has brought into celestial mechanics and which are today recognized and applied in practical astronomy are due to the circumstance that he undertook to treat anew that difficult problem and to approach nearer a solution.

The two last mentioned problems—that of Fermat and the problem of the three bodies—seem to us almost like opposite poles—the former a free invention of pure reason, belonging to the region of abstract number theory, the latter forced upon us by astronomy and necessary to an understanding of the simplest fundamental phenomena of nature.

But it often happens also that the same special problem finds application in the most unlike branches of mathematical knowledge. So, for example, the problem of the shortest line plays a chief and historically important part in the foundations of geometry, in the theory of curved lines and surfaces, in mechanics and in the calculus of variations. And how convincingly has F. Klein, in his work on the icosahedron, pictured the significance which attaches to the problem of the regular polyhedra in elementary geometry, in group theory, in the theory of equations and in that of linear differential equations.

In order to throw light on the importance of certain problems, I may also refer to Weierstrass, who spoke of it as his happy fortune that he found at the outset of his scientific career a problem so important as Jacobi's problem of inversion on which to work.

Having now recalled to mind the general importance of problems in mathematics, let us turn to the question from what sources this science derives its problems. Surely the first and oldest problems in every branch of mathematics spring from experience and are suggested by the world of external phenomena. Even the rules of calculation with integers must have been discovered in this fashion in a lower stage of human civilization, just as the child of today learns the application of these laws by empirical methods. The same is true of the first problems of geometry, the problems bequeathed us by antiquity, such as the duplication of the cube, the squaring of the circle; also the oldest problems in the theory of the solution of numerical equations, in the theory of curves and the differential and integral calculus, in the calculus of variations, the theory of Fourier series and the theory of potential—to say nothing of the further abundance of problems properly belonging to mechanics, astronomy and physics.

But in the further development of a branch of mathematics, the human mind, encouraged by the success of its solutions, becomes conscious of its independence. It evolves from itself alone, often without appreciable influence from without, by means of logical combination, generalization, specialization, by separating and collecting ideas in fortunate ways, new and fruitful problems, and appears then itself as the real questioner. Thus arose the problem of prime numbers and the other problems of number theory, Galois's theory of equations, the theory of algebraic invariants, the theory of abelian and automorphic functions; indeed almost all the nicer questions of modern arithmetic and function theory arise in this way.

In the meantime, while the creative power of pure reason is at work, the outer world again comes into play, forces upon us new questions from actual experience, opens up new branches of mathematics, and while we seek to conquer these new fields of knowledge for the realm of pure thought, we often find the answers to old unsolved problems and thus at the same time advance most successfully the old theories. And it seems to me that the numerous and surprising analogies and that apparently prearranged harmony which the mathematician so often perceives in the questions, methods and ideas of the various branches of his science, have their origin in this ever-recurring interplay between thought and experience.

It remains to discuss briefly what general requirements may be justly laid down for the solution of a mathematical problem. I should say first of all, this: that it shall be possible to establish the correctness of the solution by means of a finite number of steps based upon a finite number of hypotheses which are implied in the statement of the problem and which must always be exactly formulated. This requirement of logical deduction by means of a finite number of processes is simply the requirement of rigor in reasoning. Indeed the requirement of rigor, which has become proverbial in mathematics, corresponds to a universal philosophical necessity of our understanding; and, on the other hand, only by satisfying this requirement do the thought content and the suggestiveness of the problem attain their full effect. A new problem, especially when it comes from the world of outer experience, is like a young twig, which thrives and bears fruit only when it is grafted carefully and in accordance with strict horticultural rules upon the old stem, the established achievements of our mathematical science.

Besides it is an error to believe that rigor in the proof is the enemy of simplicity. On the contrary we find it confirmed by numerous examples that the rigorous method is at the same time the simpler and the more easily comprehended. The very effort for rigor forces us to find out simpler methods of proof. It also frequently leads the way to methods which are more capable of development than the old methods of less rigor. Thus the theory of algebraic curves experienced a considerable simplification and attained greater unity by means of the more rigorous function-theoretical methods and the consistent introduction of transcendental devices. Further, the proof that the power series permits the application of the four elementary arithmetical operations as well as the term by term differentiation and integration, and the recognition of the utility of the power series depending upon this proof contributed materially to the simplification of all analysis, particularly of the theory of elimination and the theory of differential equations, and also of the existence proofs demanded in those theories. But the most striking example for my statement is the calculus of variations. The treatment of the first and second variations of definite integrals required in part extremely complicated calculations, and the processes applied by the old mathematicians had not the needful rigor. Weierstrass showed us the way to a new and sure foundation of the calculus of variations. By the examples of the simple and double integral I will show briefly, at the close of my lecture, how this way leads at once to a surprising simplification of the calculus of variations. For in the demonstration of the necessary and sufficient criteria for the occurrence of a maximum and minimum, the calculation of the second variation and in part, indeed, the wearisome reasoning connected with the first variation may be completely dispensed with—to say nothing of the advance which is involved in the removal of the restriction to variations for which the differential coefficients of the function vary but slightly.

While insisting on rigor in the proof as a requirement for a perfect solution of a problem, I should like, on the other hand, to oppose the opinion that only the concepts of analysis, or even those of arithmetic alone, are susceptible of a fully rigorous treatment. This opinion, occasionally advocated by eminent men, I consider entirely erroneous. Such a one-sided interpretation of the requirement of rigor would soon lead to the ignoring of all concepts arising from geometry, mechanics and physics, to a stoppage of the flow of new material from the outside world, and finally, indeed, as a last consequence, to the rejection of the ideas of the continuum and of the irrational number. But what an important nerve, vital to mathematical science, would be cut by the extirpation of geometry and mathematical physics! On the contrary I think that wherever, from the side of the theory of knowledge or in geometry, or from the theories of natural or physical science, mathematical ideas come up, the problem arises for mathematical science to investigate the principles underlying these ideas and so to establish them upon a simple and complete system of axioms, that the exactness of the new ideas and their applicability to deduction shall be in no respect inferior to those of the old arithmetical concepts.

To new concepts correspond, necessarily, new signs. These we choose in such a way that they remind us of the phenomena which were the occasion for the formation of the new concepts. So the geometrical figures are signs or mnemonic symbols of space intuition and are used as such by all mathematicians. Who does not always use along with the double inequality $a>b>c$ the picture of three points following one another on a straight line as the geometrical picture of the idea "between"? Who does not make use of drawings of segments and rectangles enclosed in one another, when it is required to prove with perfect rigor a difficult theorem on the continuity of functions or the existence of points of condensation? Who could dispense with the figure of the triangle, the circle with its center, or with the cross of three perpendicular axes? Or who would give up the representation of the vector field, or the picture of a family of curves or surfaces with its envelope which plays so important a part in differential geometry, in the theory of differential equations, in the foundation of the calculus of variations and in other purely mathematical sciences?

The arithmetical symbols are written diagrams and the geometrical figures are graphic formulas; and no mathematician could spare these graphic formulas, any more than in calculation the insertion and removal of parentheses or the use of other analytical signs.

The use of geometrical signs as a means of strict proof presupposes the exact knowledge and complete mastery of the axioms which underlie those figures; and in order that these geometrical figures may be incorporated in the general treasure of mathematical signs, there is necessary a rigorous axiomatic investigation of their conceptual content. Just as in adding two numbers, one must place the digits under each other in the right order, so that only the rules of calculation, *i.e.*, the axioms of arithmetic, determine the correct use of the digits, so the use of geometrical signs is determined by the axioms of geometrical concepts and their combinations.

The agreement between geometrical and arithmetical thought is shown also in that we do not habitually follow the chain of reasoning back to the axioms in arithmetical, any more than in geometrical discussions. On the contrary we apply, especially in first attacking a problem, a rapid, unconscious, not absolutely sure combination, trusting to a certain arithmetical feeling for the behavior of the arithmetical symbols, which we could dispense with as little in arithmetic as with the geometrical imagination in geometry. As an example of an arithmetical theory operating rigorously with geometrical ideas and signs, I may mention Minkowski's work, *Die Geometrie der Zahlen*.

Some remarks upon the difficulties which mathematical problems may offer, and the means of surmounting them, may be in place here.

If we do not succeed in solving a mathematical problem, the reason frequently consists in our failure to recognize the more general standpoint from which the problem before us appears only as a single link in a chain of related problems. After finding this standpoint, not only is this problem frequently more accessible to our investigation, but at the same time we come into possession of a method which is applicable also to related problems. The introduction of complex paths of integration by Cauchy and of the notion of the ideals in number theory by Kummer may serve as examples. This way for finding general methods is certainly the most practicable and the most certain; for he who seeks for methods without having a definite problem in mind seeks for the most part in vain.

In dealing with mathematical problems, specialization plays, as I believe, a still more important part than generalization. Perhaps in most cases where we seek in vain the answer to a question, the cause of the failure lies in the fact that problems simpler and easier than the one in hand have been either not at all or incompletely solved. All depends, then, on finding out these easier problems, and on solving them by means of devices as perfect as possible and of concepts capable of generalization. This rule is one of the most important levers for overcoming mathematical difficulties and it seems to me that it is used almost always, though perhaps unconsciously.

Occasionally it happens that we seek the solution under insufficient hypotheses or in an incorrect sense, and for this reason do not succeed. The problem then arises: to show the impossibility of the solution under the given hypotheses, or in the sense contemplated. Such proofs of impossibility were effected by the ancients, for instance when they showed that the ratio of the hypotenuse to the side of an isosceles right triangle is irrational. In later mathematics, the question as to the impossibility of certain solutions plays a preeminent part, and we perceive in this way that old and difficult problems, such as the proof of the axiom of parallels, the squaring of the circle, or the solution of equations of the fifth degree by radicals have finally found fully satisfactory and rigorous solutions, although in another sense than that originally intended. It is probably this important fact along with other philosophical reasons that gives rise to the conviction (which every mathematician shares, but which no one has as yet supported by a proof) that every definite mathematical problem must necessarily be susceptible of an exact settlement, either in the form of an actual answer to the question asked, or by the proof of the impossibility of its solution and therewith the necessary failure of all attempts. Take any definite unsolved problem, such as the question as to the irrationality of the Euler-Mascheroni constant C, or the existence of an infinite number of prime numbers of the form $2^n + 1$. However unapproachable these problems may seem to us and however helpless we stand before them, we have, nevertheless, the firm conviction that their solution must follow by a finite number of purely logical processes.

Is this axiom of the solvability of every problem a peculiarity characteristic of mathematical thought alone, or is it possibly a general law inherent in the nature of the mind, that all questions which it asks must be answerable? For in other sciences also one meets old problems which have been settled in a manner most satisfactory and most useful to science by the proof of their impossibility. I instance the problem of perpetual motion. After seeking in vain for the construction of a perpetual motion machine, the relations were investigated which must subsist between the forces of nature if such a machine is to be impossible; and this inverted question led to the discovery of the law of the conservation of energy, which, again, explained the impossibility of perpetual motion in the sense originally intended.

This conviction of the solvability of every mathematical problem is a powerful incentive to the worker. We hear within us the perpetual call: There is the problem. Seek its solution. You can find it by pure reason, for in mathematics there is no *ignorabimus*.

The supply of problems in mathematics is inexhaustible, and as soon as one problem is solved numerous others come forth in its place. Permit me tentatively as it were, to mention particular definite problems, drawn from various branches of mathematics, from the discussion of which an advancement of science may be expected.

Let us look at the principles of analysis and geometry. The most suggestive and notable achievements of the last century in this field are, as it seems to me, the arithmetical formulation of the concept of the continuum in the works of Cauchy, Bolzano and Cantor, and the discovery of non-Euclidean geometry by Gauss, Bolyai, and Lobachevsky. . . .

How Mathematicians Develop a Branch of Pure Mathematics

Harriet F. Montague and Mabel D. Montgomery

Harriet Frances Montague received her Ph.D. in mathematics in 1935 from Cornell University. She has been with SUNY Buffalo since 1927 and is currently professor emeritus. She was the acting chairman of the mathematics department from 1961 to 1964 and directed summer NSF Institute programs from 1957 to 1970.

Mabel D. Montgomery received her Ph.D in mathematics in 1953 from the University of Buffalo. She taught in the public schools and then joined the faculty of SUNY Buffalo where she has been a full professor since 1962.

If a picture is indeed worth 10^3 words, then for a mathematician an example is worth 10^n words (n a large integer). In previous essays much has been written on how mathematical ideas are developed and on the philosophy of such development. The growth of mathematics through history, the trauma of creation for one individual, the sociology of the subject—all of these have been treated. But the single best way to gain some understanding of the creative process in mathematics is to follow step by step the development of a single idea. And that is exactly what this essay attempts to do. Montague and Montgomery present some simple but valid mathematical systems. They explore the nature of these systems by proving theorems, and as the theorems evolve the nature of the system and its potential for expression become clearer. Are these systems of any value? Will they help us to better deal with and understand the universe? Almost surely not. But they might further the reader's understanding of how real mathematics is done and help to convince the skeptic that mathematics is indeed a creative art.

Introduction to the Axiomatic Method

In this [article] we study the *axiomatic method.* Those who have studied geometry have seen this method used by the authors of their textbooks. The first planned use of this method in mathematics dates back to the time of Euclid (300 B.C.). Since the emphasis in mathematics at that time was on geometry, it is not surprising that the emphasis on the axiomatic method in the schools has traditionally been in the field of geometry. Today's students are being introduced to the axiomatic method in algebra as well as in geometry.

The axiomatic method is the framework within which all the "If—then" statements of mathematics are formed. The process of arriving at such a

Source: Excerpts from Harriet F. Montague and Mabel D. Montgomery, "How Mathematicians Develop a Branch of Pure Mathematics," in *The Significance of Mathematics,* 1963, pp. 119–126, 133–135. Reprinted with permission of Charles E. Merrill Books.

statement involves not only the hypothesis and the conclusion of the "If—then" statement itself; but also the laws of logic and the use of previously proved theorems, definitions, and, ultimately, certain basic statements called variously *axioms, postulates, assumptions*. To have a starting place, it is necessary to have such assumptions (axioms, postulates) and to accept them without attempts at proof. If we insisted on a proof for every statement we made, we would be involved in an unending regression. This framework developed from a set of basic assumptions is not peculiar to mathematics. Our everyday conduct is based on certain assumptions. Each of us behaves in a certain way because each of us abides by convictions or self-determined rules. A student does not cheat if he has a conviction against dishonesty. A person pays his bills because he has agreed to certain laws which compel him to pay or to suffer punishment, and he does not wish to be punished. A community makes certain laws which form the basis for the conduct of community affairs.

Consider the statement "If I do not pay my federal income tax, then I am liable to fine and/or imprisonment." Why do we make such statements? Because our national economy is based on certain laws agreed on by authorized agencies, and penalties for noncompliance have also been agreed on. In addition, there are factors involving our obligations as citizens of the United States. Behind the statement, then, there are basic assumptions acting as "rules of the game." Consider another statement: "If I go out in the rain without proper protection, then I am liable to catch cold." This statement is based on other statements such as "Rain is wet," "Being wet causes changes in body temperature, " "Certain conditions of rapid changes in body temperature, combined with other factors, provide a favorable climate for virus infections," etc.

In these nonmathematical situations we have:

1. The presence of basic assumptions
2. The acceptance of these basic assumptions

An individual's code of ethics is an example of a set of basic assumptions. It forms for him a set of axioms from which he derives the theorems determining his conduct. The Bill of Rights in our Constitution can be thought of as a set of postulates on which the laws of our country are based. An understanding of this concept of basic assumptions would do much to eliminate futile arguments between individuals and between groups of persons. Did you ever argue with another person and not come to a common meeting place? Did you ever stop to analyze why you never agreed? What were your basic assumptions? What were his? If they were contradictory sets of assumptions, then you should not have expected agreement. Even though each of you used valid reasoning in your arguments, you should have expected disagreement. Until equivalent basic assumptions are made by both parties, agreement cannot be expected. This concept applies to world affairs as well as to the affairs of individuals. Hence, we see that the conflict of political ideologies can be expected to continue as long as basic assumptions are contradictory.

In mathematics, the "cold war" of conflicting sets of assumptions has proved fruitful and exciting. Mathematicians realize that the conflicting sets of assumptions are incompatible, but they are willing to let the proponents of the conflicting sets develop the theorems which follow from each set of assumptions. Mathematician A may start with his set of assumptions, or axioms, and mathematician B may start with his set of axioms. A's axioms lead to some theorems. B's axioms lead to theorems. A does not quarrel with B's theorems unless B's reasoning is incorrect. B is tolerant of A's theorems as long as A has used valid reasoning. A and B both realize that if their sets of axioms are not in agreement, then they must expect conflicting theorems. *Neither* A nor B would deny the propriety or possible usefulness of the other's theorems. This freedom now enjoyed by mathematicians to develop theorems from many sets of axioms is a recent development in mathematics. . . .

Consistency of a Set of Axioms

Our starting place in developing a branch of pure mathematics is a set of axioms. (Alternate words for axioms are *postulates, assumptions*. We shall use the word axioms in this work.) This set of axioms forms the set of rules of the game and the rules are to be accepted without question. If the rules are changed, a different game results. In like manner, if the set of axioms is changed, a different branch of mathematics is developed.

The set of axioms must be chosen so that it contains no contradictions. This insures that no contradictory theorems will be derived from the set of axioms. We would not want, for example, to have one axiom saying "A circle is round" and another axiom saying "A circle is square" in the same set of axioms. A set of axioms which contains no contradictions is said to be *consistent*. Consistency is a required property of any set of axioms.

The axioms are statements and statements involve words. Just as it is impossible to prove every statement, so is it impossible to define every word. As a result, some words must remain undefined. This is necessary in order to avoid an infinite regression of definitions. These undefined words in the axioms form the undefined concepts in the branch of mathematics being developed. Such concepts in geometry are "point" and "line." They remain undefined. Other words in the axioms are words of ordinary language such as "and" and "exactly" and they are given the same meanings as they have in ordinary language.

Perhaps it seems strange to you for us to say that words like "point" and "line" are undefined, since most people think of a point as a dot made by a pencil or a piece of chalk, and a line as something drawn by a pencil against the side of a ruler. This reflects our desire to interpret geometric concepts in terms of the world around us. The first people who developed geometry had this same desire and for centuries no one thought of point and line in any other way. But together with the freedom to choose various sets of axioms came the freedom from being bound to the world of objects we see around us. Points and lines can have many interpretations, as we shall see in working with various sets of axioms.

The problem of showing that a given set of axioms is consistent is met by the use of *models*. A model for consistency is a concrete, physical interpretation of the undefined concepts, constructed in such a way that all the axioms of a given set of axioms are satisfied. By *satisfied*, we mean that the truth value of each one of the axioms is T. If we can construct such a model, the axioms are said to be *consistent*. In most cases, many models can be constructed if one can be. One of the appealing features of testing for consistency is this very large potential supply of models. Anyone with a creative urge usually enjoys the construction of these models.

The first set of axioms we shall examine for consistency is a set of three statements about the undefined concepts *ogg* and *uff*. We use these nonsense words purposely so that no one will have any preconceived notions about them. Also it must be remembered that we are accepting these statements without proof, if they are to serve as axioms. Let us call this set of axioms the axiom set A. The axiom set A is as follows:

AXIOM 1. There are exactly three oggs.

AXIOM 2. There is at least one uff.

AXIOM 3. There are exactly two oggs on each uff.

It is our task to test the consistency of this set of axioms. The undefined concepts ogg and uff can be interpreted in any way we wish, but we will try to interpret ogg and uff in such a way that we can construct a model of oggs and uffs satisfying the three axioms.

Interpret an ogg to be a house and represent it in our model by a cross (X). Interpret an uff to be a street and represent it in our model by a line segment (________). Figure 1 shows a model with these interpretations of ogg and uff.

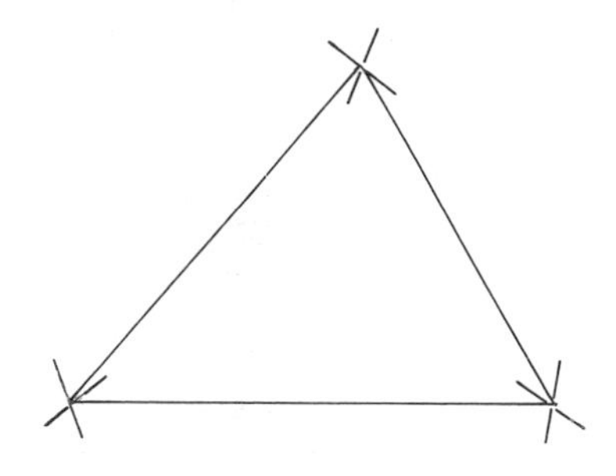

Figure 1

Is axiom 1 satisfied by this model? Yes, there are exactly three crosses, or houses—no more, no less.

Is axiom 2 satisfied? Yes, there is at least one street. The fact that there are three streets in the model does not make any difference in testing this axiom—all we need is one street in order to have the axiom satisfied.

Is axiom 3 satisfied? Yes, there are exactly two houses on each street.

We have, then, a consistent set of axioms in axiom set *A*.

Consider next axiom set *B*, in which "point" and "line" are the undefined terms:

AXIOM 1. There are exactly three points.

AXIOM 2. There is at least one line.

AXIOM 3. There are exactly two points on each line.

Instead of interpreting point as a dot and line as something drawn by using a ruler, we shall interpret a point as a captial letter such as *M*, *X*, or *T*, and line as a succession of capital letters such as *MRT*, *XYZW*, etc. If a capital letter appears in such a succession, the point designated by the letter shall be interpreted to be on the line represented by the succession of letters. For example, the point *R* is on the line *MRT*, as are the points *M* and *T*. Now for our model to show the consistency of axiom set *B*, we display the lines *PM*, *MO*, and *OP*. In this model there are exactly three points *P*, *M*, and *O*, since there are just these three capital letters used in the model. Also there is at least one line (there are, in fact, three lines). Finally, there are exactly two points on each line: on line *PM* there are two points *P* and *M*, on line *MO* there are the two points *M* and *O*, on line *OP* there are two points *O* and *P*.

The model just used to show that axiom set *B* is consistent demonstrates that not all models need to be geometric. Moreover, the undefined concepts can be interpreted in any way our fancy dictates. What we must remember is that our interpretations of the undefined concepts are supposed to help us to build models which satisfy the axioms. Thus, some measure of restraint must be exercised.

Perhaps you have observed a relation between axiom set *A* and axiom set *B*. If the word "ogg" is replaced by "point" and the word "uff" by "line," axiom set *A* becomes axiom set *B*.

Now consider axiom set *C*, in which "bird" and "nest" are undefined:

AXIOM 1. Every bird lives in a nest.

AXIOM 2. Each bird lives in a nest with exactly one other bird.

AXIOM 3. Not all birds live in the same nest.

AXIOM 4. No bird lives in more than one nest.

In building a consistency model for axiom set *C*, we interpret a bird to be a dot (.) and a nest to be what we ordinarily think of as a line. Let us examine a model made up of two parallel lines, each with two dots on it (Fig. 2).

Axiom 1 of axiom set *C* is satisfied, since each dot is on a line. Axiom 2 is satisfied, since each dot is on a line with just one other dot. Not all dots are on the same line, so axiom 3 is satisfied. No dot is on more than one line in this model, so axiom 4 is satisfied. Axiom set *C* is consistent, since the model satisfies all the axioms, with the interpretations of bird and nest as given.

A second model for axiom set *C* can be constructed by interpreting bird as a capital letter and nest as a sequence of capital letters in much the same way as we did for axiom set *B*. This new model is made up of the nests *PT* and *RS*. It can be verified that this simple model satisfies all the axioms of axiom set *C*.

Of course we wish to do more with axiom sets than to build models to show consistency. What the mathematician does after he forms a consistent set of axioms is to try to discover theorems; *i.e.*, other statements which can be derived from that set of axioms. If we used axiom set *A*, we would want some theorems about oggs and uffs.

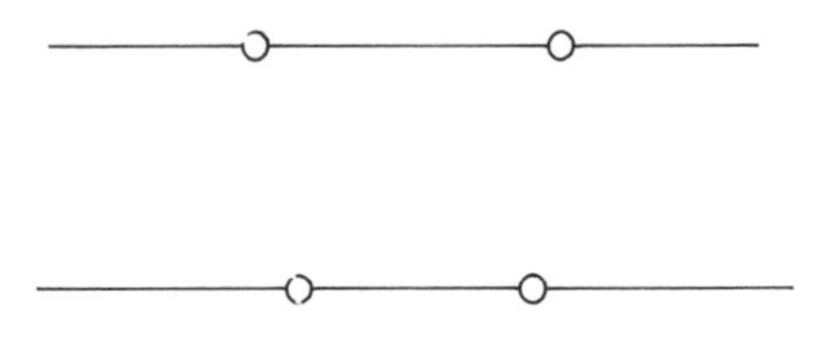

Figure 2

Axiom set C should give us some theorems about birds and nests. Sometimes hunches about possible theorems are obtained from consistency models. We might have a hunch about possible theorems about birds and nests from our consistency model for axiom set C. We might be tempted to state:

POTENTIAL THEOREM 1. There are exactly four birds.

POTENTIAL THEOREM 2. There are exactly two nests.

This is a legitimate way to get ideas for theorems. Unfortunately, theorems are usually harder to come by. Another consistency model for axiom set C will show that potential theorems 1 and 2 can never be theorems. Figure 3 shows this other consistency model for axiom set C.

The property of consistency, as we have pointed out, is a "must" property for any set of axioms. It precludes contradictory statements within the set of axioms itself and, consequently, contradictory theorems in the body of theorems derived from the axioms. We must remember that it is not uncommon to find contradictory theorems derived from two different sets of axioms. This is bound to happen if one of the two sets of axioms contains a statement contradictory to a statement or combination of statements in the other set of axioms. This is exactly the situation we find in theorems obtained from using, for one set of axioms, those axioms for Euclidean geometry and, for the other set of axioms, those for one of the non-Euclidean geometries. A theorem of Euclidean plane geometry says that the sum of the measures of the angles of a triangle is 180; a theorem from one of the non-Euclidean geometries says that the sum of the measures of the angles of a triangle is less than 180. Consistency, however, is a term used with reference to a particular set of axioms. Our attention is fixed on that one set, and we want to determine consistency within that set alone.

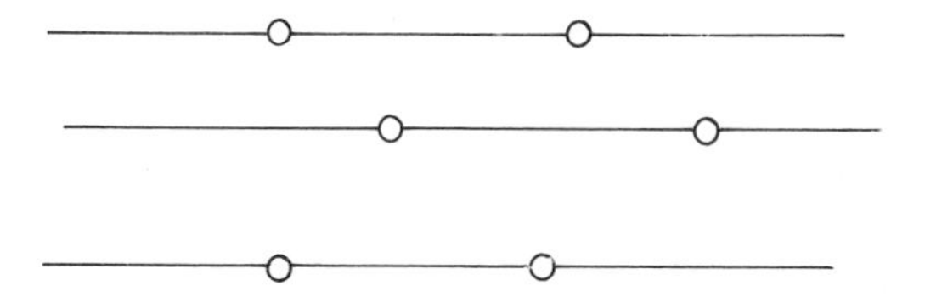

Figure 3

[The reader may wish to] test the consistency of the following set of axioms:

AXIOM 1. Each ogg is on an uff.

AXIOM 2. No ogg is on an uff by itself.

AXIOM 3. There are exactly four oggs.

AXIOM 4. Not all oggs are on the same uff.

AXIOM 5. There are exactly six uffs.

AXIOM 6. No ogg is on more than three uffs.

Independence of a Set of Axioms

Whereas consistency is an essential property for a set of axioms, there is another property, called *independence,* which is desirable from the point of view of mathematicians, but not essential. A set of axioms is said to be independent if no theorem is included in the set. In other words, no axiom can be derived from the other axioms if the set of axioms is independent.

In many elementary geometry textbooks, especially secondary school textbooks, the sets of axioms given are not independent. They often contain statements which could be proved as theorems by someone well acquainted with the subject, but which would be very difficult for the beginner to prove. There is nothing wrong in this procedure, and there are pedagogical arguments in favor of it. From the point of view of a pure mathematician, however, an independent set of axioms possesses a nicety not found in a set of axioms lacking this property. In a sense, it makes a mathematician "start from scratch" when he has an independent axiom set. He seems to derive an aesthetic satisfaction from the realization that he is using the fewest assumptions possible as a basis for the branch of mathematics with which he is concerned.

How do we test a set of axioms for independence? We try to find for each axiom a model in which all of the axioms except that one are satis-

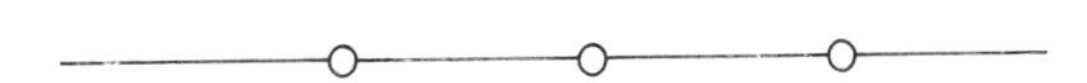

Figure 4

fied. Thus if we were testing a set of five axioms, we would need five different models. One model would not satisfy axiom 1, but would satisfy axioms 2, 3, 4, 5. The next model would not satisfy axiom 2, but would satisfy axioms 1, 3, 4, 5; and so on until the fifth model would not satisfy axiom 5, but would satisfy axioms 1, 2, 3, 4. Each one of these five models shows that one particular axiom is independent of (does not follow from) the other axioms. Therefore, if we display all five models we can declare that the set of axioms is independent.

To illustrate independence models, we display one for axiom 3 of axiom set B. We use the interpretation "dot" for point and the ordinary interpretation for line. In Fig. 4, axiom 3 is not satisfied. There are three points on the line, whereas axiom 3 states that there are exactly two points on a line. However, axiom 1 is satisfied since there are exactly three points, and axiom 2 is satisfied since there is at least one line.

Consider next axiom set C. Interpret bird as a dot and nest as an ordinary line. Axiom 2 is independent of the rest of the axioms because we can display the model of Fig. 5. All the axioms of axiom set C are satisfied except axiom 2. In this model one bird lives all by himself in a nest, contrary to axiom 2.

So far we have shown illustrations of the independence of a single axiom within a given axiom set. Now we wish to show that a complete axiom set is independent. We introduce a new set of axioms called axiom set D:

Axiom 1. There are exactly three distinct houses.

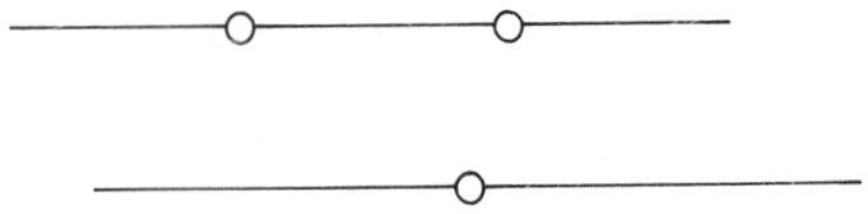

Figure 5

Axiom 2. Any two distinct houses are on a street together.

Axiom 3. Not all houses are on the same street.

Axiom 4. Two distinct streets have at least one house in common.

To prove independence, we need four models, in each of which one of the four axioms is false while the other three are satisfied. While we are making these models, we might as well look for a consistency model also. This means we will have five models. To be systematic, we display a chart (Fig. 6) reminding us of what we want to find. We mark the top of the chart with numbers to designate the axioms. We use T to indicate that an axiom is satisfied, F to indicate that an axiom is not satisfied. The consistency model must correspond to the $T\ T\ T\ T$ row. The row $F\ T\ T\ T$ corresponds to the independence of axiom 1, and so on for the rest of the rows. . . .

	1	2	3	4	
Consistency model	T	T	T	T	(a)
Independence of axiom 1	F	T	T	T	(b)
Independence of axiom 2	T	F	T	T	(c)
Independence of axiom 3	T	T	F	T	(d)
Independence of axiom 4	T	T	T	F	(e)

Figure 6

Before we look for models *(a)*, *(b)*, *(c)*, *(d)*, and *(e)*, we must remind ourselves that not all sets of axioms are independent, and that we may not be able to find all these independence models. If you look back at axiom set C about birds and nests, you can see that axiom 1 is really a theorem. It follows from the first part of axiom 2. The first part of axiom 2 is the same statement as is found in axiom 1. Therefore axiom set C is not an independent set, and we could never find a model where axiom 1 was not satisfied when the other axioms were satisfied. Also, just because one person cannot seem to find independence models does not mean that the set of axioms is not independent. Perhaps that person did not work as hard or have as much imagination as was needed. It is sometimes difficult to determine independence or to prove dependence. Axiom set D is an independent set, we assure you, and independence examples can be found.

To proceed then with axiom set D: we interpret house by a dot and street by an ordinary line. The models *(a)*, *(b)*, *(c)*, *(d)*, *(e)* in Fig. 7 show consistency and independence of the set.

[The reader may wish to] test the following set of axioms for consistency and independence:

AXIOM 1. There is at least one street.

AXIOM 2. There are exactly two houses on each street.

AXIOM 3. For each pair of houses there is one and only one street containing them.

AXIOM 4. Corresponding to each street there is exactly one other street which has no house in common with it.

Direct and Indirect Proof

Before proceeding to a discussion of the derivation of theorems from a set of axioms, we digress a bit to consider briefly some pertinent questions: (1) When may a statement be called a theorem? (2) What is a proof? (3) What are some methods (types) of proof used in mathematics?

We have touched on question (1) in [the first section]. There we referred to a theorem as a statement which could be derived from a set of axioms. Thus, when we are considering a particular set of axioms which we accept as true, any statement which we are forced to accept as true is called a theorem. We also noted . . . that a statement which may be a theorem with reference to

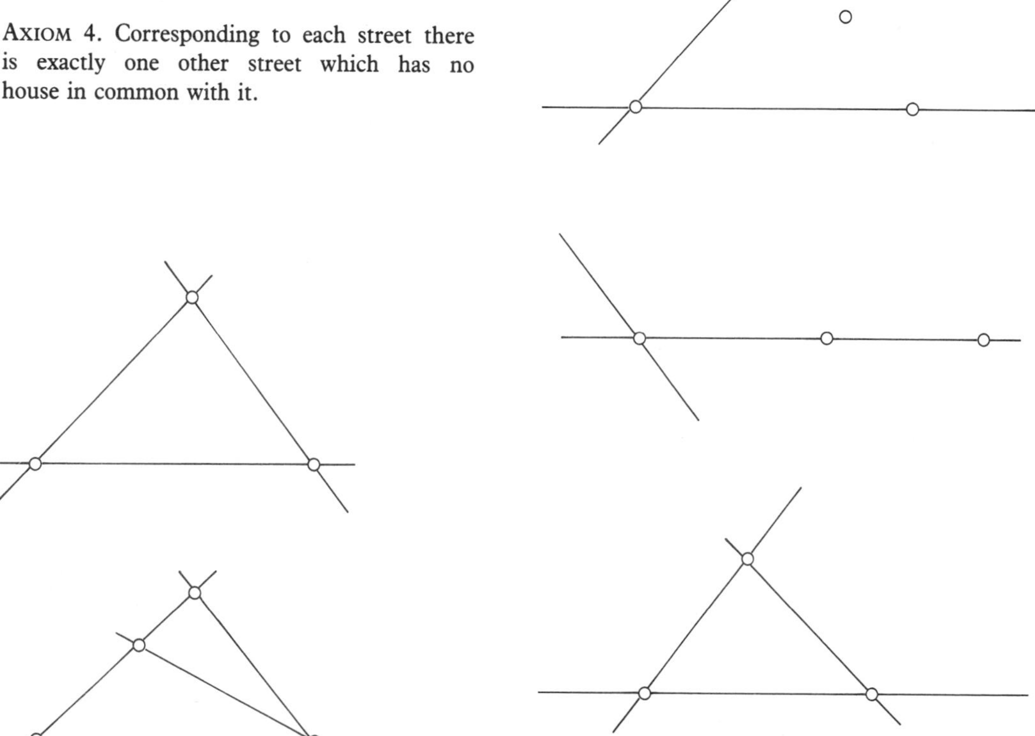

Figure 7

one set of axioms may not be a theorem with reference to a different set of axioms.

A complete answer to question (2) would require of the reader considerable background in logic and would not be appropriate in this discussion. For our purposes, it will suffice to say that a *proof* is the *procedure* by which we show that we are forced to accept a certain statement as true if we accept a particular set of axioms as true.

Two of the most common methods of proof used in mathematics, which are familiar to everyone who has studied secondary school geometry, are *direct* proof and *indirect* proof. As the name indicates, a direct proof is probably the most straightforward type. The following outline of a simple direct proof will illustrate the method. Suppose we wish to show that the statement "If H, then C" is a true statement.

1. Assume H is true.
2. For the statement "If H, then C" to be true when H is true, C must be true. Thus our goal is to show that C is true.
3. From a previous theorem, or from one of the axioms, or from a definition, we know that "If H, then P" is true. Since H was assumed true, we conclude that P is true.
4. Similarly, we know that "If P, then C" is true. Since, from step (3), P is true, it follows that C is true.
5. Therefore we state that "If H, then C" is a true statement.

A direct proof may become very long and complicated, because numerous steps such as (3) and (4) are necessary. It also may be very difficult (even for a mathematician) to see how to get started on such a proof; *i.e.*, how to get beyond step (2). In such cases, an alternative is to try an indirect proof. Although an indirect proof contains within it a direct proof, it is often true that the steps are more obvious when the indirect approach is used.

An indirect proof of a theorem of the form "If H, then C" follows a certain pattern. Using this pattern to prove "If H, then C" is a true statement, on the basis that H is true, we first assume that the statement "If H, then C" is false. Examination of the truth table for an implication shows that this is equivalent to saying that C is false and H is true, or, in other words $\sim C$ is true and H is true. We then proceed to show by a direct proof that if H and $\sim C$ are true, it follows that some statement S and its negation $\sim S$ are both true. In symbols, we then have $[H \wedge (\sim C)] \rightarrow [S \wedge (\sim S)]$. But $[S \wedge (\sim S)]$ is a false statement. Thus we have here an implication for which the reasoning is valid and the conclusion is false. . . . Our hypothesis $[H \wedge (\sim C)]$ must be false, from which it follows that either H or $\sim C$ is false. Since H is true, it must be $(\sim C)$ which is false. But if $\sim C$ is false, then C must be true. Hence we have shown that "If H, then C" is a true statement.

A few illustrations of indirect proof will clarify the idea of this type of proof. First, let us consider the following theorem: If n is an integer such that n^2 is divisible by 2, then n is divisible by 2.

To try an indirect proof of this theorem, we assume that the statement is false; *i.e.*, that n is an integer such that n^2 is divisible by 2, and n is *not* divisible by 2. It then follows that n must be of the form $2k + 1$, where k is some integer. Hence $n^2 = (2k + 1)^2$, or $n^2 = 4k^2 + 4k + 1$. But we can rewrite this as $n^2 = 2(2k^2 + 2k) + 1$, where $2k^2 + 2k$ is an integer, thus showing that n^2 is not divisible by 2. We now have arrived at a contradiction: n^2 is not divisible by 2 and n^2 is divisible by 2. Therefore, it follows that our assumption that n is not divisible by 2 is false, and we conclude that it is true that n is divisible by 2. Thus the theorem is proved.

As a second illustration, suppose we wish to prove the theorem: If $x^2 = 2$, then x is not a rational number.

We do not know how to proceed with a direct proof, so we assume that our statement is false; *i.e.*, that $x^2 = 2$ and x is a rational number, and see where that leads us.

If x is a rational number, then x can be written as p/q, where p and q are integers, $q \neq 0$, and the fraction p/q is in lowest terms. This means that p and q have no common factors.

Then it follows that

$$(p/q)^2 = 2$$
$$p^2/q^2 = 2$$
$$p^2 = 2q^2.$$

This means that p^2 is divisible by 2, and consequently p is divisible by 2. Then p can be written as $2r$, where r is an integer, and we can write

$$(2r)^2 = 2q^2$$
$$4r^2 = 2q^2$$
$$2r^2 = q^2.$$

Thus we see that q^2 is divisible by 2; hence q is divisible by 2.

Therefore both p and q are divisible by 2, which means they have a common factor equal to 2. But now we have arrived at a contradiction, for we have p and q with no common factors and at the same time p and q with 2 as a common factor. Our hypothesis that x is a rational number is false, and the statement that x is not a rational number is true. Thus the theorem is proved.

Another illustration (old and famous) is the proof of the theorem: The number of prime numbers is inexhaustible.

Suppose the number of prime numbers is exhaustible. Then there will be a largest prime number. Call it P. Now form the number $1 + (p_1 \cdot p_2 \cdot p_3 \cdot \ldots \cdot p_n \cdot P)$, where $p_1, p_2, p_3, \ldots, p_n$ are all the primes less than P.

This number $Q = 1 + (p_1 \cdot p_2 \cdot p_3 \cdot \ldots \cdot p_n \cdot P)$ is not divisible by any of the primes $p_1, p_2, p_3, \ldots, p_n, P$ since there will be a remainder of 1 whenever any one of them is used as a possible divisor of Q. Hence Q is a prime number. Also Q is greater than P. Thus we have found a prime number Q greater than P. This is a contradiction to our assumption that P was the largest prime number. This proves the theorem that the number of prime numbers is inexhaustible.

Consider, finally, some examples of the use of indirect proof in ordinary, nonmathematical situations.

1. Attorney for the defense says, "My client is innocent."
   Prosecuting attorney says, "Your client is not innocent."
   Attorney for the defense says, "O.K., assume my client guilty. If he were guilty, he would have to have been at the scene of the crime. But we can prove that he was somewhere else at the time of the crime. You will have to grant that he cannot be at two different places at the same time. It is impossible. Therefore my client is innocent."
2. First person: "It is raining."
   Second person: "No, it is not."
   First person: "If it is not raining, no one would be carrying open umbrellas which are dripping water. But I see a dozen people out there carrying umbrellas dripping water off them. Therefore it is raining."

Theorems Derived from a Set of Axioms

Let us now turn to the discovery and proofs of theorems derived from a set of axioms. We shall use a new set called axiom set E:

AXIOM 1. Each ogg is on an uff.

AXIOM 2. Not all oggs are on the same uff.

AXIOM 3. No ogg is on an uff by itself.

AXIOM 4. There are exactly four oggs.

If we interpret an ogg to be a dot and an uff to be an arc (⌒), a consistency model for axiom set E is exhibited in Fig. 8.

Notice that axiom 4 says that there must be exactly four oggs. This must be true in *every* consistency model for axiom set E. In the model of Fig. 8 there are also exactly four uffs. Does this mean that there are exactly four uffs in *every* consistency model? By axiom 1 we know that there is at least one uff. Is there a theorem which says: There are exactly four uffs? We can give a consis-

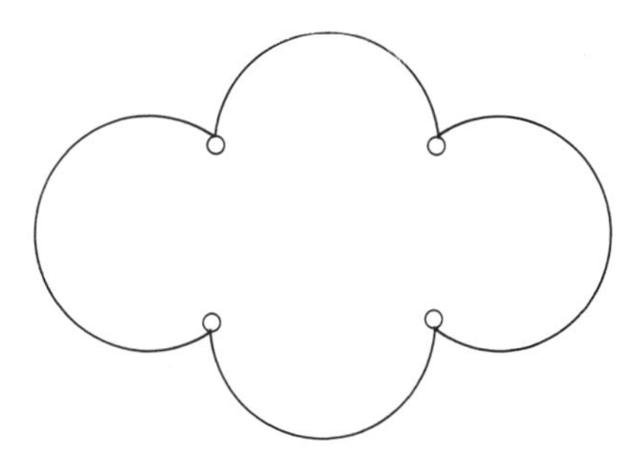

Figure 8

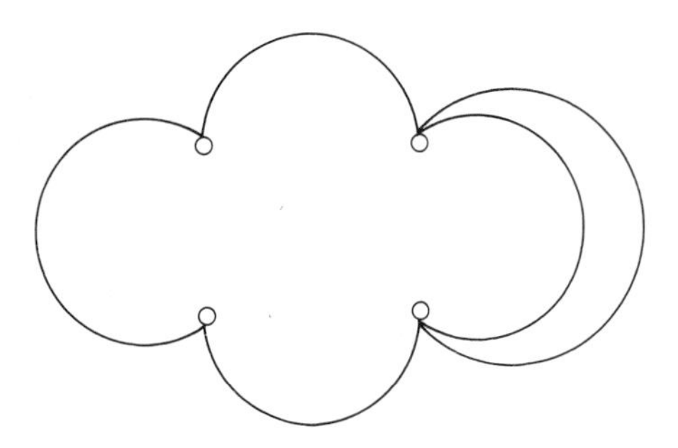

Figure 9

tency model which shows that the above statement does *not* follow from the axioms and hence is *not* a theorem. Such a consistency model is shown in Fig. 9.

All the axioms are satisfied in Fig. 9 and there are five uffs in this model. You can see that we could have six, seven, eight, or any greater number of uffs, just by adding more arcs as we did in Fig. 8 to obtain Fig. 9.

The following statement actually *is* a theorem, and we shall so label it and give the proof. You will notice that the proof follows the pattern of proofs in geometry. That is, we give a reason for each step in the proof. Each reason is based on an axiom.

THEOREM: There are at least two uffs.

PROOF: By axiom 4, there are exactly four oggs. Call them A, B, C, D.
By axiom 1, ogg A is on an uff.
By axiom 3, ogg A is on an uff with another ogg, say B.
Thus we have *one* uff AB.
By axiom 2, not all oggs are on the same uff, so there must be an ogg, say ogg C, not on uff AB.
By axiom 1, C is on an uff.
By axiom 3, C is on an uff with another ogg. This other ogg might be A. In this case we would have a second uff CA. This other ogg might be B, in which case we would have a second uff CB. This other uff might be neither A nor B, in which case it would have to be D. Then we would have a second uff CD. In every case, then, we have found a second uff. The theorem is thus proved. There are at least two uffs.

It is worth noticing that the proof given for this theorem used no diagrams, only words. It was not necessary to use any diagrams. In fact, many times students trying to prove theorems in geometry depend too much on diagrams. They are apt to think that certain statements are true because they look that way in the diagram. Again, the diagrams may illustrate special cases of the theorem and lead the student to assume statements that would not be true in general. One of the reasons we use oggs and uffs is to prevent any special meanings from being attached to them. The proof we used was "abstract" in the sense that we used no physical models in it.

We shall not go farther in obtaining theorems from sets A through E. They are hard to come by because we have so little to work with. We trust that enough has been done to give the flavor of theorem seeking and theorem proving.

Emotional Perils of Mathematics

Donald R. Weidman

Donald R. Weidman was born in the United States and received a Ph.D. in mathematics from Notre Dame University. He taught at Boston College and was a member of the Urban Institute in Washington, D.C. He died in 1980.

Who does mathematics and why? The easy answers are, naturally, that mathematicians do mathematics and they do it because they want to. But then one must ask, why do they want to do it? Most people find mathematics dreadfully dull and difficult to understand. Those who immerse themselves in it for a lifetime must be a rather curious sort of being. Still more curious is the fact that the many disappointments and discouragements of this most rigorous discipline seldom seem to dissuade its addicts. Indeed, this will seem most *curious in light of the following lament, jarringly honest in its description of the not inconsiderable emotional perils of being a mathematician.*

PEOPLE ARE TURNED aside from being mathematicians—by which I mean "pure" mathematicians—far more by temperament than by any intellectual problems. There are certain emotional difficulties which are intrinsic to the mathematical life, and only a few people are able to live with them all their lives.

First of all, the mathematician must be capable of total involvement in a specific problem. To do mathematics, you must immerse yourself completely in a situation, studying it from all aspects, toying with it day and night, and devoting every scrap of available energy to understanding it. You can permit yourself occasional breaks, and probably should; nevertheless the state of immersion must go on for somewhat extended periods, usually several days or weeks.

Second, the mathematician must risk frustration. Most of the time, in fact, he finds himself, after weeks or months of ceaseless searching, with exactly nothing: no results, no ideas, no energy. Since some of this time, at least, has been spent in total involvement, the resulting frustration is very nearly total. Certainly it seriously affects his attitude toward all other affairs. This factor is a more important hindrance than any other, I believe; to risk total frustration, and to be almost certain to lose, is a psychological problem of the first rank.

Next, even the most successful mathematician suffers from lack of appreciation. Naturally his family and his friends have no feeling for the significance of his accomplishments, but it is even worse than this. Other mathematicians don't appreciate the blood, sweat, and tears that have gone into a result that appears simple, straightforward, almost trivial. Mathematical terminology is designed to eliminate extraneous things and focus on fundamental processes, but the method of finding results is far different from these fundamental processes. Mathematical writing doesn't permit any indication of the labor behind the results.

Finally, the mathematician must face the fact

Source: Donald R. Weidman, "Emotional Perils of Mathematics," *Science* 149 (3 September 1965): 1048.

that he will almost certainly be dissatisfied with himself. This is partly because he is running head-on into problems which are too vast ever to be solved completely. More important, it is because he knows that his own contributions actually have little significance. The history of mathematics makes plain that all the general outlines and most of the major results have been obtained by a few geniuses who are not the ordinary run of mathematicians. These few big men make the long strides forward, then the lesser lights come scurrying in to fill the chinks, make generalizations, and find some new applications; meanwhile the giants are making further strides.

Furthermore, these giants always appear at an early age—most major mathematical advances have been made by people who were not yet forty—so it is hard to tell yourself that you are one of these geniuses lying undiscovered. Maybe it is important for someone to fill in the little gaps and to make the generalizations, and it is probably necessary to create an atmosphere of mathematical thought so that the geniuses can find themselves and thrive. But no run-of-the-mill mathematician expects in his heart to prove a major theorem himself.

I wonder how much of this psychological difficulty is present in other scholarly fields. I suspect that no other field suffers so acutely from all four problems. The experimental sciences in particular, I think, are pretty well preserved from the second and third difficulties. An experimentalist can perform an experiment and, at the end will have a set of data; and these data at least will indicate that such-and-such either is or is not significant. He knows before he starts the experiment that, except for equipment failure, he will finally have *something*. He is not faced with nearly certain frustration. Furthermore, publication standards permit experimentalists to describe details of procedures followed and difficulties encountered.

I also think the experimentalist has a reasonable hope for personal satisfaction. Experimental advances are frequently made by unknowns; in fact there aren't many experimentalists in history who have consistently made important discoveries, if we don't count those who have been lucky enough to head active research organizations for long periods.

Whether other speculative disciplines are immune from the four emotional problems I've outlined isn't clear to me. But I feel that differing standards of precision may ease the problem of frustration, in the sense that it is often possible in these other fields to hide the fact that you don't have anything to say. A mathematician who says nothing in an obscure manner is usually caught quickly—but alas, not always.

How Africans Count

Claudia Zaslovsky

Claudia Zaslovsky was born and educated in the United States. She has taught at a number of secondary schools and is currently on the faculty of the College of New Rochelle. She received an award from the Society of Delta Kappa Gamma for her book *Africa Counts.*

To Greek mathematicians and philosophers of the classic era mathematical concepts were the ultimate reality. Such ideas were the indivisible atoms of which existence itself was made, the very stuff of actuality. From such a viewpoint the ideas of mathematics transcend time and culture; they need no language and are the essence of human consciousness. Any beings who are aware of their own existence must understand these fundamental mathematical ideas.

The peoples and cultures of sub-Sahara Africa are geographically and culturally remote from that Hellenistic world. Yet if the fundamental mathematical ideas loved by the Greeks were to appear in so remote a setting, might one not allow that there is indeed a germ of truth in the notion of mathematics as ultimate reality? In learning how Africans count we may indeed recognize something fundamental about what we are and the universe in which we find ourselves.

Numbers One to Nine

In spite of the wide distribution of African peoples and the existence of perhaps a thousand languages on the continent, the words for two, three, and four are similar in an area covering about half of Africa. This area includes the Sudan—the region extending from the Sahara southward to the Gulf of Guinea, and from the Atlantic Ocean to the Nile River—and most of the southern part of the continent, now inhabited by Bantu-speaking peoples. The three hundred Bantu languages are classified by Greenberg as a subgroup of the Niger-Congo family, and bear a great resemblance to the languages of the western Sudanic peoples with respect to certain basic words. Linguists conclude that both categories of languages had a common origin, and that the Bantu-speaking peoples dispersed throughout the southern half of Africa from the Nigeria-Cameroons area.

The word for one exhibits great variety in African languages, as the reader will note from an examination of the lists of number words appearing in the text. Not so the words for two, three and four. "Two" is usually a form of *li* or *di*. The word for three contains the syllable *ta* or *sa*, and "four" is generally a nasal consonant, like *ne.* "Five" has a variety of forms; frequently it is the word for hand.

Dr. Schmidl introduces her discussion of Bantu numeration with these words[1] (page 168):

> When we compare the number words from one to nine in the various Bantu languages, we

Source: Excerpts from Claudia Zaslovsky, "How Africans Count," in *Africa Counts: Number and Pattern in African Culture* (Prindle Weber and Schmidt, 1973), pp. 39–51.

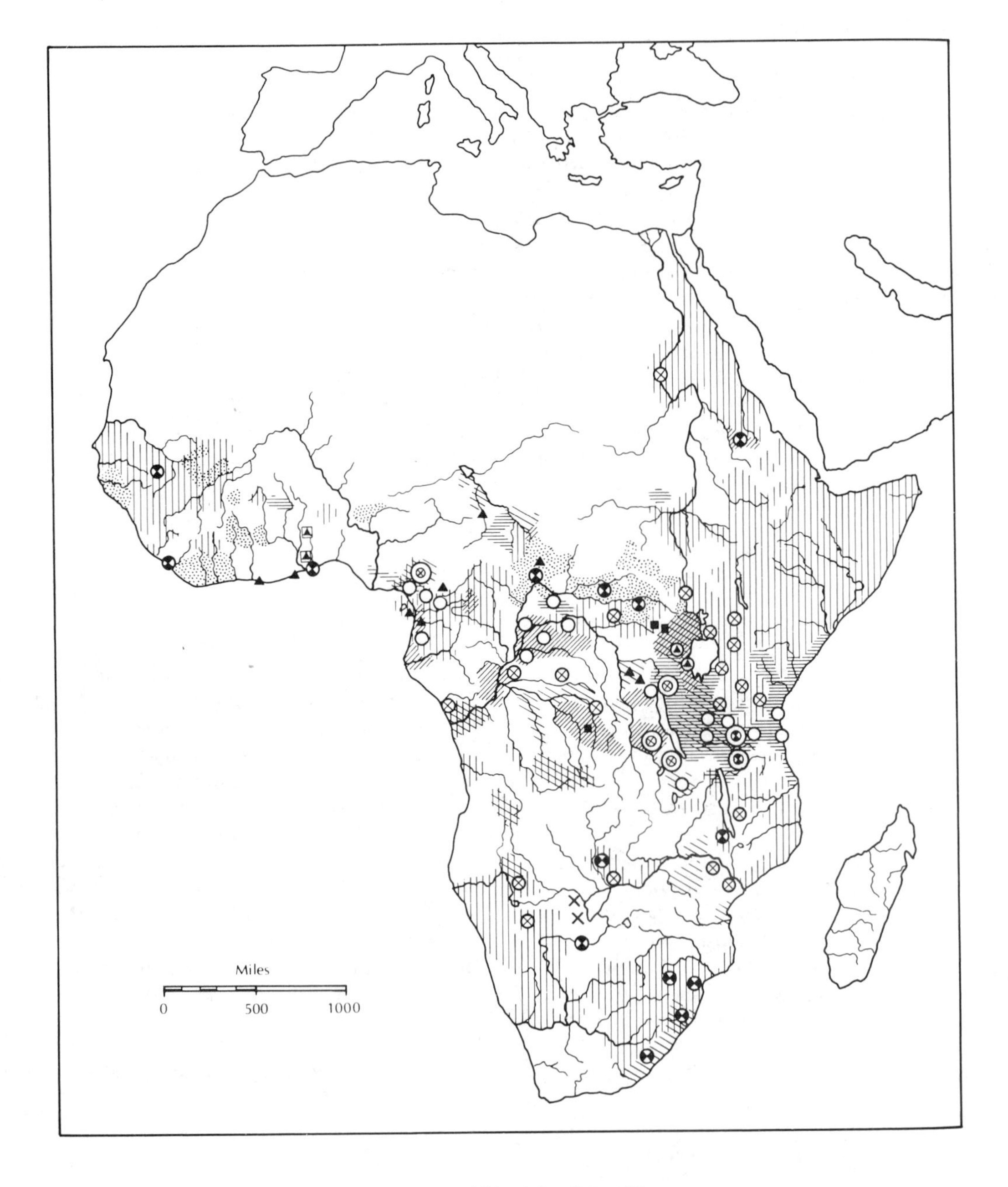

Figure 1 Distribution of numeration systems in Africa (after Schmidl).

find a similarity in the names for 2, 3, 4, and 5, while the corresponding gestures differ considerably. The basic stems are

2 *-vili* or *-vali*

3 *-tatu*

4 *-na*

5 *-tano*

. . . There are various expressions for "one," but generally they are related to *-mwe*.

In contrast, there are wide variations in the words for 6, 7, 8 and 9, and it will be necessary to deal separately with the various branches of the Bantu languages.

It is precisely the words for 6, 7, 8 and 9 which exhibit the most intriguing constructions. In some cases we find a simple addition to five; for example, in Kuanyama (southwestern Africa):

6 tano-na-mwe

7 tano-na-vali

8 tano-na-tatu

9 tano-na-ne

The composition of one with five to express "six" may not be so obvious. In the Malinke language, spoken on the upper Niger River, writes Maurice Delafosse,[2] "six" is expressed by *woro* or *wolo*, where *wo* is an abbreviation of the word for five, and *ro* or *lo* is a shortened form of "one."

In many languages the words for six through nine are derived directly from the gestures for these numbers, and are based on several different systems of gesture counting. This topic will be discussed fully later in this [article] in connection with gestures.

Remarkably, the Malinke word for nine, *kononto,* means literally "to the one of the belly," a reference to the nine months of pregnancy! This is obviously a word of the common people.

Key for Map, Figure 1.

Distribution of Numeration Systems in Africa (after Schmidl)

Code

I. Verbal numeration systems
  A. Formation of numerals for 6, 7, 8 and 9 in:
    1. Base-ten systems
      a. On the principle of two approximately equal terms
        At least two numbers
        Just one number

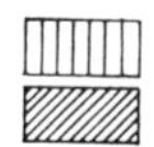

      b. By composition with five
        At least two numbers
        Just one number

    2. Base-twenty systems
      Formation of at least two numbers by composition with five
  B. Unusual bases
× 1. Evidence of base-two counting
▲ 2. Composition with six
■ 3. Composition with eight

II. Gesture counting
○ A. In the manner of the Shambaa
B. In the manner of the Hima
C. In the manner of the Zulu
⊗ D. Based on five and ten
E. Based on five twenty

Five-Ten Systems

The Sudanic and Bantu branches of the Niger-Congo language family diverge on the choice of the secondary base, the former generally using twenty, and the latter favoring ten.

Most Bantu languages use *kumi* or *longo* for ten, although that is not precisely the original meaning. The higher decades proceed smoothly as multiples of ten.

Number words for a hundred, a thousand, and higher ranks were rarely used except in association with specific objects to be counted. In the Ziba language:

100 *tsikumi,* referring to a string of 100 cowrie shells

1000 *lukumi,* a bundle of ten strings of cowrie currency

10,000 *kakumi,* a heap of ten bundles

As the need arose, these same words were applied to other objects, or new words were invented. . . .

Five-Twenty Systems

In the languages which build on twenty as a secondary base, "ten" may be an independent word, or it may mean "two hands" or "two fives." It is followed by an expression denoting "ten and one," and so on to "fifteen," which may be "two hands and one hand," or "ten and one hand," or even an independent word, as in the Dyola language of Guinea-Bissau, where the word means "to bow." Continue by adding one, two, etc. to the word for fifteen, until we reach twenty. This word in some languages means literally "man complete." In the Banda language of Central Africa the word for fifteen means "three fists," and for twenty, "take one person." The same method is then used all over again, adding to the word for twenty to form the numerals from twenty-one to thirty-nine. Forty is expressed as "two men complete," and one hundred is "five men complete"; in other words, all of the digits on the hands and feet have been counted five times. One Malinke expression for "forty" is *dibi*, "a mattress," the union of the forty digits, "since the husband and the wife lie on the same mattress and have a total of forty digits between them," to quote Delafosse[2] (page 389).

This is the basic form of a quinquavigesimal system (based on five and twenty), in which five is the primary base. It is found in many languages of the Sudan region: Dyola, Balante and Nalu in the west, Yoruba, Nupe and Igbo in Nigeria, Vai and Kru in Liberia, as well as the Nuba language of the eastern Sudan.

As a numeration system develops, special words are introduced, or words take on a new meaning. Foreign influence may bring about linguistic changes. The Malinke word *keme*, meaning "a large number," came to denote "one hundred," and was used in counting cowrie currency instead of the more cumbersome expression "five men complete."

The number words of a language may not be the same in sources from different periods. Migeod stated that the Mende say "five men finished" for one hundred, but a recent instruction book on the Mende language gives *hondo yila*—the word *hondo* is derived from the English *hundred*, and *yila* means "one."

The Igbo numeration system is based on twenty; whether five or ten is also a base is not so apparent. Thirty is expressed as the sum of twenty and ten, fifty is forty and ten, and similarly for the larger numbers (see the table that follows). In most dialects the formation of the numbers in any decade is by addition of the digits from one to nine to the appropriate word for the multiple of ten. Abraham does give an alternative method: 16, 17, 18 and 19 can be formed by deducting 4, 3, 2 and 1, respectively, from 20, and similarly for the higher decades. None of my informants could confirm this construction. Nor could I discover any basis for the statement by Thomas, a British anthropologist of the early twentieth century, that the Igbo people use 25 as a base, except for the counting of cowries.

There is a special word for the square of twenty, as one would expect in a vigesimal system—four hundred is *nnu*. The square of 400, or 160,000, is expressed as *nnu khuru nnu*, literally "400 meets 400." An "uncountably" larger number is *pughu*.

A unique system, based on twenty, is that of the Yoruba people of southwest Nigeria. Yoruba numeration illustrates an unusual subtractive principle still in effect today. The number forty-six, for example, is literally "twenty in three ways less ten less four," or $(20 \times 3) - 10 - 4$. . . .

Igbo Number Words

1	otu	30	ohu na iri (20 and 10)
2	abuo	31	ohu na iri na otu $(20 + 10 + 1)$
3	ato	40	ohu abuo (20×2)
4	ano	50	ohu abuo na iri $[(20 \times 2) + 10]$
5	iso	60	ohu ato (20×3)
6	isii	100	ohu iso (20×5)
7	asaa	200	ohu iri (20×10)
8	asato	300	ohu iri noohu ise $[(20 \times 10) + (20 \times 5)]$
9	toolu	400	nnu
10	iri		
11	iri na otu (10 and 1)		
12	iri na abuo (10 and 2)		
20	ohu		
21	ohu na otu (20 and 1)		

Along the Niger River, for several hundred miles from the coast, the subtractive method of forming number words is common. Perhaps this was the easiest method of counting cowries, or it may be that traders from one ethnic group influenced and to some extent standardized the numeration systems. Within a given language, dialects differ as to whether they compose by addition or by subtraction. Investigators give contradictory word lists, depending upon the dates and areas of their research. With the present trend toward standardization of languages, the discrepancies should be ironed out in time. Literacy and formal schooling require a uniform construction and spelling of the number words. These factors, reinforced by the mass media, tend to obliterate the various shades of local color.

Linguistic Change

The Arabs brought the new religion of Muhammed into North Africa during the seventh century. By the eleventh century there had been a fairly large immigration, and Islam spread slowly southward across the Sahara. Through immigration and commercial contacts, the influence of Islam was pervasive in many large centers of the Sudanic empires. Any person professing to practice the religion was obliged to read the Koran in the original Arabic. Thus Arabic became the language of the cultural centers, and Arabic numeration replaced or supplemented the indigenous systems. However, outside the commercial and intellectual concentrations, people lived their lives in the traditional way. This may account for the fact that in the Sudan the Soninke, the Hausa, the Fulani, and the Songhai count on the basis of tens, and even use some Arabic words, while their neighbors retain twenty as the secondary base. Further south, near the Guinea coast, we find a decimal system in use among the Ga, the Twi, and the Kpelle.

Arabic influence is apparent in Hausa, the most widely used language in northern Nigeria. Nineteenth century linguists classified the Hausa numerical system as quinary-vigesimal, but having twelve as the base for the formation of the words for thirteen through eighteen. Later sources showed a clear relationship to the decimal Arabic system with respect to the numerals starting with twenty as well as the use of the Arabic word for six. An interesting aspect of the numeration system of some eastern Hausa peoples is the use of a subtraction principle for compound numbers ending in eight and nine; e.g., $18 = 20 - 2$, and $19 = 20 - 1$. However, a recently published Hausa grammar gives a system of numeration that is pure decimal.

Migeod reported three designations for twenty in use at the same time: (1) *hauiya*, a special word applied to twenty cowrie shells, used as currency, (2) the plural of *goma*, the word for ten, and (3) the Arabic word for twenty. Some Hausa dialects had special words for a bag of 20,000 cowrie shells.

These are the Hausa number words in current use:

1	*daya*	8	*takwas*
2	*biyu*	9	*tara*
3	*uku*	10	*goma*
4	*hudu*	19	*goma sha tara*
5	*biyar*	20	*ashirin* (from Arabic)
6	*shida* (from Arabic)	30	*talatin* (from Arabic)
7	*bakwai*	100	dari

Unusual Number Bases

Most methods of number-name construction in Africa are based on five and have a secondary base of either ten or twenty. The mathematical operations of addition, subtraction and multiplication on the basic numbers give rise to names for larger numbers. Some numeration systems, however, show traces of bases other than five, ten and twenty.

As noted above, the Hausa people of northern Nigeria formerly constructed the numerals from thirteen to eighteen by addition to twelve. Some distance south of them live several technologically primitive tribes, who also use a base of twelve superimposed on a five-ten structure in which six through ten are formed by composition with five.

Although these tribes have different vocabularies up to ten, all their languages have similar words for eleven and twelve, and they all form thirteen as twelve and one, fourteen as twelve and two, etc. One would suspect a common source of the number words. A plausible theory is that these people borrowed from the Hausa, who have for centuries worked the tin mines in their area.

In the languages of the Bram and the Mankanye people of Guinea-Bissau, five is denoted by the word for "hand," ten is "two hands," nine is "hand and hand less one," and nineteen is expressed as "two hands and hand and hand less one," certainly a quinary system. But twelve is "six times two"! Twelve is formed in the same manner by the Bolan of Guinea. They carry the idea further with $24 = 6 \times 4$. Their neighbors the Balante compose with six to form the numerals from seven through twelve: $7 = 6 + 1$, $8 = 6 + 2$, etc. In the Ga language of Ghana, both seven and eight are based upon six: $7 = 6 + 1$ and $8 = 6 + 2$.

The Huku language of central Africa displays the greatest variety, basing many names on four and six:

$7 = 6 + 1$	$13 = 12 + 1$
$8 = 2 \times 4$	$16 = (2 \times 4) \times 2$
$9 = (2 \times 4) + 1$	$17 = (2 \times 4) \times 2 + 1$
10 is an independent word	$20 = 2 \times 10$

Gesture Counting

You may have ticked off on your fingers the number of guests you expected or the number of days to the next holiday. Perhaps you started with your thumb or maybe it was your little finger. Did you think about whether you used your right hand or your left? Did it matter whether you extended each finger, bent it, or tapped it with the index of the other hand?

The African would no more think of using his fingers so haphazardly than you would count: "three, one, six, eleven. . . ." To the African the finger gestures constitute as formal a method of counting as do the spoken words. Most often the two methods of expression are used simultaneously—the gestures accompany the spoken words. Go to a Hausa market and you will see people bargaining about prices, to the accompaniment of finger gestures. To indicate five, one hand is raised with the finger tips bunched together, and for ten, the hands are brought together, with all the fingertips of both hands touching.

Our two hands, a built-in calculating machine, gave rise to a decimal system of counting in many European languages. The intimate relationship between number words and finger counting is even more apparent in many African languages.

Consider the Zulu numerals, and their relation to finger counting. The finger gestures start with the left hand; the palm is up and the fingers are bent:

Numeral	*Derivation*	*Finger gesture*
1 *nye*	State of being alone	Extend left small finger
2 *bili*	Raise a separate finger	Extend left small and ring fingers
3 *thathu*	To take	Extend three outer fingers
4 *ne*	To join	Extend four fingers
5 *hlanu*	(All the fingers) united	Extend five fingers
6 *isithupa*	Take the [right] thumb	Extend right thumb
7 *isikhombisa*	Point with the forefinger of [right] hand	Extend right thumb and index finger
8 *isishiyagal-ombili*	Leave out two fingers	Extend three fingers of the right hand
9 *isishiyagal-unye*	Leave out one finger	Extend four fingers of the right hand
10 *ishumi*	Cause to stand	Extend all fingers

There is no doubt about the origin of the number words in Zulu. It is true that some words are long and cumbersome, but the efforts of the school authorities to shorten them have met with little enthusiasm.

The Sotho of southern Africa say *tselela* or *tsela* for "six," a word derived from the expression "to cross over." The reference is to the gesture language, crossing over from one hand to the other, a construction similar to that of Zulu. To continue with the Sotho numerals:

7 *supa*, "point"

8 *robeli*, "bend two"

9 *robong*, "bend one"

again similar to the Zulu expressions.

Some Bantu and Sudanic languages use words meaning "three and three" for six, "four and three" for seven, "four and four" for eight, and "five and four" for nine. Especially common in the Bantu languages is a word similar to *nana* for eight that is derived from the stem *-na*, meaning four. The origin of this compounding of words lies in the method of finger reckoning called "representation by two approximately equal terms."

The finger gestures are

6 three fingers of each hand

7 four fingers of one hand, three of the other

8 four fingers of each hand

9 five fingers of one hand, four of the other

A good example is the Ekoi language of Cameroon:

3 esa	6 *esaresa* (*esa* and *esa*, or 3 + 3)
4 eni	7 *eniresa* (*eni* and *esa*, or 4 + 3)
5 elon	8 *enireni* (*eni* and *eni*, or 4 + 4)
	9 *eloneni* (*elon* and *eni*, or 5 + 4)

This variation is widespread east of Lake Tanganyika, through the upper Congo River area, in Cameroon, near Lake Chad, and in South Kordofan. The Felup of Senegal use a combination of two methods:

$6 = 5 + 1$ (quinary)

$7 = 4 + 3$ (two approximately equal terms)

$8 = 4 + 4$

$9 = 5 + 4$

Names for the numbers from six to nine in some languages are based on the simple combination of the fingers of one hand to all five of the other. The Herero (Bantu language of southwest Africa) word *hamba* means "to change over" to the other hand when counting on the fingers. The number words are

6 *hamboumwe*, from *hamba* and *umwe* (one)

7 *hambombari*, from *hamba* and *imbari* (two)

8 *hambondatu*, from *hamba* and *indatu* (three)

9 *(hambo) muviu*, from *hamba* and *imuviu* (unexplained)

The corresponding gestures are

6 place little finger of right hand on left thumb

7 place two outer fingers of right hand on left thumb

8 place three outer fingers of right hand on left thumb

9 place four fingers of right hand on left thumb

We can now present some generalizations about finger counting. In those systems that build by addition to five, counting usually starts with the little finger of one hand and proceeds by the addition of the appropriate fingers in sequence until five is reached. This number is generally denoted by a closed fist. For six, the little finger of the other hand joins in the counting, and the fingers of the second hand are used in the same sequence as those of the first. Exceptions to this generalization are the Zulu and Sotho systems, where counting from six to nine commences with the thumb of the second hand and proceeds towards the little finger. The fingers may be extended, as with the Herero and Zulu, or bent down, as with the Malinke and Ewe of the western Sudan. This type of gesture counting is called quinary, since it is based on five.

In systems that are dominated by the principle of two approximately equal terms, counting usually starts by extending the index finger, and proceeds by using an additional finger in sequence for each number until the four fingers of one hand have been extended. The gesture for four is also based on this method of duplication, in that the first and second fingers are separated from the third and fourth. Five is again a closed fist. . . .

We have seen that number words may be derived from gesture language, as in the Zulu, Ekoi and Herero languages. By no means do the two always coincide; the contrary is often true. We may find pure quinary gestures accompanying verbal expressions based on two approximately equal terms, or vice versa. In fact, historians and anthropologists use both the oral and the gesture languages of people to trace their dispersal patterns and their relationships with other ethnic groups.

A case in point is the method of representation by two equal terms, based on the operation of doubling, or duplication. It brings to mind both the Egyptian method of multiplication and division by duplation and mediation . . . and the ancient Egyptian hieroglyphic numerals. On many monuments and temple walls we find:

||/|| (4) |||/|| (5) |||/||| (6) ||||/||| ||||/|||| (8) but |||/|||/||| (9)

The later hieratic symbols are:

[hieratic sign] or — (4) [hieratic sign] or 7 (5) [hieratic sign] (6)

[hieratic sign] or 2 (7) [hieratic sign] (8) [hieratic sign] (9)

Figure 2

These number symbols, and the fact that the Egyptians used finger gestures based on this principle (they also had quinary finger counting), confirm the prominence of multiplication by two in the development of Egyptian numeration. Going back several more thousands of years, one finds that the Ishango bone, with its sequence of notches, also suggests a notational system based on doubling: 3 followed by 6, 4 followed by 8, then 10 followed by 5 and 5. . . . It is conceivable that the system spread from Ishango northward down the valley of the Nile. Is it just a coincidence that this bone was found on the shores of Lake Edward, in the heart of the region where counting according to the principle of two approximately equal terms prevails today? Perhaps further archaeological research will give the answer.

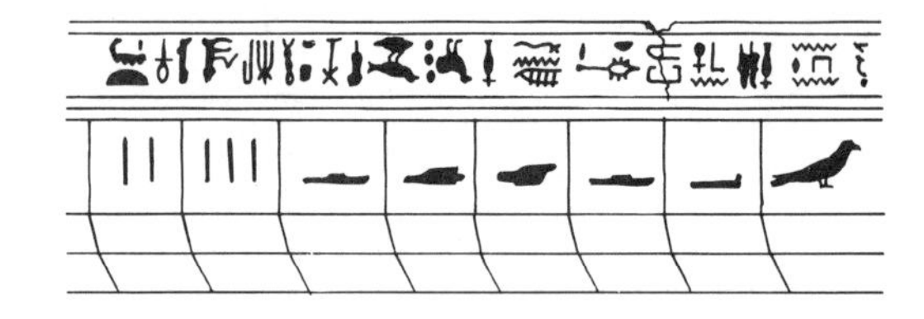

Figure 3 Egyptian carving showing the measurement of the ell, which contained twenty-eight "fingers." The first three are represented by the corresponding numbers of strokes, followed by finger gestures for four, five, six and seven. Four is a pictograph of a hand with the thumb closed over the palm, five is an open hand with thumb extended, six is a closed fist with thumb extended, but seven is not clearly portrayed.

Gesture language for the higher denominations is illustrated by the Efik, in the lower Niger Valley, who use the following elaborate system, as quoted in Schmidl's article[1] (page 191) from a work by Leonard:

> A hand clenched means 5, and along with one up to four fingers, from 6 to 9, both hands being clasped for 10. A finger is again added for 11 and so on up to 15, for which the arm is bent and the hand touches the shoulder. Twenty is signified by waving a finger in front of the body, and the reckoning proceeds as before until 30 is reached, when the hands are clapped and a finger waved. For 40, two fingers are waved, and at 50 the hands are once more clapped, and in the same way the remaining fingers up to a 100 are signalled, when the closed fist is waved and the simple sum of addition comes to an end.

The Ekoi people of Cameron are among the few African peoples who actually do count on their toes. Both their spoken and gesture languages are based on representation by two approximately equal terms, but they show "one" on the little finger, as do the base-five counters, rather than on the index finger, and their secondary base is twenty, as is common among the Sudanese. This combination of Bantu and Sudanese methods is hardly surprising, since their home is on the border of the two areas.

References

1. Schmidl, Marianne. "Zahl and Zählen in Afrika." *Mitteilungen der Anthropologischen Gesellschaft in Wien* 35 (1915), pp. 165–209.
2. Delafosse, Maurice. *La Langue Mandingue et ses Dialectes,* Vol. I. Paris: Librairie Orientaliste, Paul Guenther, 1929.
   ——. "La Numeration chez les Négres." *Africa* 1:3 (1928), pp. 387–390.

Appendix: Hilbert's 23 Problems

THE FOLLOWING IS a brief description of each of the 23 problems given in David Hilbert's address to the International Congress of Mathematicians in Paris in 1900.

The format of the presentation is as follows: first, the title of the problem in Hilbert's own words from the English translation that appeared in the *Bulletin of the American Mathematical Society*, volume 8, 1901–02; second, a brief description in the editors' words of the nature of the problem; third, a brief review of attempts at solution. The report on the status of the solutions is accurate up to 1975.

Problem 1 Cantor's Problem of the Cardinal Number of the Continuum

Cantor proved that the set of positive integers $N = \{1, 2, 3, \ldots\}$ *cannot* be placed in one-to-one correspondence with the set of real numbers. Since however, $N = \{1, 2, 3, \ldots\}$ can be made to correspond one to one with a subset of the reals (e.g., N itself), one concludes that the size (cardinality in mathematics) of the set N is less than the cardinality of the set of all real numbers. Cantor then made the hypothesis that any infinite set of real numbers has the same cardinality as that of N or of the whole set of real numbers. Thus, an infinite set of real numbers is sized like $\{1, 2, 3, \ldots\}$, or like the whole set of reals, and there is nothing in between. This assertion is now called the *continuum hypothesis*.

Hilbert tried and failed to prove this conjecture. In 1938 Gödel showed that one cannot disprove the conjecture using the usual axioms for set theory. In 1963 Cohen showed that it is also impossible to prove it. Thus, if the usual axioms for set theory are used, the continuum hypothesis is formally undecidable.

Problem 2 The Compatibility of the Arithmetical Axioms

This is rather less a problem than a suggestion that the axioms of arithmetic ought to be studied. Hilbert observed that one of the problems that ought to be investigated was the consistency of the axioms of arithmetic. By *consistency* he meant that in using the axioms of arithmetic one must not be able to prove that a statement A is *both true and false*. In 1931 Gödel showed that one can never prove that a system of axioms as complicated as those for arithmetic is consistent without going outside the given axiom system. Thus, in a sense, there is no final proof available for the consistency of the axioms of arithmetic. A great deal of work has been done on a variety of other aspects of axiom systems and their general properties.

Problem 3 The Equality of the Volumes of Two Tetrahedra of Equal Bases and Equal Altitudes

Actually, Hilbert wanted a proof that states that it is impossible to prove something using only congruent figures. The "something" in this case is the fact that two tetrahedra of equal altitudes and equal base areas have equal volumes. It is true that such tetrahedra have equal volumes, but in 1900 Dehn showed that it cannot be demonstrated by using the classic Greek method of congruent figures and nothing else.

Problem 4 Problem of the Straight Line as the Shortest Distance Between Two Points

This again is more of a program for investigation than a specific problem. Hilbert is suggesting that

Source: David Hilbert, "Mathematical Problems," ***Bulletin of the American Mathematical Society***, 1901–02, *8*, pp. 437–79. Reprinted by permission of the American Mathematical Society.

mathematicians look at various ways in which the axioms for plane geometry as given by Euclid can be altered while still keeping the "straight line is the shortest distance" property. Since 1900 this idea has been actively pursued and volumes of results obtained. The number of possible geometries is very large, and most studies deal with somewhat restricted classes of geometries.

Problem 5 Lie's Concept of a Continuous Group of Transformations Without the Assumption of the Differentiability of the Functions Defining the Group

The Norwegian mathematician Sophus Lie began to study the properties of surfaces by looking at the mathematical structure of the transformations of the surfaces. A transformation is a fairly straightforward relabeling of the points of the surface something like rotating the axes in a plane. Hilbert suggests imposing fewer restrictions on the transformations than Lie did by not demanding differentiability, for example. Since Hilbert's time some assumptions have been attached to his rather unspecific phrasing, and the problem arose, "Is every locally Euclidean group a Lie group?" In this form the question was answered in the affirmative in 1952 by Gleason, Montgomery, and Zippin.

Problem 6 Mathematical Treatment of the Axioms of Physics

Again, problem 6 is an area of study rather than a specific problem. Hilbert wanted to treat certain areas of physics, such as mechanics, mathematically, that is, by constructing a small set of axioms (rules) and obtaining the truths of mechanics deductively from these rules. Relativity and quantum mechanics rendered axiomatic treatment of mechanics somewhat pointless. However, Hilbert took account of the new developments and until the end of his life investigated various axiom schema that would be applicable to physics.

Problem 7 Irrationality and Transcendence of Certain Numbers

Hilbert wished to investigate numbers such as $2^{\sqrt{2}}$ or e^{π}. The number e was first proved to be transcendental in 1873 and π in 1882. Since these results came only about 20 years prior to Hilbert's address, it is not surprising that he suggests generalizations of results of this type. Most of the specific numbers suggested by Hilbert have had the question of their transcendental nature resolved (e.g., $2^{\sqrt{2}}$ was proved transcendental by Gel'fond in 1934). But Hilbert was clearly suggesting a general study, which continues today.

Problem 8 Problems of Prime Numbers

In this problem Hilbert is concerned with how the primes are distributed among the integers. He mentions the question of finding the zeros of the Riemann zeta function and how this would impact on prime distributions. For a more complete discussion of this subject read the article by Davis and Hersh on the Riemann Hypothesis in Volume II, Part Two.

Problem 9 Proof of the Most General Law of Reciprocity in Any Number Field

A rather technical question having to do with a general methodology for computing a type of parity function associated with number systems. In the terms in which Hilbert phrased the problem, it was solved by Takagi in 1921 and Artin in 1927.

Problem 10 Determination of the Solvability of a Diophantine Equation

A Diophantine equation is one whose solution must be found in terms of whole numbers. Hilbert wanted a general method for determining when an equation in any finite number of variables with appropriate coefficients had a solution in terms of whole numbers. In 1968 Baker produced such a method when there were only two unknowns. In 1970 Matijasevic showed that the general case was impossible.

Problem 11 Quadratic Forms with Any Algebraic Numerical Coefficients

A quadratic form Q is a polynomial in the variables $x_1, x_2, \ldots, x_m$ of the form

$$Q = a_{11}x_1x_1 + a_{12}x_1x_2 + \ldots + a_{m-1m}x_{m-1}x_m + a_{mm}x_mx_m$$

where the coefficients a_{ij} are elements of some number field. The problem that Hilbert suggests is a general study of the properties of quadratic forms where the coefficients come from any number field. Again, this question is more an area of research than a specific problem. In 1924 Hasse established a theory of quadratic forms over the rational numbers which he then extended to any field. This effort probably accomplished much of what Hilbert had in mind in proposing the problem.

Problem 12 Extension of Kronecker's Theorem on Abelian Fields to Any Algebraic Realm of Rationality

In 1881 Kronecker presented a theorem on abelian extensions of the rational numbers. This theorem characterized certain kinds of new number fields that can be constructed from the rational numbers (in essentially the way the complex numbers are created by "adding" $i = \sqrt{-1}$ to the reals). Hilbert's question is, Can the results of Kronecker for the rationals be extended to any number field? Hilbert viewed this problem and the general setting in which it occurs as being very important since solutions would probably result in tying together areas of algebra, geometry, and analysis. In its most narrow interpretation the twelfth problem was solved in 1920 by Takagi.

Problem 13 Impossibility of the Solution of the General Equation of the Seventh Degree by Means of Functions of Only Two Arguments

The general equation of the seventh degree has the form $a_7x^7 + a_6x^6 + \ldots + a_0 = 0$ where $a_7, a_6, \ldots, a_0$ are real numbers, $a_7 \neq 0$. The well-known results of Abel and Galois prove that no general solution of this equation, or indeed of any equation of degree five or greater, can be constructed using only elementary algebraic operations, i.e., addition, subtraction, multiplication, extracting roots, and so forth. However, by 1900 both the general fifth and sixth degree equations had been solved using families of curves generated by functions of two variables [$z = F(x,y)$]. Hilbert felt that even this method would not work for the seventh degree equation and proposed this as his thirteenth problem. It turned out that Hilbert's guess was wrong. The work of the Soviet mathematicians V. I. Arnold and A. N. Kolmogorov in the 1950s proved that the function of two variable methods can be extended not only to the general equation of degree 7 but to *any* finite degree n.

Problem 14 Proof of the Finiteness of Certain Complete Systems of Functions

The statement of the problem begins with a field k. This field is then extended to the field of rational functions (ratios of polynomials) in n variables over k, denoted by $k(x_1, \ldots, x_n)$. For example, when $n = 2$, $k(x_1,x_2)$ begins with all formal ratios of the form $\frac{P(x_1,x_2)}{Q(x_1,x_2)}$ where P, Q are polynomials in x_1 and x_2 with coefficients in k. Naturally, $Q \neq 0$ and there is the usual definition of equality of ratios, addition, subtraction, and so on. Now, let k be *any* subfield of this $k(x_1, \ldots, x_n)$ and let k_p be the set of all polynomials contained in k. Note that in general k_p will contain ratios of polynomials. In k_p we keep only those ratios from k that can be reduced to a polynomial $\left(\frac{P}{Q} \text{ where } Q = 1\right)$. The question is then asked, Is k_p finitely generated? That is, can every polynomial in k_p be formed from some finite subset of elements of k_p via the operations of k_p (addition, subtraction, etc.)? Hilbert felt that this may be true, but in 1959 Nagata proved otherwise. Nagata constructed a k_p that is not finitely generated and thus answered the fourteenth problem in the negative.

Problem 15 Rigorous Foundation of Schubert's Enumerative Calculus

Schubert was a nineteeth-century German mathematician who used some rather unorthodox methods to obtain a variety of constants associated with certain geometric objects. For example, Schubert claimed that the number of lines meeting four given lines in 3-space is two. A large number of other such special constants were found, some of them quite staggering, such as 666,841,048 as the number of quadric surfaces tangent to nine quadric surfaces in space. The methodology used by Schubert to get his results had been criticized. In this problem Hilbert suggests that more orthodox methodology might yield the same results in a more satisfying fashion. Much of this was done in the 1920s and 1930s by van der Waerden and others. This question is today a part of what is termed algebraic geometry.

Problem 16 Problem of the Topology of Algebraic Curves and Surfaces

In his statement of the problem Hilbert refers to a result by Harnack on the maximum number of separate branches that an *n*th order algebraic curve in the plane can possess. Hilbert's initial investigation suggested to him that these various branches must have some fixed relation to each other. Hilbert felt that while the form of these curves may vary, investigating the more fundamental topological relationship of the curves might yield useful insights. Many individuals have studied the topology of algebraic curves and surfaces but no dramatic "solution" of a type Hilbert felt might be there has emerged.

Problem 17 Expression of Definite Forms by Squares

By a *definite form* Hilbert meant a function in n-real variables $F(x_1, x_2, \ldots, x_n)$ such that F is never negative, i.e., no matter what real values are used for $x_1, x_2, \ldots x_n$, the expression $F(x_1, \ldots, x_n) \geq 0$. For example, $F(x_1,x_2) = x_1^2 + x_2^2$ is a definite form in two variables. The question is, Given a definite form $F(x_1,x_2, \ldots, x_n)$ can F always be expressed as a sum of squares of rational functions (ratios of polynomials). The answer is *yes* and the solution was first provided by Artin in 1927.

Problem 18 Building Up of Space from Congruent Polyhedra

This problem, as explained by Hilbert, asked essentially three questions.

1. Can one fill up n-dimensional Euclidean space with copies of some fundamental polyhedron whose copies are generated in a certain systematic way by transformations of the space? The answer is *yes* by Bieberbach in 1910.
2. Can one fill up n-space with a second type of polyhedra that is not produced by space transformations? The answer is *no* by Reinhardt in 1920.
3. How can you best fit equal, regular solids in space, e.g., fill space with spheres of equal radii? This has been solved for circles in the plane and a variety of other special cases but no general method is known.

Problem 19 Are the Solutions of Regular Problems in the Calculus of Variations Always Necessarily Analytic?

In this problem Hilbert is asking questions about the solutions of a certain class of partial differential equations. In particular, he wishes to know if the solutions must always be analytic (everywhere differentiable). Hilbert also suggests a general study of this class of equations and their solutions. The answer to the specific question is *yes* as Hilbert's description of the classes is usually interpreted. There were a number of solutions.

Problem 20 The General Problem of Boundary Values

Again, this is not so much a problem as the suggestion of an area of mathematics that needs investigation. A boundary value problem is a partial differential equation together with a set of side conditions (boundary conditions) that the solutions must satisfy. Hilbert suggested that such objects be studied and they have been, extensively.

Problem 21 Proof of the Existence of Linear Differential Equations Having a Prescribed Monodromic Group

This is a specific and rather technical question having to do with linear differential equations. A linear differential equation of order n is an equation of the form $y^{(n)} + c_1(x)y^{(n-1)} + \ldots + c_n(x)y = 0$ where $y^{(k)}$ represents the kth derivative of the function $y(x)$ and the $c_i(x)$ are specific functions of x. A linear differential equation is of the Fuchsian class if the $c_i(x)$ satisfy certain fairly restrictive conditions. The problem posed by Hilbert is to find a Fuchsian linear differential equation whose $c_i(x)$ satisfy yet other conditions, i.e., have a given set of singular points and a given monodromy group. It was solved in 1957 by Röhrl and others.

Problem 22 Uniformization of Analytic Relations by Means of Automorphic Functions

A parametric representation of a given algebraic relation is one in which each variable in the relation is expressed in terms of functions of a single variable called the parameter. For example, $x^2 + y^2 = 1$ may be expressed parametrically as $x = \cos h$, $y = \sin h$, $0 \leq h < 2\pi$. The points in the plane satisfying the first equation and those pairs of x and y values generated by the second pair of equations are the same. The notion of uniformization is the extension of this process of parametric representation to broad classes of very general algebraic relations. At the time of Hilbert's address Poincaré had obtained a very general uniformization result for expressions involving two variables. Hilbert wanted Poincaré's results investigated relative to the possibility of imposing some restrictions on the parametric functions. However, this is another fairly open-ended question and it has been pushed in many directions. The results of Koebe in 1907 do much of what Hilbert intended.

Problem 23 Further Development of the Methods of the Calculus of Variations

The calculus of variations treats functionals, i.e., functions of functions. For example, the arc length formula from calculus

$$L_n(f) = \int \sqrt{1 + (f')^2}$$

is a functional since the arguments of $L_n(f)$ are themselves real functions. Clearly, there are many such functionals and a standard calculus of variations problem is to maximize or minimize some functional on a given set of functions under prescribed conditions. For example, one might ask which function connecting two points in the plane has minimum arc length? Problems of the type treated by the calculus of variations were posed and solved from the very beginning of the calculus in the seventeenth century. In this problem Hilbert poses a number of specific problems in the calculus of variations and suggests other broad avenues of research. Many of the specific questions asked by Hilbert have been answered and research in the general areas continues.